Studies in the History of Civil Engineering

Volume 2

Masonry Bridges, Viaducts and Aqueducts

STUDIES IN THE HISTORY OF CIVIL ENGINEERING

General Editor: Joyce Brown

with the assistance of M.M. Chrimes, A.W. Skempton,
N.A.F. Smith and R.J.M. Sutherland

Twelve volumes are to appear from the Autumn of 1997:

1. **The Engineering of Medieval Cathedrals**
   Edited by LYNN T. COURTENAY, University of Wisconsin

2. **Masonry Bridges, Viaducts and Aqueducts**
   Edited by TED RUDDOCK, University of Edinburgh

3. **Land Drainage and Irrigation**
   Edited by SALVATORE CIRIACONO, University of Padova

4. **Dams**
   Edited by DONALD C. JACKSON, Lafayette College

5. **Water-Supply and Public Health Engineering**
   Edited by DENIS SMITH, formerly University of East London

6. **Port and Harbour Engineering**
   Edited by ADRIAN JARVIS, Merseyside Maritime Museum and Liverpool University

7. **The Civil Engineering of Canals and Railways before 1850**
   Edited by MIKE CHRIMES, Institution of Civil Engineers

8. **The Development of Timber as a Structural Material**
   Edited by DAVID T. YEOMANS, Liverpool University

9. **Structural Iron, 1750–1850**
   Edited by R.J.M. SUTHERLAND, Consultant, Harris and Sutherland, London

10. **Structural Iron and Steel, 1850–1900**
    Edited by ROBERT THORNE, Editor, *Construction History Journal*

11. **Early Reinforced Concrete**
    Edited by FRANK NEWBY, formerly F.J. Samuely and Partners, London

12. **Structural and Civil Engineering Design**
    Edited by WILLIAM ADDIS, University of Reading

Studies in the History of Civil Engineering
General Editor: Joyce Brown

Volume 2

Masonry Bridges, Viaducts and Aqueducts

edited by
Ted Ruddock

Ashgate
VARIORUM

Aldershot • Burlington USA • Singapore • Sydney

Published in the series **Studies in the History of Civil Engineering** by

Ashgate Publishing Limited
Gower House, Croft Road
Aldershot, Hampshire GU11 3HR
Great Britain

Ashgate Publishing Company
131 Main Street
Burlington, Vermont 05401–5600
USA

Ashgate website: http://www.ashgate.com

ISBN 0–86078–751–6

British Library CIP data
Masonry Bridges, Viaducts and Aqueducts.
(Studies in the History of Civil Engineering; v. 2).
1. Bridges, masonry–History. 2. Viaducts–History.
3. Aqueducts–History.
I. Ruddock, Ted.
624. 2' 09

Library of Congress Card Number: 00–103561

This book is printed on acid free paper.

Printed in Great Britain by MBC Book Manufacturers Ltd.

Contents

Acknowledgements

The chapters in this volume are taken from the sources listed below, for which the editors and publishers wish to thank their authors, original publishers or other copyright holders for permission to use their material as follows:

Chapter 1: B. Harvey, 'Testing times for arches', *New Scientist*, no. 1508 (15 May 1986), pp. 54–59. Reproduced by permission of Reed Business Information Ltd.

Chapter 2: N.A.F. Smith, 'The Roman bridge-builder: some aspects of his work', *The Structural Engineer*, 71, no. 9 (1993), pp. 160–165. Copyright © 1993 by The Institution of Structural Engineers, London.

Chapter 3: N.A.F. Smith, 'The Pont du Gard and the aqueduct of Nîmes', *Transactions of the Newcomen Society*, 62 (1990–91), pp. 53–76. Copyright © 1990–91 by the Newcomen Society, London.

Chapter 4: F.D. Prager, 'Vitruvius and the elevated aqueducts', *History of Technology*, 3 (1978), pp. 105–121. Copyright © 1978 by Mansell Information/ Publishing Limited and the Contributors. Reproduced by permission of Mansell, an imprint of Cassell, Wellington House, 125 Strand, London, WC2R 0BB.

Chapter 5: M.N. Boyer, 'Resistance to technological innovation: the history of the pile driver through the 18th century', *Technology and Culture*, 26, no. 1 (1985), pp. 56–68. Copyright © 1985 by The Society for the History of Technology. Reproduced by permission of The Johns Hopkins University Press.

Chapter 6: M.N. Boyer, 'Rebuilding the bridge at Albi, 1408–1410', *Technology and Culture*, 7, no. 1 (1966), pp. 24–37. Copyright © 1966 by The Society for the History of Technology. Reproduced by permission of The Johns Hopkins University Press.

Chapter 7: D. Calabi and P. Morachiello, 'Le Pont du Rialto: un chantier public à Venise à la fin du XVIe siècle', *Annales Économies, Sociétiés, Civilisations*, 43, no. 2 (1988), pp. 453–476. Copyright © by the École des Hautes Études en Sciences Sociales, Paris. Reproduced by permission of Librairie Armand Colin.

Chapter 8: H. Luschey, 'The Pul-i Khwājū in Isfahan: a combination of bridge, dam and water art', *Iran: Journal of Persian Studies*, 23 (1985), pp. 143–151. Copyright © by H. Luschey.

Chapter 9: J. Summerson, 'Berwick-on-Tweed Bridge', *The History of the King's Works*, 4, pt. 2 (1485–1660), ed. H.M. Colvin (London, 1982), pp. 769–778. Crown copyright. Reproduced with the permission of the Controller of Her Majesty's Stationery Office.

Chapter 10: E.C. Ruddock, 'Hollow spandrels in arch bridges: a historical study', *The Structural Engineer*, 52, no. 8 (1974), pp. 281–293. Copyright © 1974 by The Institution of Structural Engineers, London.

Chapter 11: E.I. Williams, 'Pont-y-ty-pridd: a critical examination of its history', *Transactions of the Newcomen Society*, 24 (1943–45), pp. 121–130. Copyright © 1945 by the Newcomen Society, London.

Chapter 12: E.C. Ruddock, 'William Edwards's bridge at Pontypridd', *Industrial Archaeology*, 11, no. 3 (1974), pp. 194–208. Copyright © 1974 by Bratton Publishing Limited. Reproduced by permission of Graphmitre Limited, Tavistock.

Chapter 13: E.C. Ruddock, 'The foundations of Hexham Bridge', *Géotechnique*, 27, no. 3 (1977), pp. 385–404. Copyright © 1977 by The Institution of Civil Engineers, London. Reproduced by permission of Thomas Telford Publishing.

Chapter 14: J. Manco, 'Pulteney Bridge', *Architectural History*, 38 (1995), pp. 129–145. Copyright © 1995 by The Society of Architectural Historians of Great Britain.

Chapter 15: J. Heyman and B.D. Threlfall, 'Two masonry bridges. II. Telford's bridge at Over', *Proceedings of the Institution of Civil Engineers*, 52, pt. 1 (1972), pp. 319–330. Copyright © 1972 by The Institution of Civil Engineers, London.

Chapter 16: J.B.B. Owen, 'Arch bridges', *The Works of Isambard Kingdom Brunel*, ed. Sir Alfred Pugsley (Cambridge, 1976), pp. 89–106. Copyright © 1976 by The Institution of Civil Engineers and The University of Bristol. Reproduced by permission of Thomas Telford Publishing.

Chapter 17: L.D. Lankton, 'Valley crossings on the Old Croton Aqueduct', *IA: Journal of the Society for Industrial Archeolology*, 4 (1978), pp. 27–42. Copyright © 1978 by The Society for Industrial Archeology, Houghton, Michigan.

Chapter 18: E.D. Galvin, 'The Canton Viaduct', *Railroad History*, 129 (1973), pp. 71–85. Copyright © 1973 by The Railway and Locomotive Historical Society, Inc.

Chapter 19: D. Brooke, 'The fall of Barentin Viaduct, 10 January 1846', *Journal of the Railway and Canal History Society*, 32, no. 5 (1997), pp. 363–366.

Chapter 20: J. Heyman, 'The safety of masonry arches', *International Journal of Mechanical Sciences*, 11 (1969), pp. 363–385.

***

Chapters 3 and 11 are reproduced from the *Transactions of the Newcomen Society* by permission of the Society's Council. The Newcomen Society promotes, encourages and co-ordinates the study of the history of all types of engineering and technology from the earliest times to the present day. Details from the Executive Secretary, The Newcomen Society, The Science Museum, London SW7 2DD, Tel-Fax 0171–589–1793.

Every effort has been made to trace all the copyright holders, but if any have been inadvertently overlooked the publishers will be pleased to make the necessary arrangement at the first opportunity.

General Editor's Preface

Joyce Brown

Civil engineering has a long history. Although practitioners did not begin to describe themselves as 'civil engineers' until the eighteenth century, the origins of their work lie in the construction skills of the ancient world and in the works of military engineers. Shortly after British civil engineers had formed themselves into a professional body, the Institution of Civil Engineers in London, their role was described in 1828 by Thomas Tredgold in these words:

> Civil engineering is the art of directing the great sources of power in Nature for the use and convenience of man.[1]

He thus neatly encapsulated what the civil engineer's job is about – the provision of the means to give society what it wants for its survival.

As time has gone by, the civil engineer's skills have had to diffuse over an ever-widening field and at the same time become more specialised. In all periods he has had to be the ingenious 'fixer', but more than a mere artisan – a point observed in the Institution's motto, *scientia et ingenio.*

The history of civil engineering is a fascinating one. Certainly 'the great sources of power in Nature' are not always directed without a certain amount of opposition by them, and the provision of the basic requirements – habitation, water supply, main drainage, harbours, bridges, places of worship, transport systems by land and water – has called for inventiveness and sometimes for daring and courage.

Yet interest in the history of civil engineering is not of long standing, and it seems fair to state that most of the important work has been done in the last forty years. With a few notable exceptions, most of the major books on the subject have appeared in that era; many specialist societies with their own transactions have been founded; and many of the journals in which articles might be published have come into existence. While archaeologists and architectural historians have long made their contribution to this subject, their ranks have been swelled by a large number of serious scholars – including engineers themselves – now interested in recording and evaluating the engineering achievements of earlier times. In particular, there is a lively interest in the study of our industrial past, at least in part because some of the evidences of it are already disappearing from the landscape.

Our intention in creating the series of which this volume forms part is to provide, through the reproduction of important contributions to this subject, an

---

[1] *Minutes of Council, Institution of Civil Engineers*, 4 January 1828.

invaluable reference collection for its study. The series encompasses many different branches of engineering from early times to the beginning of the twentieth century; its perspective is global; and the chosen articles have an international authorship, to the extent that this can be achieved for an essentially English-language series. The introduction to each volume has allowed the volume editor to set the pages selected in the context of the whole history of the subject and its historiography, while the provision of references and a bibliography will enable the reader to go further into the study of the topic.

This series will have succeeded if it gives civil engineers some unexpected insights into their own craft and other readers a new way of looking at engineering structures.

***

The volume editors were chosen for their particular knowledge and expertise in the field. They have worked with enthusiasm and good will to put together collections of articles intended to be both informative and stimulating, and I am grateful to them for their efforts. Where possible, they have incorporated the work of the best scholars, while at the same time giving preference to articles often obscurely published and not readily available. The learned journals are represented, as also are conference proceedings and essay chapters contributed to books.

In all of this, I have been much assisted by my Advisory Panel. Emeritus Professor Sir Alec Skempton (Imperial College) has made himself freely available to me for consultation on many aspects of this venture. Mike Chrimes (Librarian of the Institution of Civil Engineers), Dr. Norman Smith (Emeritus Reader, Imperial College) and James Sutherland (formerly a Senior Partner in Harris and Sutherland, London, consulting engineers) have all contributed valuable advice based on their extensive knowledge of particular fields. The idea for the series was developed in discussion with Dr John Smedley of Ashgate, and Publisher of the Variorum Collected Studies Series, who has given me much personal support and the benefit of his experience.

I am grateful also for the help given by the staff of several libraries, notably Kay Crooks and Susannah Parry in the library of the Department of Civil Engineering, Imperial College, and Mike Chrimes and staff of the library of the Institution of Civil Engineers, London. Many friends have given me encouragement and useful advice; in particular, I want to thank Paula Kahn and Mark Baldwin.

Joyce Brown,
Formerly Department of Civil Engineering,
Imperial College, London

Introduction

Ted Ruddock

Approaches to the History of Masonry Bridges

Historians have in general been prompted to write about the history of particular masonry bridges by one or more of three reasons:

- the bridge of which they write is of importance in the history of engineering;
- it played an important part in the general history of a community, a nation, or the wider world;
- the quality or quantity of records render it a valuable subject of study.

These factors can be seen to justify, in varying proportions, the choice of the papers which are printed as chapters in this collection. Each of the factors will now be illustrated by discussing the chapters where it is of particular importance.

Masonry Bridges in the History of Civil Engineering

Importance in the history of engineering is the most obvious reason for inclusion of a particular study here, though in the editor's view it is neither a unique, nor necessarily the strongest, reason for inclusion. That will be argued more fully later. It is the main reason for inclusion of the first and last chapters, but they both have an additional relevance for readers who are professional engineers, namely, that they contain material which can be used directly in the practice of bridge engineering today. For that reason, their descriptions of bridges and any analyses they make are consistent with the descriptions and analyses which would be given by engineers reporting present-day bridge designs and assessments.

The overt professional purpose of the final chapter by Jacques Heyman, 'The safety of masonry arches', is to develop a practical method for assessment of the stability and strength of existing arches. As part of the development, Heyman chronicles previous theoretical studies of arches from 1700 onwards,[1] and after developing his present-day method of assessment he applies it to well-known historic bridges. He thus enhances our knowledge, not, it should be noted, of their builders' thoughts, but of the structural behaviour and competence of the

[1] In several subsequent writings Heyman has repeated his survey of theoretical studies of the arch, perhaps most fully in his 1982 book, *The Masonry Arch*.

bridges.[2] He also informs us, incidentally, of some of the thoughts of early scientific writers, the antecedents of modern engineering theorists including himself. Harvey, whose opening chapter 'Testing time for arches' sets the scene for the whole collection, is another clear successor to the seventeenth- and eighteenth-century theorists and experimenters whose writings Heyman surveys. The knowledge and assessment procedures which these two authors provide are more or less independent of other historical knowledge, but the converse is not true; technical matters should not be omitted from the study of bridges within the history of communities and the wider world.

Both Harvey and Heyman are distinguished professors of civil engineering who, to their credit and to the benefit of the profession, value history as a serious component of engineering knowledge. Of the remaining eighteen chapters, by a variety of authors, of whom only a minority have practised civil engineering, four contain quantitative mathematical analysis of the bridges studied. Firstly, the arch analysis of Telford's bridge at Over given by Heyman and Threlfall in chapter 15 is made by similar methods to Heyman's analyses in chapter 20, namely, by methods of the late twentieth century. The arch analysis which forms a small part of my own study of hollow spandrels in chapter 10 applies not dissimilar concepts, but I use them as found in publications of eighteenth-century theorists such as William Emerson.[3] I make no inference, however, that such concepts were understood or used by more than a very few designers of bridges in the eighteenth century.

Owen's treatment of I.K. Brunel's arch bridges in chapter 16 reports in some detail, from Brunel's own surviving manuscripts, the mathematical theory and calculations he used in designing his bridges. They are the only authentic calculations by a historical bridge designer to be quoted in this collection. The paucity of such manuscripts is consistent with the conclusion of most historians of engineering that, until at least the first decade of the nineteenth century, masonry bridges in Great Britain were designed without the aid of calculations based on a theory of structures. The statement would not, however, be true of some of the bridges designed by engineers of the French Corps des Ponts et Chaussées in the eighteenth century.[4] I do note in chapter 10 and also in a book[5] a few *probable*

[2] 'Structural behaviour': the physical response(s) of a structure to the application to it of force or environmental conditions (such as temperature change). The most obvious response(s) are generally those visible to the naked eye, such as parts of the structure undergoing change of dimensions or shape, or movement relative to each other.

[3] Emerson, *The Principles of Mechanics*.

[4] Publications by French engineers from Belidor (1729) onwards show their knowledge and use of the theories of the academicians. As a very clear example, Perronet's mature understanding of the relevance of the works of de La Hire and Couplet ('De la poussée des voûtes', *Histoire de l'Académie des Sciences* (Paris, 1729, 1730), 79, 117) – *and of limits to their relevance* – is shown in his 'Mémoire sur le cintrement et décintrement des ponts', which he read at the

uses of such calculations in Britain in the same century, but without surviving manuscripts; and in the absence of manuscripts one cannot be certain about the type of calculations that were made or on what theoretical writings they were based.

The fourth chapter which contains mathematical analysis is my paper on the foundations of Hexham Bridge (chapter 13), where the theories adduced are of two kinds: firstly, theories of the flow of water through bridge arches, in which I first deduce from the *results quoted* in the documents – as the calculations have not survived – that John Smeaton calculated the velocities of flow by a formula known to men of science at the time, and I then, for comparison, recalculate the flows by a modern textbook method; secondly, modern theories of foundation bearing capacity are used to assess the risk of overloading of the foundations of the bridge. The second group of calculations is analogous to Heyman's assessments of arch stability, in that it attempts, as he does, to understand events more fully than men of the time of their occurrence could have done. The first group records a significant though not unique early application of contemporary hydraulic science to practical bridge engineering. The man who made the application, John Smeaton, was the first native British engineer to enter into scientific discourse. Priority in use of the hydraulic theory actually belongs to Charles Labelye,[6] but he was a Frenchman who only became naturalised as British near the end of his career.[7] Few, if any, other examples of such calculations at dates before 1800 are likely to be found in Great Britain.

Coincidentally, Smeaton was the first British man to adopt consistently the style 'civil engineer'. Labelye, who built the first Westminster Bridge between 1738 and 1750, and Smeaton, who built four large stone bridges in the years between 1763 and 1781,[8] were the first two men to bring some of the scientific knowledge taught by the members of academies of science to bear on the design and construction of masonry bridges in Britain.

Before leaving the discussion of mathematical theory and its use, attention must be drawn to a type of perverse conclusion which has been surprisingly prevalent in publications on the history of engineering. Instances are noted in N.A.F. Smith's study of the Pont du Gard (chapter 3) and the nature of the errors is spelled out both in that and in his chapter 2, 'The Roman bridge-builder'. The general mistake is to infer that, if features of a historic bridge are found

Académie des Sciences in 1773. He criticises the writers' 'formulae' as based on 'hypotheses with which they had to be satisfied', for lack of actual observations of the deformations of arches such as he had made during de-centring of the arches of three of his bridges. The 'Mémoire' is printed in his *Description des Projets...*

5 Ruddock, *Arch Bridges...*, 13, 69–70.

6 Ibid., 4–5 and 203.

7 Ibid. and Walker, *Old Westminster Bridge.*

8 Smeaton, *Reports*, Ruddock, *Arch Bridges*, 80–104, and Ruddock, 'Bridges and roads...', (1981).

satisfactory by a modern analysis, the principles on which the modern theory is based must have been clear to the designer or builder, and/or that they followed the same thought-process as the analyst. Such inferences deserve Smith's criticism and are unworthy of any serious student of history. There must, however, be a place for hypotheses about designers' intentions and about methods of construction when they are based on observation of surviving structures (including the findings of archaeological investigation) and/or authentic historical records.

Masonry Bridges in the History of Communities

The second reason suggested for historians' interest in masonry bridges is very broad: that bridges have mattered throughout recorded history for trade, for cultural exchange, for government and for war. For many centuries the building of a bridge might be considered an act of piety; while in Asian and European cities and towns a bridge was often a chief place of public assembly. Both in use and in their symbolic meaning, bridges were the links between communities large and small, and facilitators of social and commercial intercourse. These facts define the human context in which all engineers, and their forebears who with various titles built bridges in earlier centuries, have designed and constructed the bridges that history celebrates. Communities through their leaders, or monarchs, noblemen or clergy – the Pope himself being called 'the Pontiff' – conceived, ordered and paid for each bridge made by a mason, architect or engineer, and the story of any bridge must, to be complete, relate the activity of the patron or ruler well as that of the builders. The history is much more than a story of the physical structure. However, although a writer's interest may quite justifiably be focussed on the part played by bridges in the stories of communities, for his study to be a contribution to the history of civil engineering it must also describe, or in some way illuminate, the activities which eventually became the responsibility of civil engineers, namely, the making and maintenance of bridge structures.

The studies in this book have previously been available only in seventeen different books and journals published in Britain, the USA and France. Most of them have some specialised content which justified their publication in particular journals or compilations. Such are, for instance, Sir John Summerson's chapter 9 on 'Berwick-on-Tweed Bridge', taken from the official British *History of the King's Works*, and Jean Manco's chapter 14, 'Pulteney Bridge,' which, as befits a study of a bridge designed by the architect Robert Adam, was published in the journal *Architectural History*. E.I. Williams's chapter 11 on the history of the bridge at Pontypridd is an example of a study based, like Summerson's, entirely on documents, but published as history of engineering in *Transactions of the Newcomen Society*. Unusually for those *Transactions*, its conclusions involve no engineering judgement, but it provides crucial clarification of the previously muddled story of a famous bridge.

Influence of the Quality of Sources

The third broad reason which has stimulated the studies described in the collection is the existence of a satisfactory quality and/or quantity of source materials. Every chapter shows that it is based on at least adequate source documents, but some serious constraints laid on authors by gaps in the source information are also revealed, and may be important to the whole field of study. An explicit example is reported in Boyer's 'History of the pile driver through the eighteenth century' in chapter 5. (Note that 'through' is used in the American sense of 'up to and including'). Description of the means of driving piles, she tells us, was absent from manuscript sources from the time of the Assyrian empire right through to that of medieval France because, variously, (i) it was not considered important by the writers of ancient chronicles; (ii) to the architects and builders to whom Vitruvius addressed his instructive treatise in the Roman era it would have been common knowledge; and (iii) in later times it was presumably withheld as secret knowledge of the masons' lodges. There were undoubtedly other aspects of the bridge-building techniques that civil engineers inherited that had been similarly excluded from old source documents.

General Features of the History of Masonry Bridges

Each of the twenty chapters treats a specific and limited topic – twelve of the chapters deal with a single bridge – while the Bibliography points the reader to some broader histories of bridges and of civil engineering. Specific local studies do not supplant the broad histories – rather, they need each other and, at their best, both give appropriate attention to broad, contextual issues as well as the bridge or bridges they describe.

As background to the local and detailed stories in the chapters, some general comments are offered here under three headings: terminology for the parts of masonry bridges, the costliness of masonry relative to alternative materials, and the aesthetic aspect of masonry bridge design.

Terminology

The names of parts of a typical masonry bridge are shown in Fig. 1. A few other terms which occur in this introduction are explained in the footnotes, and a short glossary printed at the end of chapter 12 may also be found helpful.

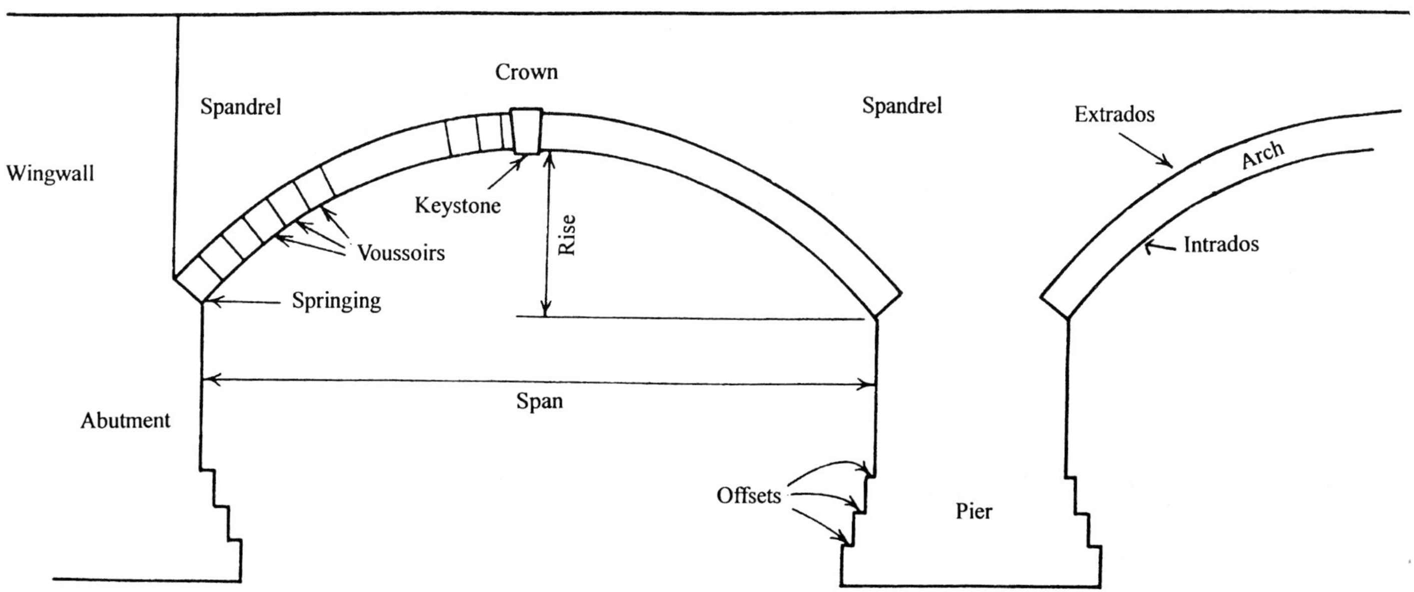

Figure 1: Parts of an arch bridge

Expense

Over the wide range of contexts portrayed in this volume, encompassing two thousand years of history, one fact is invariant, namely, that masonry bridges have always been expensive, and the men, communities or states that commissioned their building were rich and often very rich. Other types of bridge and other means of crossing waterways were always cheaper. Fords and ferries were used throughout history, wooden bridges were common in many parts of the Roman Empire and from the medieval period onwards in forested lands of Europe, Asia and North America. They were followed in the late eighteenth and nineteenth centuries by iron bridges, once the scale of production of iron had made it a cheap material. Wood and iron were also the common materials in the early years of most colonial territories, and even the building of bridges of hand-made ropes and branches following primitive practice continued in some countries. In the later stages of the Industrial Revolution, steel, followed by reinforced concrete, stole the market for bridges. Throughout those developments, as through all the centuries before they began, those who opted for masonry and rejected cheaper materials paid a high price, generally double, sometimes ten times that of the alternative. A famous example is the first Westminster Bridge, built between 1738 and 1750, which was the first modern bridge over the Thames at London and cost almost £200,000, while the wooden Fulham-to-Putney Bridge a few miles upstream, with a carriageway half as wide, had been built a few years earlier for less than £14,000.[9] What the rich authorities paid for was the durability of stone – though stone's durability varies and the stone used at Westminster was not ideal – and, in some countries and periods, of brick. In general their investment paid off, as the survival of bridges hundreds of years old, and a few which are two thousand years old, testifies.

The stories of bridges told here illustrate the variety of contexts in which the high cost of a masonry bridge could be afforded. Near the Roman city of Nîmes in the first century AD the astounding aqueduct bridge, the Pont du Gard, was erected to carry fresh water from a faraway source (chapter 3); the Muslim state of Persia could afford the Pul-i-Khwājū at Isfahan to provide water conservation, a river crossing and a place of rest for travellers (chapter 8); the city of Venice, long famed for trade, wealth and architectural excellence, spent the necessary money, after long hesitation, to build the new Rialto bridge, mainly to be its commercial centre and for civic pride (chapter 7); Pulteney Bridge in Bath was built by the first outlay of funds in a scheme of speculative town development by one of the wealthiest men in eighteenth-century England (chapter 14); the pretensions of American railroad companies prompted the building of

[9] Ruddock, *Arch Bridges*, 28–9 and references quoted there.

several stone viaducts before they were forced by a very competitive economy to adopt timber and iron for bridges (chapter 18); and the proprietors of New York's Croton water supply scheme had to build for permanence in their structures, the consequences of a loss of supply being thought disastrous (chapter 17). The constant prerequisite for building in stone is wealth; other factors of need, use and economy have varied greatly.

The expensiveness of stone masonry persists to the present day, as does the relative durability of well-chosen stone. Early enthusiasm for concrete as an 'artificial stone' proved a delusion and its visual texture, in spite of great research applied to it, has never approached that of good stonework.

Aesthetics

In any full history of masonry bridges, aesthetic qualities have an important role, though they have had little or no place in the professional training of many civil engineers in the twentieth century. Occasionally even a historian of engineering has been heard to dismiss aesthetic intentions from the serious history of bridge design. In contrast, many civic authorities throughout history have aspired to build their bridges 'noble' or 'grand' and individual owners have generally wished their bridges to be beautiful. The aesthetic motives of both owners and builders are recorded in some of the chapters here. Ascription of comparative levels of aesthetic merit to the bridges of different ages and cultures, however, can hardly be meaningful; rather, the rich variety of styles and ornament should be enjoyed.

It has actually been the case for long periods in history that no masonry elements were added to bridges simply to ornament them. The aesthetic impact of Roman bridges, for instance, lay most often in their sheer size. The overhead aqueducts that fed water to cities were of awesome length and rhythm of arches, all the more impressive when they were built of two or more tiers; and most Roman highway bridges impressed chiefly by their solidity and apparent strength. In medieval times also, plainness was sometimes favoured by the hard-headed burghers of merchant cities, for whom a safe river crossing was the paramount need; and in not a few instances the view of the bridge itself was partly or wholly obscured by buildings placed on top of it, either because the river crossing was an ideal place for commerce or to earn property taxes for the bridge's upkeep. Happily, such practice was not universal and Rome, Florence, Avignon and Prague are examples of European cities where beautiful bridges existed during the medieval centuries. With the coming of the Renaissance a new emphasis on aesthetic quality and style affected bridge designs, and conscious aesthetic intentions have been a regular feature of designs and of criticism ever since.

Sequence of Historical Developments

The remainder of this introduction is devoted to a brief history of masonry bridges from the days of the Roman Empire to the nineteenth century. One of its functions is to note some important phases of bridge-building which are not represented in the chapters of this book due to the unavailability of suitable published papers. Another function is to outline the social and administrative context of bridge-building in each succeeding period of history.

Roman Bridges and Aqueducts

The stone bridges of the Roman Empire, like its public services as a whole, have been studied constantly since they were first built. For that reason, and because chapter 2 by N.A.F. Smith provides a synoptic study of Roman bridge engineering, little more need be said here. Detailed local studies of individual Roman bridges can be found in various countries and languages, and Smith's study of the Pont du Gard in chapter 3 is one of the best, happily focussed on the acknowledged paragon of surviving Roman arched structures.[10] The third study of Roman bridges here is Prager's 'Vitruvius and the elevated aqueducts' (chapter 4). Prager seeks, by detailed scanning of the *Ten Books of Architecture* written in the first century BC by Marcus Vitruvius, and quoting everything that may relate to the elevated aqueducts that supplied the city of Rome with water, to estimate how much use was made of Vitruvius's instructions in their design and construction. In passing, a summary history of the development of the system from about 270 BC to 300 AD is given.[11]

No builder of a substantial masonry bridge in Europe for at least a millennium after the break-up of the Empire can have been unaware of the legacy of Roman bridges. Their general characteristics became accepted as rules for ideal bridge-building, a central dogma being that perfect arches were semicircular, as perfect domes must be hemispherical. The 'perfectness' was treated as a single quality, which caused some confusion in later centuries when it became desirable to think separately about the two aspects of an arch, its structural behaviour and its aesthetic quality. The separation of these two concepts later contributed to the breaking of the Roman rules and adoption of other shapes of bridge arches.

---

[10] The most useful general book on Roman bridge-building is Colin O'Connor's *Roman Bridges*, written by a professor of civil engineering and providing chapters on the road system of the empire, the builders, the technology, and all the substantial remaining bridges in each part of the empire, with extensive tabulated information and bibliography.

[11] A more-or-less contemporary description of the system, by Frontinus, is available in English (see Bibliography).

Medieval Bridges

In the great bridge of St. Bénézet at Avignon,[12] built between 1177 and 1187, of which three original arches still stand, Roman building techniques were used: each arch was formed by building four parallel arches, each of them one stone wide, using the same narrow centre four times,[13] and also the stones were shaped so perfectly that they could be laid without mortar. Such arches could be seen in the Pont du Gard not far away. At Avignon, however, the form of the arches was no longer a semicircle but a much lower curve. Such low arches came into use in other places, a common shape being one-third of a circle, or segment of 120 degrees. They reduced the necessary height of a bridge and the number of arches and piers required, as compared with bridges of semicircular arches; they also offered options for more varied architecture. Similar arch shapes, but with the whole of each arch built at once on centring of the full width and using mortar – methods which simplified the masons' task – were used in the fourteenth-century Italian bridges of Ponte Vecchio at Florence (1345–67), Ponte Coperto at Pavia (1351–56) and Ponte Scaligero at Verona (1355–58);[14] and in Newton Cap Bridge at Bishop Auckland (1388–1405) in England.[15] Of these four bridges the last was financed by a bishop, as were many others, usually built of traditional, semicircular, arches, including the fourteenth-century bridge at Glasgow and sixteenth-century bridges at Aberdeen and Dunkeld, all in Scotland.[16]

There is frequent evidence of the involvement also of abbots, monks and other churchmen in the building and maintenance of European bridges. They encouraged rich men and women to make financial contributions, often by bequests of land or buildings whose produce or rentals would contribute year by year to the maintenance of bridges. Studies of the management and use of such funds gifted to London Bridge and Rochester Bridge have been published,[17] and Boyer's chapter 6 describes the management and techniques of maintenance and rebuilding of the town bridge at Albi in southern France. Elsewhere, in an

---

[12] Celebrated in many writings. Boyer in *Medieval French Bridges*, 38–42, tells the story of its construction, identifies medieval sources and makes many other references to the bridge, which she pairs with Old London Bridge as the two most famous of medieval bridges.

[13] A *centre*, or *centring*, is a framework, almost invariably of timber, covered with boards or 'battens' forming the curved shape required for the soffit – i.e., intrados – of the arch. Centring was costly and the quantity of timber required could be reduced by the Roman method, but if enough timber was bought to make the centre of the full width of the arch, much time would be saved.

[14] Dates derived from Whitney, *Bridges*, checked against other sources and slightly amended.

[15] Date proposed by Johnson and Scott-Giles, *British Bridges*, 94, rather tentatively, but involvement of the Bishop of Durham, whose palace is nearby, seems undeniable.

[16] See Ruddock, 'Bridges and roads ...' and references quoted there.

[17] Home, *Old London Bridge*, and Becker, *Rochester Bridge*.

important paper and a book, she has cast serious doubt on the assumption by other authors that semi-religious 'bridgebuilding brotherhoods', or a single such body, operated through most of Europe.[18] Collective funding by residents of towns, districts, and sometimes religious communities, was the norm, and the way bridge-building projects were to be run was determined locally.

Boyer's other contribution, chapter 5, deals with the important technique of pile-driving which had undergone some development, but not very much, since its use by the Romans. A significant innovation in medieval bridges was the use of ribbed arches, probably resulting from the experience of ribbed vaulting in Gothic churches, but in bridge-building it represented a reversion to the Roman practice described above. Four or more parallel arch ribs were built, presumably one at a time on the same narrow centre, but now with use of mortar and with spaces between them. The gaps were bridged over with stone slabs to complete the arch, over which filling and road metal were placed. A further innovation, certainly Gothic in spirit, was that some arches were pointed. Most pointed arches were of short span, but spans of 16.5 metres were built in the Pont Valentré at Cahors (1308–55)[19] and one of 21.3 metres in the Brig o' Balgownie at Aberdeen (1320).[20]

A type of stone bridge which reached an amazing peak of development in the medieval era, but not in Europe, is the multi-span bridge of stone slabs or beams. At its simplest it amounts to no more than supporting a footway of stone slabs on stepping-stones set in a shallow stream. Such 'clapper bridges' are found in several regions of the British Isles, some of them at least two centuries old, but in Cornwall several bridges of granite slabs or beams spanning 2–3 metres carry modern motor traffic.[21] In Fukien province, China, however – a maritime province – some bridges of longer stone beams can still be seen, and records exist of viaducts built from the eleventh to the thirteenth century of many spans and stone beams up to a stated maximum of 20 metres. The largest stone beams would weigh up to 200 tonnes, but could presumably have been carried to their sites on floats or barges and set in place by the falling tide.[22]

---

[18] Boyer, 'The bridge-building brotherhoods', and *Medieval French Bridges.*

[19] Boyer, *Medieval French Bridges,* and de Dartein, *Études,* vol. 1, 7–14 and pls. 3–7.

[20] Johnson and Scott-Giles, *British Bridges* ... 432–3.

[21] One Cornish clapper bridge used by vehicles is at Bradford and another is the nearby Poley's Bridge, see C. Henderson and H. Coates, *Old Cornish Bridges and Streams* (London, 1928).

[22] The dimensions result from the extensive documentary researches of Needham's team, reported in *Science and Civilisation in China*, vol. 4, pt. 3, 152–8. Examples of the surviving bridges – of modest spans – are described and illustrated by Needham and by Fugl-Meyer, *Chinese Bridges.*

Renaissance to Seventeenth-Century Bridges

The ideals of the Romans as expressed by Vitruvius were a strong influence in Leone Battista Alberti's *Ten Books on Architecture*, which was first printed at Florence in 1485. Thereafter, knowledge of the design and construction of masonry bridges was never restricted to masons' lodges. Bridges were seen by the literate public as subjects for architectural design. Alberti's advice was about the practical form, dimensions and construction of stone bridges, but a century later in 1570 Andrea Palladio's *Four Books of Architecture* was published at Venice, with much longer treatment of rules of construction and also drawings of desirable architecture of bridges, both the general forms and the ornament. Many designs across Europe were based either on the drawings of Palladio's design for the Rialto in Venice, which had not been adopted, or on his drawing of the five-arch Roman bridge at Rimini which, with classical aedicules[23] on all the spandrels between the arches, was one of the more decorated of ancient Roman bridges.[24]

The effect of the Renaissance on bridge-building was naturally felt first in Italy. Calabi and Morachiello's study of the Rialto, a fairly late Renaissance bridge, in chapter 7, provides a more complete history of the construction of a famous bridge than any other study presented here. Though a fully chronological record is not attempted, the details of legal and administrative procedures for funding the project at various stages, resolution of disputes about acquisition of the necessary ground, and other administrative aspects, are described just as fully as the contracts for purchase of materials and the methods, labour and equipment used for all parts of the construction – including the very surprising expedient of supporting the centring for the arch on a *floating* pontoon. Unusually but logically, the last part of the chapter is devoted to description and comparison of the various interesting designs which had been offered, and from which the structure built and all its details were derived.

The Rialto was, of course, an extreme development of the medieval town bridge, as it offered no vehicular passage at all, but a street of shops for trade and financial dealings in the heart of the city, rising high over the Grand Canal to give clear passage to the water traffic. The considerable engineering problems which had been solved were those of a high arch of long span, and the making of foundations below the level of the water and on poor ground very close to existing buildings. For more of the story of the Rialto, and those of other ambitious bridges of the same period in Italy and France, the reader is referred to the writings of W.B. Parsons.[25]

---

[23] Aedicule: properly a shrine framed with two columns and an entablature and pediment, but used of any opening, such as a door or a niche, which is similarly framed,

[24] See Ruddock, *Arch Bridges,* especially chapters 1 and 8.

[25] Parsons, *Engineers and Engineering* ..., after scanning the Renaissance literature relevant to structural engineering, makes discrete studies of two Italian and four French arch bridges.

Of equal social significance and much more complex engineering functions, was the Pul-i-Khwājū bridge-dam at Isfahan, Iran, described by Luschey in chapter 8. As noted earlier, it had, and has, an intrinsic social function as a place of rest and meeting-place for travellers. Doubtless it was also a place of commerce to some degree, though without the intended concentration of business that characterised the Rialto and other town and city bridges in Europe, such as those of London, Paris and Florence. The Pul-i-Khwājū is an exotic among the subjects of study here, and as such it is the tip of an iceberg; were there studies available, the bridges of other Islamic countries, of China, Russia and possibly of the Indian sub-continent would have merited inclusion.[26]

Summerson's chapter 9, 'Berwick-on-Tweed Bridge', describes, from a good surviving archive, the construction of a large bridge during a period when few large bridges were being built in Britain. The Reformation's destruction of the wealth of the Catholic Church, and perhaps the loss of indulgences offered to benefactors of bridges, had left a gap in funding which was only overcome, for the bridge in question, by large grants from King James VI and I, who was making a political point about the union of his two kingdoms of Scotland and England by supporting the project.

Eighteenth-Century Bridges and Aqueducts

Writings about bridges underwent considerable change in the eighteenth century. For one thing, they began to provide the sort of technical information about construction that Boyer found to be absent from medieval French records. Alberti's and Palladio's books became well known, with translations printed in England in 1726 and 1715–16 respectively. Secondly, a new type of writer emerged, as mathematicians began to seek understanding of the stability of arches and to try therefrom to formulate methods for the rational design of arches. The first mathematical explanation of 'the arch', written by Philippe de la Hire in France in 1695, was, however, criticised in 1717 by the practical engineer Henri Gautier because it was expressed in an esoteric language, namely algebra, which neither tradesmen nor architects would understand.[27] Nevertheless, other mathematicians in several countries published similar theories, and a few went further and tried to compare their theories with the behaviour of real arches. For such comparisons, accurate measurements of the dimensions of the real arches were required and also reliable detailed descriptions of their movements and any collapses that had occurred. Some such data could be obtained from the experiences of architects

[26] For Chinese historic bridges, see Fugl-Meyer, *Chinese Bridges*, and Needham, *Science and Civilisation*, and for Iranian bridges Pope and Ackerman, *Survey of Persian Art*.

[27] La Hire, *Traité de Mécanique*, and Gautier, *Dissertation*, 6.

and builders, and it was occasionally also recorded in the writings of gentleman antiquarians.

The books of Alberti and Palladio were joined by others, most importantly the excellent practical treatises in French by Gautier and Belidor,[28] both used by leading British designers. However, the use of books cannot explain the most surprising fact about British bridges built in the eighteenth century, namely, that each of the first two large bridges built in Britain to the century's radically new standards was the work of a man who had no previous personal experience of bridge construction. The first was Charles Labelye, a new sort of professional in Great Britain who called himself 'engineer'. He was the named 'engineer' for both the design and construction of Westminster Bridge (1738–50), and he was obviously aware of all that Gautier had written, in Labelye's native language of course. The second was the Scotsman Robert Mylne, who performed the same functions for Blackfriars Bridge (1760–69), and had the benefit of Labelye's experience, accounts of which had been published in several pamphlets.[29] Mylne may also have read Belidor's book. No praise can be too high for their building of bridges of many large arches over the tidal reaches of the Thames – the building of a bridge in tidal water being inherently more difficult than above the tidal limit – and the confidence with which they used innovative methods of construction. These included the laying of foundations and building of the lowest courses of masonry of the piers in 'caissons'. The caisson was an open-topped wooden crate which sank as the masonry was built within it on to a bottom prepared by dredging, the crate being kept dry by pumping until the masonry was high enough to reach well above the level of low tide, when the sides would be removed and construction would continue between tides. Wooden centring of unprecedented size and strength was also devised, as were the special methods needed to permit its removal for re-use without damage when an arch had been completed.[30]

Unlike Labelye, Mylne called himself an architect and was known to others as such, but he accepted many commissions for engineering work as well. Of the group of men who formed the first professional association of civil engineers – the Society of Engineers, which began meeting in London in 1771 – only one, other than Mylne, could count bridge design and building as an important part of his practice. He was John Smeaton, who had been the favoured candidate for the Blackfriars Bridge commission for several years before Mylne's sudden arrival on the scene in 1759. Smeaton is more recognisable as father of the bridge engineers of the succeeding century than Mylne, because he looked to the developing science of mechanics to inform his decisions about the form and

[28] Gautier, *Traité des Ponts* (1714, 3rd edn. 1728), Belidor, *Architecture Hydraulique*.

[29] Labelye, *Description of Westminster Bridge*, and *idem*, A *Short Account of the Methods Made Use of in Laying the Foundations of the Piers of Westminster Bridge* (London, 1739).

[30] For details, see Ruddock, *Arch Bridges*.

economy of structures. Smeaton's bridge-building is represented here in chapters 10 and 13 and also studied in books by Skempton (1981) and myself (1979).[31]

By the end of the century the technical methods introduced by Labelye and Mylne at Westminster and Blackfriars Bridges had been further developed with better pumping and pile-driving, both already powered on some sites by steam-engines, and various uses were being made of iron bars both to reinforce traditional masonry and to accelerate its construction.[32]

A bridge whose story is equally famous and which was not the work of either engineer or architect is the single arch at Pontypridd in south Wales, completed in 1756. Its history is given here by chapters 11 and 12, the first by Williams and dealing solely, but very valuably, with printed and manuscript writings, and the second containing my analysis of early extant drawings of the bridge. Together these testify to the importance, for historical accuracy, of detailed documentary research. Jean Manco's chapter 14, on Pulteney Bridge, Bath, which was built between 1770 and 1774, provides similar detail of the story of a fairly late example of a bridge designed by a major architect, Robert Adam. It is a bridge as unique as Pontypridd but for different reasons, being built as a street over the river Avon and lined on both sides with retail shops. The placing of buildings on bridges was a notable anachronism by this date, but Adam was a free-thinking designer and also an active 'developer' as well as architect,[33] which doubtless appealed to his client and gives Manco's chapter value in the socio-economic study of bridges.

The rapid spread of canals required navigable aqueducts, the largest of which were of stone and carefully styled, but by engineers, not architects, the most notable examples being Robert Whitworth's Kelvin Aqueduct in Glasgow and John Rennie's Lune Aqueduct at Lancaster.[34] Another element of canal-building was the construction of large numbers of small bridges crossing the new waterways, all the bridges on an individual canal being commonly built to a standard design. Almost all were hump-backed and carefully shaped with curved wingwalls; they were sometimes of brick but also often of well-cut stone. They were made just wide and high enough for the passage of a barge and its towing horse, and complex shapes were devised for two special situations: (i) to permit the horse to cross from one bank to the other without unhitching it from the tow-rope, and (ii) for the roadway to cross 'on the skew', that is, at an oblique angle to the line of the canal.[35] Clever geometrical constructions were worked out for the

[31] Ibid., and A.W. Skempton, ed., *John Smeaton, FRS* (London, 1981).

[32] Ruddock, *Arch Bridges*.

[33] Ibid., 116–20.

[34] Ibid., 128–31.

[35] The first use of skew bridges was claimed by William Chapman at three sites on the Sallins-Naas branch of the Grand Canal of Ireland in 1787–89. See Chapman's article 'Oblique arches', Rees's *Cyclopaedia*, vol. 25 (London, 1813).

setting-out, centring and precise stone-cutting which was needed for both types of bridge and the geometry of the skew bridge became an important topic in handbooks of masonry design and civil engineering. Much larger skew bridges, often of several arches, were required in the nineteenth century, especially on the railways, and were designed by the same geometrical principles.[36]

Architects continued to be commissioned for the majority of large road bridges until 1790, after which most commissions went to engineers. The last two decades of the architects' dominance produced bridges of similar quality to Westminster and Blackfriars in many cities and towns of Britain, such as Richmond-on-Thames, Maidenhead, Shrewsbury, Worcester, Lancaster and Glasgow.[37] With the increase of overland traffic on the growing network of rural roads, small and large bridges were built according to need at small towns and in countryside locations. The majority of them were designed by the bridge surveyors who, in most of the counties of England, progressed from ad hoc employment as masons repairing bridges for the county to appointments as salaried officers responsible for upkeep of all their county's bridges.[38] A number of the larger bridges of the period, however, were promoted by the turnpike trusts which, working in parallel with the counties, built or improved important routes, being empowered by Parliament to recoup their capital by charging tolls for a limited period of time. Two large bridges so funded were Coldstream Bridge over the Tweed on the England/Scotland border, built between 1763 and 1767 to a design by John Smeaton, and Drygrange Bridge, also over the Tweed, designed and built between 1779 and 1780 by a mason/architect named Alexander Stevens.[39] Not until the beginning of the nineteenth century, however, did the government in London move to direct funding of large engineering developments, which included bridges, out of their general revenue.

Bridge-building by central government had started much earlier in France. The country's roads and bridges had been placed under a *Directeur-Général des Ponts et Chaussées* in 1715 and by 1747 there was a central school in Paris, the École des Ponts et Chaussées, for training of engineers to maintain and

---

[36] Buck, *Essay on Oblique Arches*, was perhaps the best publication on the subject, and written after the author had worked as resident engineer on the London and Birmingham Railway.

[37] See Ruddock, *Arch Bridges*, chapter 9.

[38] Much evidence of the work of bridge-building and maintenance is found in the quarter session records of county justices and housed in local archives throughout Great Britain. Outstanding examples are the 'Book of Bridges' dated 1752 containing survey drawings of all of the 116 bridges maintained by the West Riding of Yorkshire, and the 'Bridge Book' of *c.* 1805, with similar drawings of the 113 bridges maintained by the North Riding. The information from both books is summarised by Ruddock, *Arch Bridges*, 25–7, 114–16, 205–7. Chalklin, *English Counties* ..., 67–132, explains the administrative background and lists and discusses data from the records of many counties.

[39] Ruddock, *Arch Bridges*, and *idem.*, 'Hollow spandrels' (chapter 10).

extend the system of roads and bridges. Under the leadership of Jean-Rodolphe Perronet, first as head of the school and later as chief engineer of roads and bridges, a corps of engineers educated in both the science and the practice of bridge-building was created, and French masonry bridges reached standards of design and construction which have never been surpassed. New forms of all the parts of their bridges were tried successfully: the roadways and lines of parapets horizontal or nearly so, the arch curves either elliptical or of very low circular segments with span/rise ratios as large as 12, the span/pier thickness ratio stretched to 9 in Perronet's masterpiece, the Pont Neuilly near Paris, which could only be achieved by making separate centres for all five arches and keeping the arches supported until all were finished so that piers were never subject to unbalanced thrusts.[40] The unavailability of a study of the work of the eighteenth-century French engineers suitable for use in this collection is regretted. Their work is well recorded in French books both of their own and recent dates.[41] Similar centralisation of national transport infrastructure and training became established in other European countries, beginning with Spain in 1785.[42]

Nineteenth Century Bridges, Viaducts and Aqueducts

The British Parliament had voted modest grants towards the construction of individual bridges during the eighteenth century, responding to specific petitions, but at the turn of the century several commissions were created by Parliament to undertake more continuous involvement in projects which were seen as contributing to nation-wide development. A movement to rebuild London Bridge emerged from proposals for improvement of the Port of London and brought into consideration a number of designs for a new bridge,[43] but the desired end was not achieved until the late 1820s. A commission to consider the provision of roads and bridges in the Highlands of Scotland moved much more rapidly from a first report by Thomas Telford in 1803 to the formation of a management team and contracts for actual road- and bridge-building. A long sequence of contracts was awarded through the years 1805 to 1826. Telford was engineer for the whole

---

[40] Perronet, *Description des Projets*. From that source the French innovations were copied in British encyclopaedias: see, for instance, Telford, 'Bridge: History', and 'Practice', also Ruddock, *Arch Bridges*, 175–7. On the thrust of arches, see note 47 below.

[41] See, for instance, Belidor, Gauthey, Perronet, de Dartein and Mesqui, in the Bibliography.

[42] M. Chrimes, *Civil Engineering, 1839–1889: A Photographic History* (London, 1991), 18–19.

[43] All the designs, some of masonry and others of iron, were printed with the reports of the Select Committee on the Improvement of the Port of London, in *Commons Reports, 1715–1801* (1803), 14, 461–635. See also Ruddock, *Arch Bridges*, 155–8.

project and designed many substantial bridges, while hundreds of smaller ones were built within the scope of road contracts.[44]

Later, Telford was appointed engineer for major road improvements and new bridges along the whole route from London to Holyhead, which had political importance as the link between London and the satellite administration of Ireland in Dublin. In the years after the battle of Waterloo, when serious hardship of disbanded soldiers and others prompted more spending by government on capital works to create employment, Telford became engineering adviser to the Exchequer Loan Commissioners, who dispensed loans to various major bridge projects.[45] As a sign of the times, the longest spans required on the Holyhead road and in projects supported by the Loan Commissioners were bridged by iron structures, but many modest-sized masonry bridges were still being built, and one which set a milestone in bridge history. Grosvenor Bridge at Chester, built between 1827 and 1833, has the longest-spanning stone arch ever built in Great Britain, 200 feet in span. It was designed, at least in the architectural sense, by the elderly architect Thomas Harrison, but the structural design was validated by George Rennie and its construction was supervised by Jesse Hartley, both engineers.[46] There was little further involvement of architects in masonry bridge design in subsequent years.

British engineers of the nineteenth century designed and built a huge number of masonry bridges and multispan viaducts for railway companies. Many were built of brick, but many also of stone; some used both materials, either brick for arches and stone for walls or some other combination. Surprisingly, although descriptions of many railways and at least some of their viaducts were published in engineering journals of their own time, they have not attracted the attention of historians of civil engineering in this century, and the making of British railway structures is therefore represented in this collection by only two chapters: chapter 14 by Owen on I.K. Brunel's arch bridges and chapter 19 by Brooke on the collapse of the Barentin Viaduct in France, a viaduct of which both the designer and contractor were Englishmen. The Barentin story is a short classic in the genre of bridge collapse tales. The fall of twenty-eight tall piers in what is today called a 'progressive collapse'[47] evinced a quick acceptance

---

[44] The fullest modern history of the Commission's work is A.R.B. Haldane, *New Ways through the Glens* (Edinburgh, 1962), quoting their many printed reports and extensive manuscript papers in Edinburgh and London. See also Sir Alexander Gibb, *The Story of Telford* (London, 1935), and L.T.C. Rolt, *Thomas Telford* (London, 1958).

[45] Telford's work on the Holyhead road and for the Exchequer Loan Commissioners is dealt with by Gibb and Rolt, see note 44.

[46] For Grosvenor Bridge, see Ruddock, *Arch Bridges*, 186–8, and Woodward, 'The Brunels ...'.

[47] The term 'progressive collapse' describes a structural accident in which the initial collapse of one element of the structure removes necessary support from one or more other elements which

of blame by the contractor, equally quick decision about design changes, and an almost immediate start on rebuilding – all of which would seem incredible in any but the Victorian era, but they are parallelled by the phlegmatic behaviour in the face of apparent threats of disaster and loss of reputation reported of I.K. Brunel by Owen. Mention of the very extensive export of British engineering skill and of British capital during Victoria's reign to build railways in European countries and in South America, India and other Asian countries, and Australia, only emphasises the great dearth of historical study of nineteenth-century viaduct-building.[48]

In the United States, where masonry bridges were contemplated only in the very early years of railway development, and used for only a few prestigious viaducts, there is interest in and care for most of the few examples.[49] In chapter 18 Galvin describes the Canton Viaduct on the Boston and Providence Railroad, which is of quite unique design. It looks like a normal sequence of high arches, but its main load-carrying members are actually two continuous longitudinal walls standing the whole height of the structure from ground to rail-bed. A second study, of a much later series of stone bridges, is chapter 17 by Lankton, describing the 'Valley crossings on the Old Croton Aqueduct'. All these structures are conventionally arched and not unlike bridges built for similar purposes in Victorian Britain. The skills and techniques in use seem not to have changed from those imported with the emigrants of one or two centuries earlier, although the manner of commissioning of public works had changed a lot.

So much has been written of historic wooden and iron bridges in the United States that the use of masonry for bridges of ordinary size and functions is neglected by almost all bridge historians. A string of well-built arch bridges of local sandstone were, however, built in the first road development to be funded directly by the US Congress – what became known as the National Road. It was the subject of very careful survey and planning and was constructed from Cumberland, Maryland, to Wheeling on the Ohio River, 130 miles long over the Alleghany Mountains, between 1811 and 1820. Extensions in the 1820s and early 1830s took the road westward to Vandalia in Ohio State and more stone bridges

then collapse. In a viaduct each individual pier experiences equal and opposite horizontal thrusts from the two arches which rest on it, and the removal of one of those thrusts may cause the thrust from the other side to tilt the pier, causing collapse of the second arch and exposing the next pier to unbalanced thrust; repetitions of the tilting and collapse then ensue, pier by pier, to the end of the structure.

[48] M. Chrimes, *Civil Engineering, 1839–1889* (see note 42), in its later chapters, and A. Burton, *The Railway Empire* (London, 1994) sketch the rapid spread of British railway building in the Victorian years, but with only occasional specific reference to masonry viaducts.

[49] Plowden, *Bridges: The Spans of North America*, notes that only four masonry viaducts were built on American railroads before 1850. He prints beautiful photographs of those and some earlier canal aqueducts, with good historical narrative.

were built there.[50] Moreover, there are thousands of masonry bridges, most but not all of small size, in the eastern states and the Mid-West, and they are now receiving some attention from at least a few local historians.[51]

Bibliography

Works Covering Several Centuries or All Centuries

General and National Histories

Casado, Carlos Fernandez, *Historia del Puente en España* (Madrid, n.d.).

Civil Engineering Heritage. A series of handbooks to surviving historic works of civil engineering in Great Britain and Ireland. Many masonry bridges are included. At the time of writing, five regional volumes have been published and two more are to follow. Published by Thomas Telford Ltd. for the Institution of Civil Engineers.

de Dartein, F., *Études sur les ponts en pierre remarquables par leur décoration antérieurs au XIXe siècle*, 4 vols. (Paris, 1907–12).

de Maré, Eric, *The Bridges of Britain* (London, 1954).

Emerson, W., and G. Gromort, *Old Bridges of France* (New York, 1925).

Fugl-Meyer, H., *Chinese Bridges* (Shanghai, Hong Kong and Singapore, 1937).

Hopkins, H.J., *A Span of Bridges* (Newton Abbot, 1970).

Jervoise, E., *The Ancient Bridges of the South of England* (London, 1930).

—, *The Ancient Bridges of the North of England* (London, 1931).

—, *The Ancient Bridges of Mid and Eastern England* (London, 1932).

—, *The Ancient Bridges of Wales and Western England* (London, 1936).

Johnson, S.M., and C.W. Scott-Giles, eds., *British Bridges: An Illustrated Technical and Historical Record* (London, 1933).

Mesqui, Jean, 'Les ponts en maçonnerie jusqu'à la fin du XVIIIe siècle', in G. Grattesat, ed., *Ponts de France* (Paris, 1982), 39–96.

Needham, Joseph et al., *Science and Civilisation in China*, vol. 4, pt. 3 (Cambridge, 1971), 145–210.

---

[50] See Harley J. McKee, 'Original bridges on the National Road in Eastern Ohio', *Ohio History*, 81, no. 2 (Spring 1972) 131–9 and Billie Joe Peyton, 'To make the crooked ways straight and the rough ways smooth: building the Cumberland Road, our nation's first interstate highway', *Proceedings of an International Conference on Historic Bridges...* (Morgantown, West Virginia, 1999), 1–13.

[51] See, for instance, Thomas E. Boothby, Cecilia J. Rusnak and Linda Black Weber, 'Hunterdon County's [Pennsylvania] stone-arch bridges: history, technology and preservation', *APT Bulletin (Journal of Preservation Technology)*, 30 (1999), nos. 2–3, 26–31, one of several papers by these authors.

O'Keeffe, Peter, and Tom Simington, *Irish Stone Bridges* (Blackrock, 1991).
Plowden, David, *Bridges: the Spans of North America* (New York, 1974), 9–32.
Pope, A.U., and P. Ackerman, eds., *Survey of Persian Art*, 7 vols. (London, 1938–58), 570–72 and 1226–41.
Robins, F.W., *The Story of the Bridge* (Birmingham, n.d., 1940s).
Séjourné, Paul, *Grandes voûtes,* 6 vols. (Bourges, 1913–16).
Smith, H. Shirley, *The World's Great Bridges* (revised edn., London, 1964).
Troyano, Leonardo Fernández, *Tierra Sobre el Agua: Visión histórica universal de los puentes* (Madrid, 1999), 110–164.
Whitney, C.S., *Bridges: A Study in their Art, Science and Evolution* (New York, 1929).

Histories of Single Bridges

Fraser, G.M., *The Bridge of Dee* (Aberdeen, 1913).
Home, Gordon, *Old London Bridge* (London, 1931).
Morris, James A., *The Brig of Ayr and Something of its Story* (Ayr, 1912).
Thomson, R. (signed 'An antiquary'), *Chronicles of London Bridge* (London, 1827).

Works Concerned with Specific Periods of Time

Roman

Frontinus, S.J., *De aquaeductibus urbis Romae* (c.100 A.D.). Translation by C. Herschel as *The Two Books on the Water Supply of the City of Rome* (Boston, Mass., 1899).
O'Connor, Colin, *Roman Bridges* (Cambridge, 1993).

Medieval

Becker, M.Janet, *Rochester Bridge, 1387–1856: A History of its Early Years* (London, 1930).
Brierley, J., 'The medieval Exe Bridge', *Proceedings of the Institution of Civil Engineers*, 66, pt. 1 (1979), 127–39.
Boyer, Marjorie Nice, 'The bridge-building brotherhoods,' *Speculum*, 39, no.4 (1964), 635–50.
—, 'Moving ahead with the fifteenth century: new ideas in bridge construction at Orleans', *History of Technology*, 6 (1971), 1–22.
—, *Medieval French Bridges: A History* (Cambridge, Mass., 1976).

—, 'A fourteenth century pile driver: the *engin* of the bridge of Orleans', *History of Technology*, 9 (1984), 31–41.

—, 'Working at the bridge site in late medieval France', in *Artistes, artisans et production artistique au Moyen Age (Actes du Colloque édités par X. Barral*, vol. 1, *Les Hommes* (Paris, 1987).

Renaissance to c. *1700*

Alberti, L.B., *De re aedificatoria* (Florence, 1485). English translation by J. Leoni as *The Architecture of Leone Battista Alberti* (1726), edited by J. Rykwert as *Ten Books on Architecture by Leone Battista Alberti* (London, 1955).

Palladio, Andrea, *I quattro libri dell' archittetura* (Venice, 1570). English translation by Isaac Ware as *Four Books of Architecture* (1738), republished in facsimile (New York, 1965).

Parsons, W.B., *Engineers and Engineering in the Renaissance* (revised edn., Cambridge, Mass., 1968).

Ruddock, Ted, 'Bridges and roads in Scotland, 1400–1750', in A. Fenton and G. Stell, *Loads and Roads in Scotland and Beyond* (Edinburgh, 1984).

Post-1700

Belidor, B.F. de, *Architecture hydraulique*, 4 vols., especially vol. 4 (Paris, 1737–52), 441–64.

Brees, S.C., *Railway Practice: A Collection of Working Plans ... of the Most Celebrated Engineers,* 2 vols. (2nd edn., London, 1838).

Carson, P.M., 'Provision and administration of bridges over the lower Thames, 1701–1801, with special reference to Westminster and Blackfriars'. Thesis submitted for the degree of Master of Arts: University of London (1954).

Chalklin, Christopher, *English Counties and Public Building, 1650–1830* (London, 1998), 76–132.

Day, Tom, 'Did Telford rely, in Northern Scotland, on vigilant inspectors or competent contractors?', *Construction History*, 13 (1997), 3–15.

Gauthey, E.-M., *Traité de la construction des ponts*, 2 vols. (Paris, 1808 and 1813).

Gautier, H., *Traité des ponts*, (3rd edn., Paris, 1728).

Labelye, Charles, *A Description of Westminster Bridge* (London, 1751).

Perronet, J.-R., *Description des projets et de la construction des Ponts de Neuilly ....et autres...*, 2 vols. (Paris, 1782–83).

Revely, Georges, 'Les ponts en maçonnerie au XIXe siècle', in G. Grattesat, ed., *Ponts de France* (Paris, 1982), 97–116.

Rickman, J., ed., *The Life of Thomas Telford, Civil Engineer, written by Himself*, 1 vol. text and 'Atlas' of plates (London, 1838).

Ruddock, E.C., 'The building of North Bridge, Edinburgh, 1763–75', *Transactions of the Newcomen Society*, 47 (1974–76), 9–33.

Ruddock, Ted, *Arch Bridges and their Builders, 1735–1835* (Cambridge, 1979).

—, 'Bridges', in A.W. Skempton, ed., *John Smeaton, FRS* (London, 1981).

Semple, George, *A Treatise on Building in Water* (Dublin, 1776).

Sheppard, Francis, 'Henley Bridge and its architect', *Architectural History*, 27 (1984), 320–30.

Smeaton, J., *Reports of the late John Smeaton, FRS*, 3 vols. (London, 1812).

Smith, Martin, *British Railway Bridges and Viaducts* (Shepperton,1994).

Telford, Thomas, 'Bridge: History,' and 'Practice', *The Edinburgh Encyclopaedia,* vol. 4, 479–89 and 519–44 (Edinburgh, 1830). Volume 4 was issued in 1812.

Walker, R.J.B., *Old Westminster Bridge: The Bridge of Fools* (Newton Abbot, 1979).

Weale, J., ed., *The Architecture of Bridges*, 2 vols. text, 3 vols. plates (London, 1843).

Woodward, G., 'The Brunels and the Grosvenor Bridge, Chester', *Transactions of the Newcomen Society*, 69, pt. 1 (1997–98), 129–45.

Theoretical Studies

Belidor, B.F. de, *La science des ingénieurs*, (6 vols.), vol. 2 (Paris, 1729).

Buck, G.W., *A Practical and Theoretical Essay on Oblique Arches* (London, 1839).

Emerson, W., *The Principles of Mechanics* (2nd edn., London, 1758).

Gautier, H., *Dissertation sur les culées, voussoirs, piles et poussées des ponts* (Paris, 1717).

Heyman, J., and C.J. Padfield, 'Two masonry bridges: I. Clare College Bridge', *Proceedings of the Institution of Civil Engineers*, vol. 52, pt. 1 (1972), 305–18.

Heyman, Jacques, *The Masonry Arch* (Chichester, 1982).

Hutton, Charles, *The Principles of Bridges* (Newcastle-upon-Tyne, 1772). New edn. extended, with other papers on bridges, in *Tracts on Mathematical and Philosophical Subjects*, (3 vols.), vol. 1 (London, 1812), 1–166.

La Hire, P. de, *Traité de Mécanique* (Paris, 1695).

Nimmo, Alexander, 'Bridge: Theory', *The Edinburgh Enclyclopaedia*, vol. 4 (Edinburgh, 1830), 489–519. Volume 4 was issued in 1812.

Robison, J., 'Arch', in *Supplement to the Encyclopaedia Britannica*, 3rd edn., vol. 1 (Edinburgh, 1801), 14–31.

1
Testing times for arches

B. Harvey

ARCHED BRIDGES are naturally elegant structures. Preservation orders protect many arched bridges, and very few defile their surroundings. Of the hundred thousand bridges that connect roads in Britain, about seventy thousand are arches. Thousands more arches carry railway lines and canals. Some of the biggest arches—Thomas Telford's Over Bridge at Gloucester and the railway viaducts at Ballochmyle and Ribblehead, for example—inspire awe. The smallest may span only 2 or 3 metres across a stream, but even they have considerable value.

It might cost a hundred thousand pounds to replace a modest bridge on a minor road, so we can value the arches that carry Britain's roads at £7000 million or more. Some bridges are more valuable than others. Ribblehead Viaduct would cost about £4 million to replace, according to a recent estimate. Other bridges are important because they carry a disproportionate amount of traffic, or because they are tourist attractions in their own right. Peter Colechurch's Old London Bridge, built around 1176, earned so much money from tolls before it was replaced in 1831 that the invested capital has paid for all four bridges in the City—London, Tower, Southwark and Blackfriars—since then. The bridge across the River Tay in Aberfeldy, designed by William Adam and built for General George Wade in 1733, draws crowds all summer and brings income to the local community.

If a bridge closes, or new regulations restrict the weight of the vehicles using it, local businesses have to live with the commercial strain. On the other hand, people living near a bridge often have a vested interest in weight restrictions, which they might disguise as concern for the bridge. The engineers, employed by the council to care for old bridges and design new ones, must consider politics and conservation as well as structural soundness when assessing whether a bridge is safe to carry traffic.

Money for maintaining arches is hard to come by, for the simple reason that they rarely fall down. People often regard them as truly fixed assets. On the whole engineers subscribe to this popular view, though recent changes in the weights of vehicles have encouraged them to think more deeply. Until quite a short time ago, the weight of the vehicles passing over a bridge was negligible compared with the weight of the bridge itself. Now, however, vehicles with five axles weighing 38 tonnes gross can go anywhere on Britain's roads. Some of these vehicles are greatly overloaded, yet they use bridges that were built to carry horses.

Every bridge is regularly inspected: its strength is reassessed at least every six years. Engineers base that assessment on a mixture of science and judgment. The complexity of the problem limits the value of the science, however, as no two arches are alike. The shape may be semicircular, segmental, elliptical or even pointed. The arch might be of dressed stone voussoirs (wedge-shaped blocks), bricks or rough stone. The road may be high on a bank of earth over the arch, or it may rest directly on the back of the ring. The road may be level, sloping or humped. The space between the arch and the road—the spandrel—may be filled with earth, stone or concrete. Some spandrels are hollow.

The science of arches developed slowly. The Romans seem to have believed that force followed the ring of the arch. (The ring comprises the blocks that line the underside of the arch.) The Romans thought, as a result, that semicircular arches exerted only vertical forces on their supports. In fact, the concrete that filled the spandrels in Roman arches carried the neglected horizontal forces.

Throughout the Renaissance, engineers and architects struggled to understand what happened to the forces inside an arch. In 1675, Robert Hooke hit upon the idea that an arch was simply the inverse of a loaded chain hanging across the same span (Box 1). The line of the chain represents the line of thrust—a line joining the points of greatest stress in each block of the arch's ring (Box 2).

Hooke was an inspired thinker and an experimentalist, but he lived in an age dominated by Isaac Newton and theory. When Hooke hinted at his ideas on arches, the Secretary of the Royal Society wanted to give Newton or Christopher Wren the credit. Hooke finally presented his theory, which he had proved by experiment, as a Latin anagram at the end of a paper describing helioscopes. A translation of the solution to the anagram reads: "As hangs a flexible cable, so, inverted, stand the touching pieces of an arch." Hooke could claim precedence, but the dispute dragged on and, although many scientists knew of the solution, he never formally published it.

Unlike many architects and engineers in the 17th century, Thomas Young clearly understood all the implications when he wrote his article "Bridge" for the *Encyclopaedia Britannica*

TESTING TIMES FOR ARCHES 3

1: Hooke's experimental approach to the arch

Brian Shuel

HOOKE used two types of model to show how arches work—flexible cables, and rods joined together with pins. Many different shapes of cable can support the same loads in the same positions. In each case, as in the three examples on the far left, the tension in the cable is different. If you displace the cable, it will swing back to its initial position. This form of stability is similar to that of a ball sitting in a dish: it will always roll back to the centre if you move it.

If you use jointed rods to form a shape similar to that of a cable supporting different weights, and invert it, it will stand. Unlike the cable, the structure is unstable (although friction in the joints of Hooke's models allowed these structures to stand up). Once displaced, the pinned rods will collapse, just as a ball will roll off an upturned dish.

The line of the chain or the cable in Hooke's models represents the line of thrust. In a real arch built with deep voussoirs (wedge-shaped blocks), many lines of thrust will fit, as the diagram on the immediate left shows. The arch is thus overstable: it will take a large force to displace it. Once you remove that force, the arch will snap back into position. The structure's stability is now similar to that of a ball sitting in the bottom of a conical cup.

An arch with a much thinner ring allows only one line of thrust to fit within the arch and support the loads. The arch is still stable—but only just. □

2: How an arched bridge responds to stress

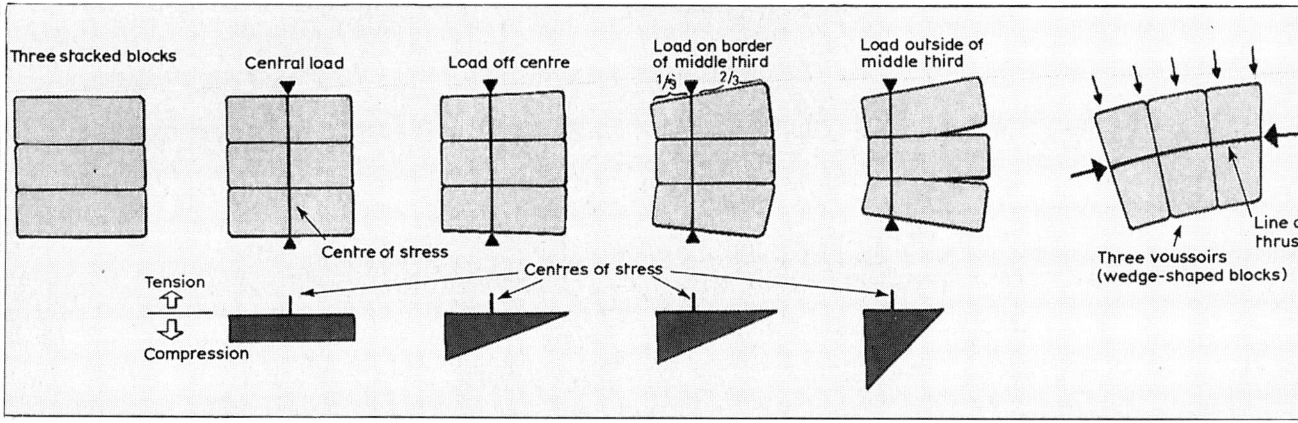

Load affects the distribution of stress. Cracks form between the blocks once the centre of stress moves outside the blocks' middle third

STONE deforms when it is under stress. The diagrams show three blocks of stone, stacked together without mortar. Stress (pressure) develops between the stones to resist any applied load. The centre of stress—the point of greatest stress—is always directly in line with the load. As the stress increases, it compresses the stone. There can be no tensile stress because the stones would simply move apart.

If the load moves away from the centre of the blocks, the stress increases at the edge towards which the load has moved, allowing the centre of stress to stay in line with the load. When the load is one-third of the way across the blocks, the stress falls to zero at the edge farthest from the load. The maximum stress is then double what it was when the load was at the centre of the blocks.

Moving the load further in the same direction causes cracks to open. The distribution of stress is still triangular, and the centre of stress is still at the point at the base of the triangle. Where the stress is zero, the blocks are not squashed, so the sections of the blocks between the cracks are the same shape as they were originally.

In a real arch, tapered blocks, called voussoirs, often make up its ring. The diagram above on the far right shows the forces imposed on three voussoirs in the ring of an arch.

The forces in the ring represent the effect of stress on the sloping faces of the stones. Other forces, applied on the extrados (the longer curved edge of the ring), cause the force to turn through an angle as it passes through the blocks. In this case, engineers call the line through the centres of stress the line of thrust.

The diagram below shows the line of thrust through a real arch at the point of collapse. Wherever the line passes outside the middle third of the voussoirs, cracks should develop on the unloaded face. Mechanism analysis (Box 3) assumes that one wide crack will open at each of four points. In practice, many narrow cracks will often appear, although they may be difficult to see. □

TESTING TIMES FOR ARCHES 5

Mark Entwisle

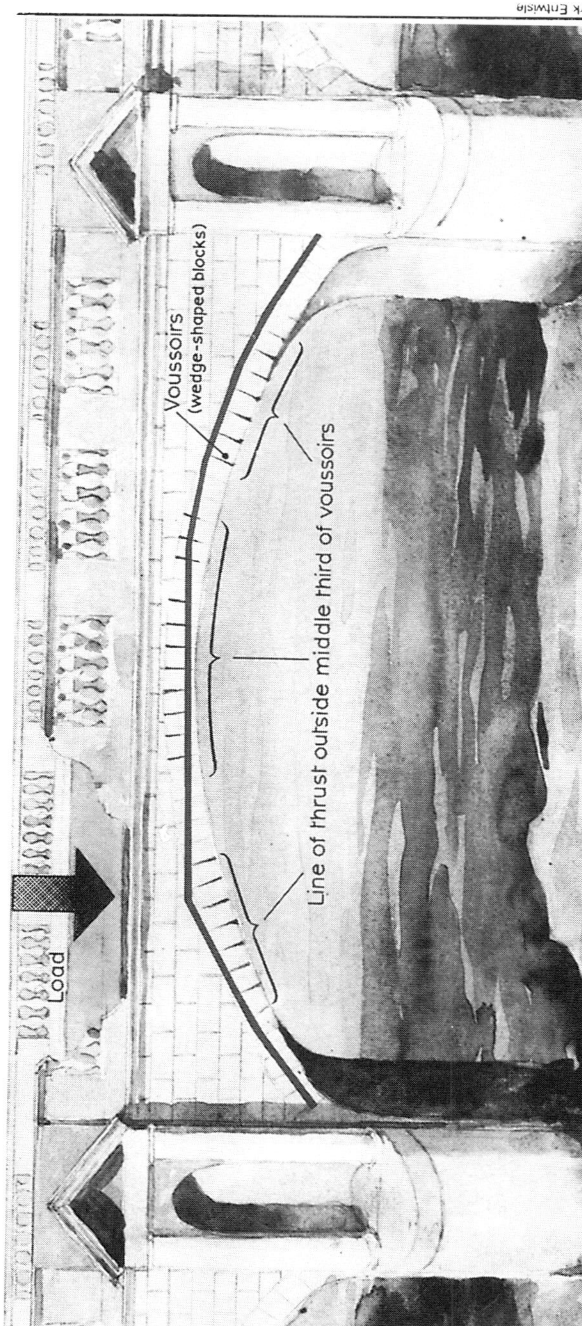

The typical line of thrust in a real bridge at the point of collapse. The loads on the bridge include the arch's own weight, pressure from the material filling the spandrels and the load from the vehicle's wheels. Loads distributed over a wide area cause the line of thrust to turn smoothly. Concentrated loads cause it to turn sharply. If you increase the load any further, the line of thrust will move outside the arch's ring, and the bridge will fall down

in 1817. He explained that a very thin arch will stand up if it is exactly the right shape for the loads imposed on it. Deepening the blocks which make up the ring of the arch allows the arch to withstand the complex pattern of loads produced as a vehicle crosses the bridge. The ring must be wide enough to accommodate all the different lines of thrust that result.

In France, in 1730, Pierre Couplet studied the way in which an arch collapses when it is overloaded. He found that models folded up with very little deformation of the arch between the "folds". It is much easier to calculate the load at which collapse will occur (Box 3) than it is to compute the line of thrust generated by a lesser load—a "safe working load", for example—but engineers did not appreciate the value of this approach until recently. They were concerned with how things stood up, not with how they fell down.

Throughout the 19th century, many practising engineers developed methods for computing the line of thrust of an arch. Most of them realised that it was sufficient to find any line of thrust which would support the required loads. However, the engineers still pursued a method that would provide them with a single solution to the problem.

In 1879, Alberto Castigliano, an engineer in Turin, Italy, produced such a method. He calculated how the arch would deform under a load. He could then work out the pattern of stress through the ring of an arch and thus the line of thrust. The calculation was complex and tedious, but it did lead to a single solution. Unfortunately, his analysis considered only the arch and assumed that the supporting abutments remained rigid. In fact, small movements of the supports produce large changes in the distribution of stress within an arch. As most foundations move, engineers now view Castigliano's results with caution.

Towards the end of the 19th century, engineers in France and Austria resorted to testing large-scale models to help them to design arches with longer spans than those that already existed. This method of proving designs is good, provided that the scale of the model is large enough, but it is very expensive. Some of the Austrian models had a span of 23 metres and a rise of only 4·6 metres. These arches were so strong that only explosives would induce them to collapse. In France, engineers built a model with a span of 36 metres and a rise of only 2·1 metres. That arch proved equally resilient. The French never built the intended bridge, however, which would have had a span of 120 metres. The bridge with the longest span ever constructed, some 90 metres, stands in Plauen, in the German Democratic Republic.

The advent of cheap steel and concrete meant that the

TESTING TIMES FOR ARCHES 7

3: Mechanism analysis—or how bridges collapse

IF YOU overload a model arch, it collapses by folding up. Individual voussoirs—the wedge-shaped blocks that form the ring of the arch—deform negligibly because the stresses on them are so low (Box 2). But cracks form at four points, alternately on the inner and outer curves of the arch's ring—the intrados and extrados. We can regard the points of contact between the sections as hinges.

If we slowly increase the load until the fourth crack just forms, the bridge reaches a state of unstable equilibrium, almost collapsing. The line of thrust passes through the root of each hinge. Each section of the arch A, B and C is just in balance, rotating neither to the left nor to the right.

The turning effect of the forces acting on each section must then be in balance, so we can write an equation describing that balance for each section, or for combinations of sections. Thus at crack 3, the weight of section C and the horizontal force at crack 4 are trying to turn section C clockwise, while the vertical force at crack 4 is trying to turn this section anticlockwise. At hinge 2, the weights of sections B and C, plus the horizontal force, all act one way, resisted by the vertical force. At hinge 1, the weight of section A and the load are also involved.

We can, of course, measure the dimensions of the arch and calculate the weights of each section of arch, so we can form three simultaneous equations with three unknowns. Solving these equations gives us approximate values for the horizontal and vertical forces and for the load which will just cause the arch to collapse. The values are only approximate because we have had to estimate the position of the hinges.

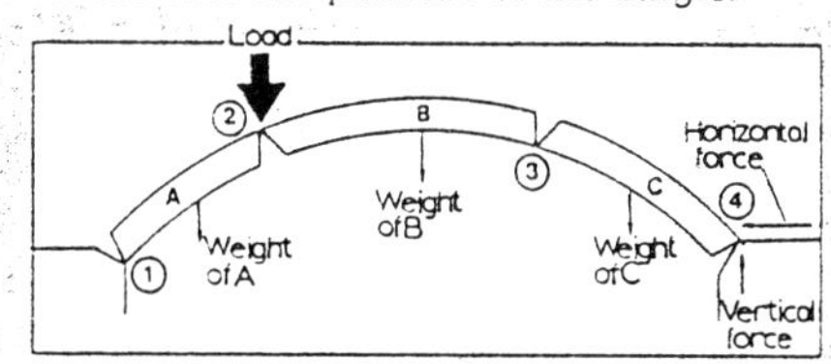

We can represent the approximate values for the horizontal and vertical forces as two sides of a triangle whose lengths correspond to the ratio between the two values. The slope of the triangle's hypotenuse gives us the slope of the line of thrust at hinge 4. We can plot this line back through the arch: if our calculations are wrong, the line of thrust will pass outside the arch. In this case, we choose new positions for the hinges and repeat the process until the line of thrust stays within the ring. We then use the new values for the horizontal and vertical forces to give us a more accurate value for the load that will cause the bridge to collapse.

Real arches also contain filling to bring the road up to the required level. The weight of the filling acts downwards like the weight of the arch itself, but the filling can also push sideways on the arch. The amount of sideways force is difficult to assess, but neglecting this effect nearly always leads to a low result for the calculated load at which collapse will occur. □

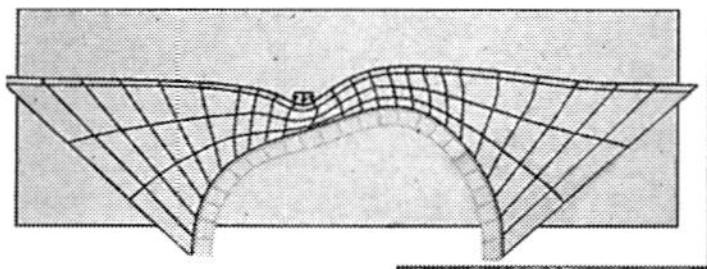

This semicircular arch at Bridgehouse, near Bargower, failed (its masonry crushed) at a higher load than computerised analysis (above) predicted. Its filling of stone made it stronger

popularity of arched bridges waned early this century. However, interest in arches—particularly in how to maintain them—began to revive in the 1930s. Although most vehicles were still relatively light, military appliances were rapidly increasing in weight and the army needed to be certain that a bridge would carry them. The problem of how to assess existing bridges presented itself for the first time. Alfred Pippard and his research assistant, Letitia Chitty, began a new study of models in 1936. They cooperated with Norman Davey of the Building Research Station, near Watford, on some full-scale tests. Their efforts resulted in the Military Engineering Experimental Establishment's (MEXE) method of assessment, which, modified by the Ministry of Transport, is still in use today.

Pippard used Castigliano's method to calculate the value of a load at a central point which should cause cracking. This load would take the line of thrust just outside the middle third of the depth of the ring of the arch (Box 2). Yet when Davey tested the bridges, he could see no cracks even at much higher loads. Indeed, there were still no cracks at the load that Pippard calculated should take the line of thrust outside the middle half of the ring. In theory, this load should induce cracks through one-quarter of the ring's depth. The lack of cracks in the real arches probably resulted partly from the effect of horizontal forces from the material filling the spandrel, which Pippard ignored, and partly from the effect of the mortar joints in the arch. The theory predicted cracks quite accurately in the model bridge, which was made of closely fitted steel blocks and bore only vertical loads. Eventually, Pippard's team decided to set the maximum load that the

TESTING TIMES FOR ARCHES 9

arch should carry on the basis of how much compressive stress they calculated the blocks of the ring could stand, and ignoring the implied cracking.

Once Pippard's team had established suitable limits, it was a straightforward matter to calculate the capacity of any bridge. The army, however, needed to produce answers in minutes, not hours. MEXE produced a chart based on parabolic arches with a rise of a quarter of the span. This chart, called a nomogram, allows someone to read off, for a bridge of a given span and construction depth, a provisional load for each of two axles which the arch could safely carry. Most bridges have circular curves, making it necessary to apply a correction factor for the shape. The engineer also has to modify the value that the nomogram supplies to take account of other factors such as different span-to-rise ratios and building materials, and the general condition of the arch. It is possible to measure or calculate some of these factors; others rely on the judgment of the engineer. In any case, the method is useful only for spans of between 1·5 and 18 metres, as the tests that confirmed it were all on bridges of this size.

Some bridges carried heavy loads for years before engineers assessed their strength and imposed weight limits on them. It is certain that the MEXE method underestimates the capacity of most bridges, but the people who developed it did, after all, intend it to be safe. Recent increases in the weights of vehicles permitted on bridges prompted a survey of arched bridges. The results suggested that 12 per cent of the sample were not strong enough. Many of these arches, however, were already carrying the new loads without apparent distress. Restrictions on the use of bridges that appear to be safe are both politically and economically undesirable. We need a more generous method of assessment. The two most useful approaches are Couplet's mechanism method and finite element analysis.

Professor Jacques Heyman of Cambridge University revived Couplet's work in 1980. He combined many of the old theories on arches with the latest thinking on structural safety to produce a method, called mechanism analysis, based on simple calculation. In his calculations, he used the true shape of the arch, but took account of only half of the ring's true thickness. He applied the loads from the bridge's own weight to this theoretical arch, and then calculated the additional axle loads which would cause it to collapse. Drawing the line of thrust provides a check on the calculations (Box 3). As the line of thrust supports the true loads on the bridge, we can assume that it lies in the middle half of the actual ring. Pippard discarded this assumption—the rule of the middle half—as too onerous, but the factors in the

MEXE method which allow for different shapes of arch are certainly fixed with caution. Mechanism analysis often indicates that bridges are stronger than the MEXE method suggests. It can also cope with strange-shaped arches and large spans, both of which the MEXE approach specifically excludes.

Like Pippard, Heyman chose to ignore the effect of horizontal forces on the arch. To do so seems to be safe, and to include them takes the method beyond calculation by hand. However, ignoring these forces means unnecessary weight restrictions on some high-rise arches. It is possible to get round this problem by using a microcomputer to carry out the calculations. A program written at Dundee University allows for horizontal forces and stresses. It allows the computer to complete a mechanism analysis and plot the line of thrust on the screen in three seconds.

The improvement in the apparent capacity of bridges that stems from Heyman's method is quite small, so the Transport and Road Research Laboratory in Berkshire and the Science and Engineering Research Council are jointly sponsoring a large research programme. There are plans for 10 full-scale tests; four have already been carried out. Engineers will use the results of these experiments, and of tests on models at Edinburgh and Dundee universities and Bolton Institute of Higher Education, to calibrate analytical work at the Transport and Road Research Laboratory and Dundee.

The first of the full-scale tests involved a shallow segmental arch with a span of 18·3 metres, at Bridgemill, near Girvan in Strathclyde. A team of researchers from Edinburgh University, directed by Professor Arnold Hendry, carried out the test for the Transport and Road Research Laboratory. The bridge did not collapse because the jacks that the researchers used to exert force on the bridge ran out of movement. Cracks formed as predicted by the mechanism analysis, but the maximum load was much less than that calculated. As the load increased, the arch bent into a smooth hollow under it, rather like the exaggerated curve shown in the computerised mesh diagram on page 57.

We need a much more complex analysis in order to explain this behaviour. In addition to a mechanism analysis, Mike Crisfield at the Transport and Road Research Laboratory is developing a computer program which analyses the bridge by considering the deformation of small parts of it. Crisfield divides the arch into about 20 parts, called finite elements to distinguish them from the infinitesimal elements used in integration. Other elements represent the filling. Each element must model the way in which the material of the real

TESTING TIMES FOR ARCHES 11

At Preston upon the Weald Moors, crushing of the stone again caused the bridge to fail. This bridge would have carried very little load without the help of its stone backing

Bill Harvey

bridge deforms right up to the point at which it fails. The rate of deformation changes with increasing load, so we call these elements "non-linear finite elements".

Within the computer, the assembly of elements forms a mathematical model of the bridge. If we apply a theoretical load to this model, it will deform. If the model is a good one, this deformation will match that of the real bridge. The geometry of the bridge changes as the load is applied, so we must perform the analysis in stages, applying slowly increasing loads to a slowly deforming bridge. The calculation takes about four hours—comparable to the time it takes to carry out a full-scale test. The cost is much less, however, and this method does not involve knocking the bridge down! The results from this program are encouraging, and each new test provides further data against which to calibrate the program.

The researchers working on the joint project have made scale models of each bridge that they plan to test. The models do not respond to loads in the same way that the real bridges do. A possible reason is that the materials that the engineers

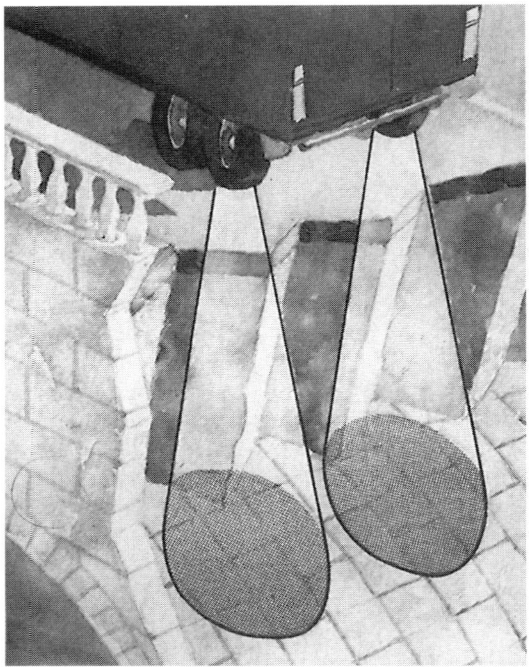

The pressure from the tyre of a passing vehicle spreads out and down, becoming weaker on the way. By the time it reaches the ring of the arch, it covers a much wider area than that between the tyre and the road surface

use for the models are similar to those in the bridge, while the stresses are reduced in proportion to the scale. It is important to develop the analyses until they can predict results from scale models. We can then scale up the results.

The second test involved a semicircular arch with a span of 10·5 metres at Bridgehouse near Bargower, also in the Strathclyde region. This arch failed by crushing of the stone at a much higher load than any analysis had predicted. After the test, it became clear that large blocks of stone partly filled the bridge, considerably enhancing its strength.

The effect of such a filling was even more apparent on the elliptical arch with a 5-metre span that the researchers tested next. This bridge, over a disused canal at Preston upon the Weald Moors in Shropshire, would have carried very little load without the help of its stone backing.

Another canal bridge of a similar span, but segmental, formed the raw material for the fourth test. The Transport and Road Research Laboratory has not yet published full details of the results of this test, but the bridge failed in the classic way predicted by mechanism analysis.

TESTING TIMES FOR ARCHES 13

I have not yet mentioned the importance of the width of the bridge. The pressure under the tyres of a vehicle spreads and weakens as it penetrates downwards. By the time the load reaches the arch's ring, it covers a patch many times larger than the area of contact between tyre and road. This loaded area tries to move, but the surrounding arch restrains it, forcing an even larger area to carry load. Each stage of distribution—the spread of the tyre on the road, the spread of the load through the road surfacing, through the material filling the spandrel and through the arch's ring—reduces the effect of the load on the arch. It is safe, if uneconomic, to ignore distribution in the direction of the span, but it is important to use a realistic measure of the width of the arch.

In skewed bridges, where the span is not perpendicular to the supports, the problems are greater because stress will concentrate in the obtuse corners. Testing using models may not be accurate, and comprehensive full-scale tests are very expensive. Extending non-linear finite element analysis to three dimensions demands vast computing power. For the time being, engineers will have to investigate these problems using established programs based on linear finite elements.

But eventually it should be possible for researchers to build up a computerised database of information about bridges, which engineers can feed into a program whenever they need to check a route for a heavy load.

The value of existing masonry bridges is beyond doubt. Because many have endured for hundreds and even thousands of years, we should probably consider the arch as a design for new bridges too. Many modern bridges, on the motorways around Birmingham for example, have deteriorated rapidly. The cost of maintenance after 20 years already approaches the original cost of the bridge. Skilled masons are expensive, but modern machinery would mean that we would need far fewer of them to build an arch than in the past. Modern scaffolding would also dramatically reduce the cost of supporting the arch during its construction.

Recently, a student doing a project at Dundee University came up with a design for an arch, which impressed Jack Maxwell, Tayside Regional Council's bridge engineer. A local contractor priced the design at only slightly more than a modern alternative. In 1987, the department of civil engineering at Dundee University will hold a competition to design a new bridge. Arches may yet prove to be the bridges of the future.

2

The Roman bridge-builder: some aspects of his work

N.A.F. Smith

To judge from the literature on Roman engineering, there was a time when the history of bridge building was a prominent theme closely associated with a parallel and equally well-developed interest in Roman roads. Recently, as a result of a variety of new approaches to archaeological, technical and social themes, the emphasis has moved to aspects of hydraulic engineering and, in particular, problems of water power and water supply. Of course, fashions in engineering history, Roman or later, are bound to change from time to time. That is understandable and nothing new. On the other hand, any overall view of Roman civil engineering needs to integrate the various approaches, especially in so far as they affect one another. Bridge building, after all, is bridge building, whether the structures are for roads or aqueducts and, in this lecture, both functions will be considered.

However, before turning to engineering historical matters, it is important to notice, even briefly, a few points about method that are fundamental to comprehending the ancient past and especially its engineering. It is noticeable how frequently engineers who write history are disposed to imagine that any failure to comprehend early engineering is essentially a failure of modern intelligence. Too rarely is it realised that the real problem is a failure of the ancient evidence to reach the minimum level that comprehension and conclusions require. Nor, contrary to a number of recent exercises, is it a foregone conclusion that resort to modern analytical devices and computation is a means of making good gaps and faults in the evidence or illuminating problems of interpretation. What is more, and what is often worse, is that retrospective analysis of all manner of ancient technologies – machines, ships or metallurgical processes, never mind structures – can often become a debate about the validity of analytical methods *per se* and the historical issues are lost sight of. When it comes to trying to comprehend the Roman engineer's thoughts and ideas about design and construction, one must never forget that the evidence from the structures themselves is no substitute for the written word – and that commodity is always rare and frequently ambiguous.

A second proviso is a function of survival. Those Roman structures on which we are encouraged to base our historical and structural interpretations are not necessarily typical – and they might have come down to us for particular reasons. Perhaps they were well constructed of particularly durable materials in good conditions; they may have occupied an especially strategic location which encouraged sustained and thorough maintenance over the centuries; or, quite the opposite, survival may have been ensured by isolation. Some of the gems of Roman bridge-building, however small in size or infrequently reported, are in the middle of nowhere and they survive – often unspoiled and original – simply because they have hardly been used at all.

Thirdly, it is important, in these historical assessments, to remember that any bridge which, in 2000 years, has been used by traffic, abused by armies, and misused by the ignorant, will surely have had to be maintained and repaired, altered and rebuilt – in some cases very considerably. How much of a surviving Roman bridge is actually Roman can often be a very good question; one is reminded of what purports to be an original W. G. Grace cricket bat with its two new blades and three new handles.

Road bridges were built by the Roman Government. Their function, above all, was to meet the needs of military transport and the administration of an empire. It is true that they were used for commerce as well but that, on the whole, was not their purpose. The freight traffic of the Roman world was moved far more efficiently by boat, on the sea and rivers and, here and there, on canals. Bridges for aqueducts, on the other hand, were built by towns and cities; the Imperial Government was not involved usually, although famous names are associated with the sponsorship of many aqueducts. Privately-built bridges appear to have been undertaken but rarely, the only certain example being in the mountains above Aosta. Il Pondel (as it is called) is a curious structure, apparently built as an aqueduct bridge at the outset, quaintly modified during its life to serve as a fortification, and then becoming a road bridge – a function that it continues to meet, albeit for a very narrow track. Unusually, it can be dated with precision, to the year 3BC according to an inscription that survives above the keystone.

Even though we have a record of bridges being built by the Assyrians, Etruscans and Greeks, it is with the Romans that bridge building first became a widely practised aspect of civil engineering, so fundamental was its contribution to the management of an empire and the life of its towns and cities. Moreover, bridge building was an activity that typifies those three features of Roman engineering which exemplify its character and even anticipate modern times – arcuate construction, the use of concrete, and the concept of public works. Let us now see if something of the nature of Roman bridge-building can be portrayed by considering the sequence of building a bridge, the choices which had to be made, and the techniques adopted.

At the outset, some surveying might be necessary. In the case of aqueduct bridges, significant requirements had to be met because these structures not only had to carry their water channels at the appropriate slope, or on the level, but it was necessary also to match them with sufficient accuracy to the falling gradient of the channels upstream and down. Moreover, substantial and last-minute adjustments to levels in the manner that might be allowable – however infrequently required – with a road was not necessarily a practical expedient in the case of an aqueduct. However, aqueduct surveying is a complex subject, appropriate to mention here but too involved for consideration in its own right.

How much surveying went into the setting-out of a typical road bridge is impossible to determine. Surveyors' marks on one pier of the Narni Bridge (Fig 1) appear to record distances measured from the pier, but whether they refer to inter-pier distances or route mileage is unknown. It is also worth noticing that, in the *Corpus Agrimensorum*, is a work entitled *Fluminis Varatio* which includes a description of a geometrical technique for

Fig 1. One arch spanning 20m survives in the bridge at Narni on the Via Flaminia

determining the width of a river[1]. One can imagine how such a method might have been used to establish certain basic features of a bridge's construction, but we do not know that it was ever done.

Surveying the layout of a bridge may not have been an especially significant procedure; examining foundation conditions probably was because experience would have told the Roman engineer what the historical record tells us – that ancient bridges were most vulnerable in their foundations and that was where failure was most likely to be initiated. Had the typical Roman bridge been built on dry land, the generally massive construction would have been more than adequate to resist everything except, perhaps, military action or seismic activity. And this contrast between the fragility of the piers below and the security of the arches, trusses or girders above is evident in bridge building right through to the 18th century.

One solution to the problem was to dispense with piers altogether. The bridge of boats is an idea at least as old as Xerxes' crossing of the Struma, while the best-known Roman example is depicted in a finely preserved mosaic in Ostia. It shows the boat bridge at Arles, the only Roman bridge across the Rhône below Lyon and so compactly built that it prevented navigation further up-river by seagoing vessels. Vestiges of the bridge's abutments survive.

Alternatively, bridge foundations might be built well above water level, at least normal water-level (the situation that might prevail in the bridging of a gorge, for example), and a number of examples are known. The concomitant, of course, was often the need for a considerable span.

Another recourse to assist foundation construction was to confine work to the dry season, so that as many piers as possible were easily accessible. It must have been a technique of considerable advantage at such places as Mérida, across the Guadiana (Fig 2), or Salamanca, across the Tormes (Fig 3), provided that work proceeded fast enough, or in well-judged phases, so that the huge floods that threaten every year did not wash out unfinished or unprotected construction.

We have little evidence that rivers were ever diverted in order to make foundations accessible. It has been conjectured that some sort of diversion was achieved by Trajan's engineers on the Danube – although the evidence and the arguments on which it is based are not convincing – and an interesting variation has been suggested for the Pont du Gard (Fig 4)[2]. Under this aqueduct bridge, the rocky bed of the river appears – one can put it no stronger – to have been artificially deepened to confine the flow, at least when the Gard was running normally. There is, moreover, some evidence that Roman engineers occasionally resorted to rudimentary methods of diverting, or at least containing, rivers in order to build dams, and it is possible that comparable techniques were adopted for bridge building.

Ultimtely, of course, *the* bridge-building problem arose in deep and wide perennial rivers such as the Thames, the Rhine or the Rhône.

Fig 2. The half-mile-long bridge across the Guadiana at Mérida, often damaged by the river and wars and frequently rebuilt therefore

Temporary military bridges are a type to be included here because one of the most celebrated of all Roman bridges was the one thrown, as they say, across the Rhine by Julius Caesar in an operation described by him in *De Bello Gallico*[3]. This structure and others comparably large and strategic must have been routine for legionary engineers. What the technique required, essentially, was the driving of sets of vertical and angled piles to create across a river a series of parallel trestles which were then spanned one to the next by beams and balustrades. With no great difficulty, presumably, the piles

Fig 3. This part of the bridge at Salamanca is still essentially Roman; the remainder, to the south, was considerably restored in 1677

Fig 4. The Pont du Gard, an aqueduct bridge, possibly Augustan, and notable for being very largely original

of such a bridge were readily loosened and floated away for further use as the army moved on.

Permanent all-wooden bridges were undoubtedly resorted to as well, even though their inability to endure has left us little evidence of their existence and even less as to how they were constructed. One supposes that they were common in northern Europe and, in fact, modern opinion tends to the view that, along with other Roman bridges in Britain, the first London Bridge was entirely of wood. Interestingly enough, one can make a case for sequences of bridges along the Tyrrhenian coast of Italy also being made of wood[4].

Pile driving was fundamental to Roman bridge-building because, even when the temporary piers of a military bridge or the permanent piers of a wooden one were not the overriding issue, piles were needed to create cofferdams in which to establish the base foundation for masonry construction above. The history of pile driving, its machinery and techniques, is not yet a well-researched field and what is well understood dates from periods well after the Roman. However, given the crudity of even early modern pile-driving (never mind that of the Middle Ages), and assuming that Roman engineers were no better equipped, we can reasonably extrapolate back from the later evidence. Early pile-driving utilised a barge, or two moored side-by-side, from which the piles were driven by brute force, gangs of workmen manually raising a weight – by ropes if a heavy impact was intended – which fell repeatedly on to the pile. Hard, tedious and slow it must have been, but we must never underestimate the ancients' capacity to achieve results by the massive application of brute and labour-intensive force.

Cofferdams were formed of piles driven close together, and there is some evidence that they were interconnected in certain circumstances, either as an aid to positioning, or to improve watertightness, or both. At Trier, archaeological evidence suggests a double row of piles with clay packed between to seal the cofferdam[5]. Cofferdams must have been drained and maintained dry with relatively simple and compact machines such as the chain-of-pots or bucket devices. Anything more elaborate would have been too bulky when one remembers how closely pier construction had to be fitted to the cofferdam wall. It was, after all, difficult enough to drive piles in the first place, never mind pull them afterwards, so that leaving them in place around the finished pier to act as 'starlings' was expedient . That such protection was worthwhile is clear from the regularity with which, over the centuries, bridge piers have required their starlings to be reconstructed – so much so that, in some cases, veritable islands have been created.

If good bedrock could not be found within the completed cofferdam, a foundation of piles would be driven. Here and there the archaeological record is able to indicate something of what could be achieved. Great arrays of piles 2-3m long – as much as 8m occasionally – and up to 50cm in diameter have been discovered in the foundations of some major Roman bridges. Even a minor bridge such as the one at Aldwincle in Northamptonshire featured 5m piles with a cross-section of 50cm x 50cm. Iron tips of some size appear to have been a standard accessory. As Villard d'Honnecourt shows us in a medieval drawing, once driven the piles were cut to a level and a grillage was laid as a basis for pier construction, and doubtless the Roman technique was the same.

The typical masonry pier comprised an outer wall of cut masonry, the blocks often bound together with iron clamps set in lead. Some bridge piers consisted of cut masonry throughout; in other cases, rubble concrete was placed within the masonry outer wall. In itself either technique could be relied on to give very strong and stable support for the road structure above, as indeed

Fig 5. The piers of the Ponte Leproso at Benevento are protected by a platform; the superstructure is a mixture of workmanship

was still the case 2000 years later when the same principles of masonry bridge-building continued to prevail. However, against the ravages of rivers in full spate the typical Roman bridge pier was often found to have insufficient strength. The reasons, in proportions which varied from one bridge to another, were inadequate underwater construction and very defective hydraulic performance; in short, the risk was of being undermined. Any number of Roman bridges featured piers of totally adequate dimensions as such but, being furnished with cut-waters only on the upstream side, the downstream face being built flat, their susceptibility to erosion and scour was considerable and sometimes fatal. Curiously enough, there are exceptions. The first bridge at Chesters[6] had cut-waters up-and downstream and there is evidence from elsewhere that doubly pointed cut-waters were utilised occasionally.

Some Roman bridges featured platforms, or aprons, which surrounded the feet of the piers and usually covered the whole of the riverbed in the vicinity of the bridge. Such platforms may be original (which appears to be the case at Piercebridge on the River Tees) or, equally, they may be remedial, which is probably true of the bridges at Benevento (Fig 5) and Cordoba (Fig 6). Either way, here was a measure designed to resist scour and erosion. Also, it is noticeable throughout Europe that medieval weirs, built for a variety of primary purposes, were often located *downstream* of Roman bridges, and very probably the intention was to allow the bridges' piers the security of still water. It is, moreover, the case that the cut-waters we see today on many Roman bridges, up-and downstream, are not Roman in origin but are later additions or remodellings. My earlier question about how Roman is a Roman bridge is nowhere more pertinent than in these matters of piers, starlings and cut-waters.

Fig 6. In this downstream view, many and various styles of construction are visible in the piers of the Córdoba Bridge

Fig 7. The large tunnels in the spandrels of the Pons Milvius (109BC) were probably intended to assist the passage of the Tiber in flood

One other feature of Roman bridges may indicate a concern about waterflows: it is the provision of tunnels in the spandrels. Conceivably, this was a device to ease the problems of construction and lighten the structure's weight. But equally, these tunnels would have acted as flood arches and such may have been the intention (Figs 7 and 8). Interestingly, however, what was usual in the early days became less commonly practised later on, and how far this might reflect the evolution of other aspects of bridge design is at the moment unclear.

In its superstructure, the traditional Roman masonry bridge used the full semicircular arch. A bridge such as Salamanca (Fig 3) shows the style to perfection. Fundamental to modern interpretations of this form is the belief that the combination of semicircular arches and heavy piers allowed construction to proceed arch-by-arch from one end to the other or both ends towards the middle without risk of collapse at any stage. And, indeed, this may be a proper conclusion. Moreover, it is interesting to notice that the fine and famous Alcántara Bridge (Fig 9) in western Spain may, in an unexpected way, confirm this view. Because of its strategic importance, the Alcántara Bridge has, over the centuries, frequently been breached in one or other of its spans – and yet it has never failed, which indicates that even so relatively tall and slender a structure would stand securely even when denied its structural integrity: in which case the low multispan bridge was hardly likely to have been a bigger problem. Indeed, it is not fanciful to postulate, given the Alcántara evidence, that even high bridges were built from end-to-end or ends-to-middle, one span at a time, rather than integrally from

Fig 8. The Pons Fabricius (62BC), another Tiber bridge with provision to cope with deep water by means of a central flood-relief tunnel

foundations upwards with all the arches, or at least their falsework, being completed simultaneously. Variations in the dates and styles of successive piers in the Narni Bridge (Fig 1) also seem more consistent with a constructional sequence from end-to-end rather than one proceeding from bottom-to-top. Just as significant, too, is the fact that, in modern times, one single, large span of the Narni Bridge has survived alone, like a triumphal

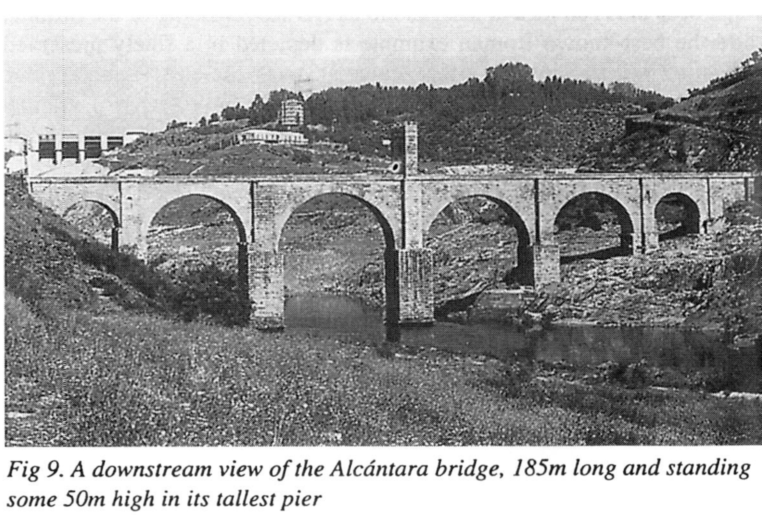

Fig 9. A downstream view of the Alcántara bridge, 185m long and standing some 50m high in its tallest pier

Fig 10. The Alcoñetar Bridge whose ruins were not only restored recently but, in addition, were moved from their original location as the waters of a new reservoir began to rise

arch, and other ruined Roman bridges exhibit structural stability in one or a pair of surviving arches, even though all else has fallen.

Contrary to a frequently encountered opinion, the Roman bridge-builder was not restricted to semicircular arches, not even in masonry and certainly not in wood. The remains of the Roman bridge in Narbonne, nowadays all but lost between later bridges and under modern buildings, indicates the use of segmental arches, as does the recently relocated Alcoñetar Bridge (Fig 10) in Spain[7].

Precisely how bridge arches were formed is not always clear and, in any case, methods no doubt varied. All the same, many road and aqueduct bridges display traces of the slots and projections that were obviously used to support wooden scaffolding and falsework during construction, not to mention subsequently for maintenance and repairs when those were necessary. Such falsework might be built to the full width of the bridge – sufficient, in other words, to support the construction of a complete arch. Voussoir rings might be of masonry, possibly clamped with metal, or brick and tile with infill above of rubble concrete into which, as noted above, flood relief arches might be incorporated using further falsework. In the Pont du Gard an interesting variation seems to have been preferred. The voussoir rings apparently were erected one after another, using the same piece of centering in successive positions, three or four, across the width of the bridge at each of its three levels. In the large and single span of the Ponte sul Lys at Pont-St-Martin (Fig 11), east of Aosta, the same technique appears to have been used, although here the difficulty of distinguishing Roman work from medieval is considerable. In the medieval Pont d'Avignon, there is further evidence of the same procedure.

Masonry piers were not necessarily spanned with voussoirs of masonry, brick or tile. Where it was economic and expedient to do so, wood was used to construct a superstructure, a deck of beams where spans were suitably short but, more interestingly, trussed arches for longer spans. It is conventional, and convincing, to reconstruct, in both models and the mind's eye, the bridges of Roman Britain as utilising wooden beams or arches while, in the case of Apollodorus' bridge over the Danube (built for Trajan in AD 105/106), we can be sure of the use of wooden trussed arches from the unique evidence on Trajan's Column[8]. Attempts to decide the constructional details of such arches and how they might have been assembled are essentially futile; it is the general principle of Trajan's bridge which is so important, particularly when it is remembered that this was a bridge of considerable length, perhaps 3500 ft, across a big river into whose bed some 20 piers in masonry were sunk. Fragments of them discovered in modern times verify aspects of the contemporary descriptions. Dio Cassius says the structure had spans of 170ft and piers 60ft wide, in which case arched trusses in wood do seem appropriate. Apollodorus' familiarity with this style of construction, and his confidence in it, possibly led to its being used elsewhere, and the most frequently claimed example is Iberian. The emperor Trajan was responsible for a substantial rebuilding of roads in Spain, including the Via de la Plata from Astorga to Mérida. The bridge of Alconetar (Fig 10), which carried that road over the Tagus, is conceivably by Apollodorus, or at least reflects his influence. Certainly, the bridge's proportions are broadly those of the Danube bridge, and the ruins indicate the use of segmental rather than semicircular arches, but almost certainly in masonry rather than wood.

Fig 11. The Ponte sul Lys on the road to Aosta with a span of 37m

Various details of Roman bridge-building having been referred to, it will be useful now to generalise with regard to questions of types and styles and their distribution. The masonry-piered bridge was adapted to suit valley profiles, as one would expect. Wide and flat river valleys were crossed with long multispanned structures using semicircular arches. The choice between wood and masonry for the superstructure would have depended on the economics of availability and skill, as well as considerations of span length in the conditions. For example, it seems clear that bridges across rivers as wide as the Danube or Rhine featured wooden arches and decks. In deep and broad valleys, fewer, but larger, spans were often the style, as at Alcántara or Narni for roads or the Pont du Gard for an aqueduct. If a deep valley was also narrow (in short, a gorge), a masonry bridge with one suitably large span might have been the solution; Vulci and Ascoli Piceno are examples. A fourth type was what, in the Middle Ages, was sometimes known as a 'devil's bridge' – a high, single arch to clear a river in one go but with steeply graded approaches if the banks were at a relatively low level (Fig 12). Bridges wholly of wood were either military structures, quickly built and just as quickly dismantled, or else they were used where economy and resources favoured nothing else – in northern Gaul or Britain, say.

Particular routes often display a characteristic structural style for which we have evidence or can make the appropriate inference. Along the Via Flaminia there was a sequence of important masonry bridges of which the Ponte d'Augusto at Narni and the Ponte di Tiberio at Rimini (Fig 13) are the significant, but not the only, surviving examples. On the other hand, although we know of roads along the western coast of Italy, their crossing

Fig 12. The Pont Julien near Apt in Provence, as thin at the crown as any Roman arch anywhere

Fig 13. The Ponte di Tiberio at the end of the Via Flaminia in Rimini, often damaged, as frequently repaired, and still in use

points of major rivers – the Arno, say, or the Garigliano – have left us no bridge remains. As noted earlier, it is probable that, over those wide and deep alluvial rivers, it was wooden bridges, long since washed away or burned down, which predominated, a proposition borne out by recent work at Minturnae[4].

Until the Augustan period, wooden bridges also predominated along the Via Domitia in southern Gaul. Their replacements were well-built masonry structures, some of which survive, either intact and in use (e.g. at Narbonne and Sommières (Fig 14) or else as intriguing and impressive

Fig 14. A downstream view of the bridge at Sommières **(only seven arches of the original bridge still span the river, the others having become *caves* and garages in the town to the right of the picture)**

Fig 15. One arch of the Pont d'Ambroix refuses to yield to the Vidourle

remains like those of the Pont d'Ambroix at Gallargues (Fig 15). Less well-known are the numerous places along the western side of the River Rhône where the northbound road had to cross such tributaries as the Escoutay, Ardèche and Doux. Above la Voulte-sur-Rhône, one can still admire the abutments of a mighty bridge which once spanned the Eyrieux with a single arch of some 30m.

Aqueduct bridges were a special case of bridge building, and this led to distinctive styles tailored to purpose. The essential point here, as we have seen, was that these bridges, in addition to carrying the rather trivial loading of a water channel and its contents, had to maintain a suitable hydraulic gradient as well – and, not infrequently, the latter was essentially the main purpose. It is this requirement which imparts such a striking appearance to the beautiful bridge in Segovia (Fig 16) with its tall, slender piers closely spaced in a structure much narrower than any road bridge could have been, or the majestic arcades that stride across the Campagna to Rome (Fig 17). So massive an edifice as the Pont du Gard merely maintains the slope of a not very large aqueduct. One can observe here, incidentally, how much more susceptible were such tall, narrow structures to lateral displacement and, in fact, in some aqueduct bridges the Roman engineer provided an element of buttressing; Fréjus and Mérida are examples and the same technique appears in the upper part of the Alcántara road bridge.

How Roman bridges, and other structures, were designed – and the term 'design' must be used with due care and caution – remains very obscure so far as detail is concerned, even though, in broad terms, one or two aspects of what might have been involved can be postulated. Simultaneously, we can reject those recent inquiries, based firmly on modern theory, hindsight and wishful thinking, which have encouraged some modern writers to

Fig 16. One of the finest of all Roman structures, the 1st century AD aqueduct bridge which dominates Segovia

attribute to ancient – and, for that matter, medieval – engineers a ridiculously advanced understanding of design concepts, techniques and calculation. All the same, some process of structural design, however rudimentary, must have been employed; it is perfectly absurd to imagine that, one morning, gangs of workmen with materials and equipment turned up at a site and simply got busy. There must have been a 'design' of some sort to work to. And indeed the ancient sources do indicate something of the nature

Fig 17. The Aqua Claudia, of large masonry blocks, carries on top the Aqua Anio Novus, of brick, across the Campagna

of the rules of proportion, and the readiness to *rely* on proportion, that was fundamental to ancient design. For later periods, the forms of the structural design rules by then in use – and no less interesting their variety – are prominent in a number of bridge designers' work. From as late as the 18th century we can study the rules used by John Smeaton to relate arch thickness, span and the degrees of arc to the width of pier and notice, moreover, that he more than once changed his mind. Overall, it is probably not fanciful to extrapolate bridge-building rules of this type, or some anticipation of them, however simple, back through the Middle Ages to the Roman epoch.

If Roman engineers had learnt to proportion a bridge on the basis of some assumed, or derived, relationship between (shall we say) arch span, voussoir thickness, pier width and possibly rise (and a similarly appropriate group of parameters for a truss), one might suppose that the statistical analysis of sufficient bridges ought to reveal the rule, or rules, used. The problem, however, is that, even if a rule that met all the requirements could be established, it would by no means follow that it was *the* rule. It is rather like trying to recover the clues to a crossword from the finished puzzle: it is easy to arrive at an answer, but impossible to know whether it is correct. Similarly, we should be cautious in considering the claim by J. B. Ward-Perkins that the Pont du Gard's design is based on the ratios 4:3:1:6 for the proportions of the central, lateral and upper arches and the overall height. In fact, such a 'design' is unconvincing because, in reality, the proportions claimed are very approximate, to the nearest whole number when measured against the bridge itself. For design proportions of this sort to be really persuasive, they need to be exact.

Let me now add a fourth qualification to the three outlined at the beginning. Much that is written about Roman engineering – and probably because, more often than not, the authors are following an established tradition – advances the view, or at least presupposes, that any field of technical activity was essentially an entity throughout the empire. Whether it was mining, irrigation, swordmaking or shipbuilding, it is not acknowledged in many studies that, from place to place in the Roman world, there was probably more than one way of doing things. And, as here discussed, in relation to bridges, some facets of regional variations can be put down to the materials available, to constructional necessities characteristic of certain localities, and to specific structural problems. There are at least a couple of other factors to consider.

How far Roman engineering was derivative from earlier societies, and how far it was indigenously Roman, is a complex matter and overall the balance between the two must have varied greatly with technologies and localities. That Greek influence in building was powerful is well established and, for the early period in Italy, so probably was Etruscan. In northern Europe, it seems clear that traditional techniques and craftsmanship inevitably were adopted from captive peoples. Whether or not bridge building was thus absorbed is impossible to determine, particularly because it was an activity whose extent in the pre-Roman period is, in any case, very uncertain.

Secondly, we ought to notice that some variations in Roman bridge-building techniques may well have been matters of opinion among designers and builders. Conceivably, the differences detectable in arch forms, pier shapes and the size of flood arches were due to the application of theories, novel or routine, of individual engineers or embodied within the expertise of a practising group such as a legion – a 'school of thought', in other words. In reality, it is exceptionally difficult to draw even the outlines of the lives of more than a handful of Roman architects and engineers, never mind scrutinise their opinions and ideas. Moreover, within the ranks of army engineers and their staffs, varying experience and familiarity with particular techniques in different circumstances and places – legions did, after all, move about – might have served to diffuse and distribute practices in a manner that led to a degree of standardisation but might, depending on a legion's travels, have led to quite the opposite. It is not a matter that has received much study.

Overall, these observations on the nature of Roman engineering design and associated concepts and influences allow a final thought.

It is customary, and not unnatural, to present developments in Roman bridge-building, and indeed other branches of structures, as assuming that the same sequence occurred everywhere at the same time. This leads some writers to date structures and place them within a progression of developments and events according to how their features fit the generalised chronology. In fact, it might not be so straighforward, if only in the sense that, if the take-up of techniques from one place to another was subject to a timelag, the chronology would be displaced, so to speak. And, in any case, the dating of a sequence of bridges can *appear* incompatible even when the opposite is known to be the case. Augustus undertook the rebuilding of the bridges along the Via Flaminia in 27BC. The fact that the new set of bridges dates from the one period does not preclude a great variety in methods of construction brought about not least by the expediency of rebuilding existing structures. Reconstruction of older edifices is capable of undermining dating as tboroughly as any other single factor. Questions of style, technique and chronology are easy to propose but much less easy to resolve, and this shows how far we are from a definitive account of Roman bridge-building.

References

1. Dilke, O. A. W.: *The Roman land surveyors*, Newton Abbot, David & Charles, 1971, pp59-61
2. Lassalle, V.: 'Le Pont du Gard et l'Aqueduc de Nîmes', *Dossiers de l'Archéologie*, **36** , 1979, p57

3. Bundgärd, J. A.: 'Caesar's bridges over the Rhine', *Acta Archaeologica,* **XXXVI,** 1965, pp87-103
4. Brookes, A. C.: 'Minturnae: the Via Appia Bridge', *American Journal of Archaeology*, **78**, 1974, pp41-8
5. Mesqui, J.: *Le pont en France avant les temps des ingénieurs*, Paris, Picard, 1986, p243
6. Bidwell, P. T., Holbrook, N.: *Hadrian's Wall bridges*, London, English Heritage, 1989, pp15-18
7. Fernandez Casado , C.: *Historia del puente en España*, Madrid, Instituto Eduardo Torroja, 1980, chap. 3
8. Lepper, F., Frere, S.: *Trajan's Column*, Gloucester, Alan Sutton, 1988, pp147-152

3
The Pont du Gard and the aqueduct of Nîmes

N.A.F. Smith

Olwen Brogan calls[1] it the most beautiful of all the monuments of Roman Gaul and on any short-list of the finest historic structures in Europe the Pont du Gard (Fig. 1) would almost certainly figure; for many people it would probably rank in the first half-dozen. By any standards it is a mighty monument; it is old, it is large, and, quite its most impressive feature, it is notably complete. For at least two hundred years, from the time when Europeans were first excited by their antiquities, the bridge has been visited, admired, illustrated and described. Indeed an account of the history of Pont du Gard studies would undoubtedly make a paper in its own right. For all I know it may have been done somewhere already.

Fig. 1. The Pont du Gard viewed from the south.

THE PONT DU GARD AND THE AQUEDUCT OF NÎMES

One of the first to set down his admiration is the one most often quoted, Jean-Jacques Rousseau:[2]

> . . . It was the first work of the Romans that I had seen. I hoped to see a monument worthy of the hands that had raised it; and for the first time in my life my expectations were surpassed by the reality. It belonged only to the Romans to produce such an effect. The aspect of this simple and noble work struck me all the more that it was in the midst of a desert where the silence and solitude increased its effect . . . I walked along the three stages of this superb construction, with a respect that made me almost shrink from treading on it. The echo of my footsteps under the immense arches made me think I could hear the strong voices of the men who had built it. I felt lost like an insect in the immensity of the work. I felt, along with the sense of my own littleness, something nevertheless which seemed to elevate my soul; I said to myself with a sigh: "Oh! that I had been born a Roman!" . . . I remained several hours in this rapture of contemplation. I came away from it in a kind of dream, and this reverie was not favourable to Madame de Larnage. She had been careful to warn me against the attractions of the young women of Montpellier—but not against the Pont du Gard.

For a man who was seeing an example for the first time, Rousseau is remarkably sure of his view that this was truly Roman work.

Only recently I learned of another famous visitor, and a rather unexpected one, namely Michael Faraday who, in the company of Humphry Davy, passed by on Friday 7 January 1814 on his way to Montpellier.[3]

Later on (1835) came Prosper Mérimée who was impressed by not only the bridge but the river it spanned:[4]

> The Gardon, swollen by heavy rain, was overflowing and the colour of coffee as it roared under the arches of the aqueduct. The sky was full of thunder and shafts of sunshine gilded the monument which sparkled in the light; this wild site, the solitude of the place, the noise of the raging river gave a poetic splendour to the mighty structure before my eyes.

The power of the Gard in full spate is a point we shall return to. Alexandre Dumas refers[5] obliquely to his visit in an incident in *The Count of Monte Cristo* while later on Henry James came more in the spirit of a tourist and was willing to claim that he "remained there an hour and got a complete impression".[6] He was certainly correct when he observed "You are very near it before you see it,"[7] something which always heightens the drama of a visit, and, like Rousseau, James was aware of the bridge's Roman qualities:[8]

> Over the valley, from side to side and ever so high in the air, stretch the three tiers of the tremendous bridge. They are unspeakably imposing, and nothing could well be more Roman. The hugeness, the solidity, the unexpectedness, the monumental rectitude of the whole thing leave you nothing to say—at the time—and make you stand gazing. You simply feel that it is noble and perfect, that it has the quality of greatness.

And yet surprisingly Henry James was inclined to qualify this view when he says:[9]

> It came to pass that at the same time I discovered in it a certain stupidity, a vague brutality. That element is rarely absent from great Roman work, which is wanting in the nice adaptation of the means to the end. The means are always exaggerated; the end is so much more than attained.

And then later on he adds:[10]

> The preservation of the thing is extraordinary; nothing has crumbled or collapsed; every feature remains, and the huge blocks of stone, of a brownish-yellow (as if they had been baked by the Provençal sun for eighteen centuries), pile themselves, without mortar or cement, as evenly as the day they were laid together. All this to carry the water of a couple of springs to a little provincial city!

Oddly enough I know what Henry James means and some elaboration of this point will emerge later in the paper.

THE PONT DU GARD AND THE AQUEDUCT OF NÎMES

The Pont du Gard is equally well recorded pictorially[11] and the Musée de Vieux Nîmes houses a fine collection of prints and engravings. Some of the most appealing record the beginnings of tourism which Henry James represents and which has now reached epidemic proportions to the extent that visitors threaten the structure as surely as the structure threatens visitors when they mass on the narrow, top-most level. It is said that after Versailles and Notre Dame the Pont du Gard is the most visited historic monument in France. The pictorial record is important as a source of certain information, for example the fact that, in order to use the structure as a road bridge, the bases of the middle range of piers were once cut back sufficiently to create a road-way rather more than 2.5 m wide. This modification appears in a drawing of 1557 by Poldo d'Albenas[12] and is subsequently depicted by other artists until the piers were at last re-established early in the eighteenth century. Almost immediately Henri Pitot, in the years between 1743 and 1747, added the extant road-bridge to the downstream side of the Pont du Gard. And then, in the 1850s, the interest of Napoleon III led to a very thorough restoration of the whole structure which brought it to essentially the condition in which it can be admired and studied today.[13]

The Pont du Gard has produced a not inconsiderable family of look-alikes. Henri Pitot surely had it in mind in designing his own imposing aqueduct bridge of St. Clement which dominates the centre of Montpellier[14] while across the Rhône, near Aix-en-Provence, the engineer de Montricher, in building a water supply for Marseilles in the 1840's, imitated the Pont du Gard's form so successfully that his Roquefavour aqueduct bridge[15] is regularly reported as itself being Roman and is known to have been called the Pont du Gard!

By any standards then the Pont du Gard has been well recorded, endlessly admired and frequently studied and one might be disposed to believe that interpretations of it, historically and structurally, are well established. And yet it is not so. Much about the bridge continues to be problematical and when the enquiry is extended to the whole of the aqueduct, of which the bridge is after all only a fraction, however distinguished, the position is even more puzzling, misunderstood or not studied at all. In what follows some attempt is made to comment on our knowledge of the Pont du Gard and the aqueduct of Nîmes, to raise some questions, suggest a few answers and even admit that here and there we do not know and, the nature of the evidence being what it is, probably never will. Let us begin with the most basic question of all, the date of the Pont du Gard's construction.

THE DATE

The truth is we do not know when the bridge and the aqueduct were built.[16] It is conventional to date the works to the reign of Augustus and the designs of Marcus Agrippa. But the evidence, however persuasive, is no more than circumstantial. Nîmes was an Augustan colonial foundation of about 28 B.C. Coins of Nîmes dating from then until 15 B.C. show a crocodile chained to a palm tree (as the emblem of modern Nîmes still does) and this has been taken to indicate that the Emperor settled veterans from Egypt in the city after the civil war. From an inscription of 16 or 15 B.C. it is known that Augustus gave the colony gates and walls of which the surviving Porte d'Auguste was a part. Agrippa appears to have been in Nîmes in the winter of 20/19 B.C. and it was then that a number of buildings were planned and even begun.[17] It is generally accepted, for example, that the Maison Carrée was one of these and possibly there were baths as well. By comparison the aqueduct cannot be positively associated with either Augustus or Agrippa. That the attribution is made is based on two convictions. The first is that a structure of the grandeur of the Pont du Gard must be a work of a personality as powerful as Augustus and an engineer as bold as Agrippa who to be sure was an hydraulic engineer of repute and experience having already built Rome's Aqua Julia and being involved with the Aqua Virgo at the very time he was in Nîmes.[18] And secondly, it is argued that water-supply was so important to a Roman community that its provision was bound to be realised at an early date. Doubts have been expressed by those who advance social and economic arguments to point out that Roman towns and cities did not normally have the means to build lavish water-supply systems at an early stage of urban development. The essence of this view has been put recently by A. F. L. Rivet when he writes:[19]

> Beyond this, while on the Pont du Gard, as in the amphitheatre, the inscriptions (here mainly recording architectural instructions) include the name of a man (Veranius), he is otherwise unknown and the possibility that such a structure could have been completed in the Augustan period has recently been questioned. Finally, even leaving aside the intricacy of the work, the building of so large an aqueduct so early in this area seems highly improbable and a date no earlier than the late first century AD (in the time of Frontinus) or a little later seems much more likely.

It can also be noticed that Nîmes' impressive amphitheatre is a contemporary of the one at Arles and both date from the later first century A.D. The Temple (so-called) of Diana is probably even later, some say Hadrianic, and in the second century A.D. Nîmes was especially flourishing in the period of Antoninus Pius (A.D. 138–161) so why not attribute a mighty aqueduct to one of these later eras? The Russian archaeologist Nicolayev was inclined to do this because it was his opinion that apart from all else the style and structural refinement, as he judged it, of the Pont du Gard must be the result of some centuries of development and could not possibly be an early work, an achievement of beginners so to speak.[20]

The lack of a confirmed date for the Pont du Gard is very unsatisfactory. It is unsatisfactory in the first place because one feels that a structure of such significance ought to be dated—the famous only tantalise when they refuse to disclose their date of birth. Even more unsatisfactory is the consequent difficulty of placing the bridge within the evolution of Roman building, or relating the Pont du Gard to its predecessors and its deratives, if such there were. Instead, the temptation is to adopt a sequential argument first and to try to fix a date from it. In other words, by looking at the evolution of engineering technique, is it possible to position the Pont du Gard accurately enough in the developmental sequence to approximate its date? I believe that this is a possibility as we shall see.

The date of the Nîmes aqueduct then is a puzzle, but the route is not and here we can proceed with some confidence.

THE ROUTE

The source of the Nîmes aqueduct was at Uzès, 20 km to the north. The remains of a cylindrical masonry tank, 4 m in diameter, found in 1922 have been taken by some to indicate the manner in which local springs were captured and directed into the conduit, but as so often seems to be the case with Roman aqueducts the headworks have now disappeared and nothing is known with certainty.

The route of the aqueduct[21] did not follow a straight line to Nîmes; instead it took a great loop to the east in order to find a practical point at which to cross the Gard, some 3 km above Remoulins (Fig. 2). Herein lies the essential engineering problem of the Nîmes aqueduct: not only was the choice of water sources exceptionally limited but the route from the only viable one had to swing far to the east if excessive tunnelling and bridging were to be avoided. The Gard had to be crossed at some point and the best place proved to be well off the direct route.

In fact, from Uzès to the river the route of the aqueduct, north of the modern D981 and parallel to it, is easy going, only a modicum of bridging and tunnelling being required of which some interesting fragments survive. From Vers to the river the channel is carried on increasingly more substantial elevated structures before it crosses the river on the Pont du Gard. From here, for a few kilometres, the going is very sinuous as the conduit follows the contours of one minor

THE PONT DU GARD AND THE AQUEDUCT OF NÎMES

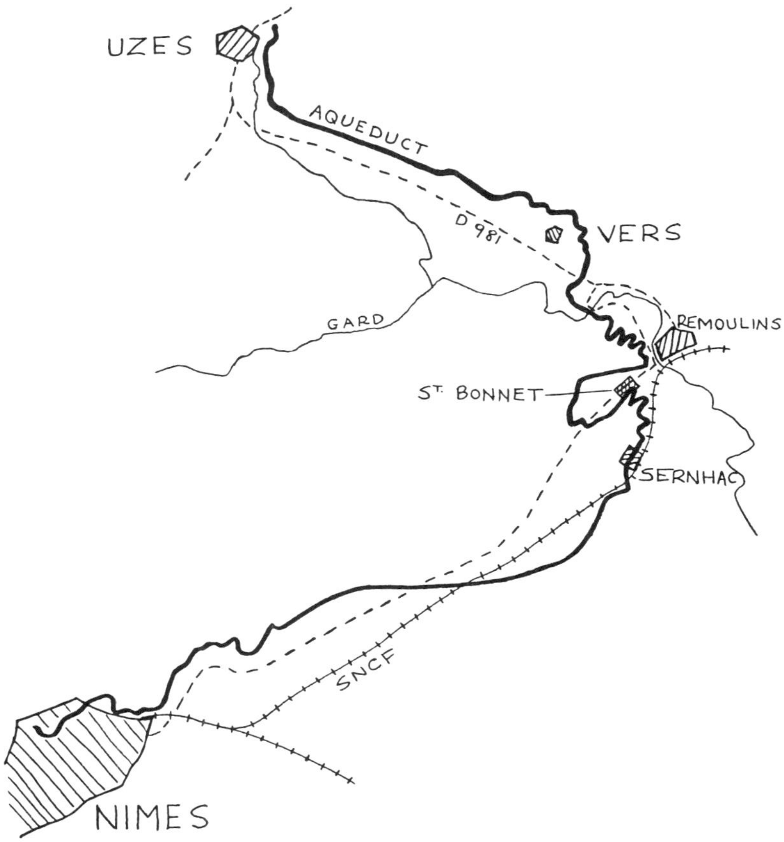

Fig. 2. The route of the Nîmes aqueduct from Uzès to Nîmes showing the geographical relationship to modern towns, roads, railway and the River Gard.

valley after another until a long loop brings it to St. Bonnet from where it turns south through two tunnels, large in section but short in length, (Fig. 3) to Sernhac after which an interesting cross-section of tunnel is exposed where a cutting on the Nîmes-Avignon railway line was dug in the nineteenth century. From here the route, so far as it can be identified, was relatively straight and straightforward until the final section into Nîmes which required some tunnelling not all of which has been located by archaeologists. One priceless component has come to light, however, and that is the *castellum divisorium* half-way up the modern Rue de la Lampèze where it was unearthed in 1844 during building works. The *castellum* will be discussed later; suffice to say here that it was the point at which Nîmes' water ended the aqueduct phase of its journey and was set up for distribution within the city below. That process of distribution is essentially another story—and a brief one in that little is known of the network of pipes and channels within Nîmes.

For the main course of the aqueduct one feature is especially notable and that is the amazingly shallow gradient; (Fig. 4) in a total length of channel of 49.75 km the fall is a mere 17 m so that the average gradient is 1 in 2926 and over one section it is no more than 1 in 14,285.

THE PONT DU GARD AND THE AQUEDUCT OF NÎMES

Fig. 3. Views of the square-cut tunnels near Sernhac. In cross-sectional area these parts of the aqueduct are much larger than elsewhere on the route.

THE PONT DU GARD AND THE AQUEDUCT OF NÎMES

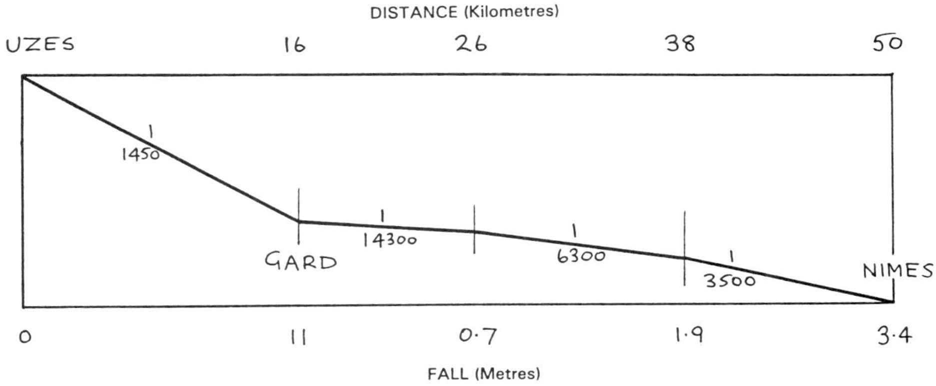

Fig. 4. Simplified profile of the aqueduct's route. Slopes are shown as fractions along the gradient line.

From the Pont du Gard to Nîmes the aqueduct falls only 6 m in 34 km. These are gentle gradients by any standards and this feature of the Nîmes aqueduct leads inevitably to a consideration of how an aqueduct was surveyed and what order of slope was intended, and could be attained.

THE SURVEYING

For engineering historians Roman surveying presents significant problems. Certainly some aspects of ancient surveying have been thoroughly studied and are well worked out. As much as is likely to be done on the textual evidence has probably been done and from it we know a good deal about ancient instruments and the surveying geometry which served them in their applications. Based on studies of Vitruvius, Hero and the *Corpus Agrimensorum* there has evolved the view that Roman engineers depended on three instruments: the *groma* for setting out, the *chorobates* for levelling, and the *dioptra* for either. The use of these instruments for such purposes as measuring and allocating land, setting up boundaries, map-making, cadastral surveys, centuriation and town-planning forms the principal focus of studies of Roman surveying to date and Professor Dilke has written the standard work on the subject.[22] Now all of this is essentially agrimensorial work, land surveying, and as William Kriesle has recently commented:[23]

> A major mis-understanding as to the role of the engineering surveyor versus that of the land surveyor has evolved and has been carried on by both surveyors, with no background in engineering, and by engineers with little knowledge of engineering history or enough field experience to appreciate the role of the land surveyor.

And it is true, compared with land surveying, really very little seems to have been established about Roman engineering surveying, aqueduct building being a typical case and tunnelling possibly the only exception.

Gradients have always been a pre-occupation with writers on Roman aqueducts, one suspects because it is so obviously easy to measure what was built and compare it with the advice of the various authorities whose recommendations are: Vitruvius, not less than 1 in 4800 or 1 in 200 depending on which manuscript is consulted; Pliny, who agrees with Vitruvius at 1 in 4800; Faventinus, 1 in 40 to 1 in 67; and Palladius who agrees with Faventinus because he copied him.[24]

Frankly these numbers are of limited value as a guide to how the ancients sloped their aqueducts. Perhaps they might be taken to represent the limits at either extreme but that means little enough in the day-to-day design of practical systems. Roman engineers had no more control over the topography than a twentieth century road constructor, a nineteenth century railway engineer or an eighteenth century canal builder. Bringing water from source to city was a matter of adapting differences in level to the distance between and not to trying to demonstrate that such-and-such a gradient was the one to use. Thus the Nîmes aqueduct has a notably shallow gradient (say 1 in 3000), the aqueducts at Rome from the Sabine Hills are of the order of 1 in 700, while the Yzeron aqueduct of Lyon has a ruling grade of 1 in 60, so steep that a reasonable slope of the open-channel sections was achieved only by the expedient of building *chutes*, or cascades, in between.[25]

What, by comparison, *was* a critical surveying problem and one that engineering historians persistently overlook, is how a water source for city or town, garrison or villa, was located in the first place and its elevation above the delivery point established. After all, everything else depended on this initial measurement being accurate. To some extent the lie of the land could be judged by eye. At Tivoli, for example, a key point on the routes of the Anio Vetus, Marcia, Claudia and Anio Novus, it is obvious that Rome lies at a lower level across the Campagna and that to the east the steeply graded Aniene must be a likely source. So much can be established without recourse to instrument surveying and the same is true at Segovia, for example, where the elevation of the Guadarrama Mountains is manifest. But in most places it was not so simple; where the lie of the land was complex, practical aqueduct gradients were anything but obvious so that real surveying, not visual inspection, must have been required to determine whether or not a potential source had sufficient elevation, however that was judged. Route selection would follow with accuracy of levelling once again the priority, in effect the first survey—the one to find the source—being conducted in reverse, back to base so to speak.[26]

Unfortunately we do not really know how it was done. Most opinion selects the *chorobates* as the instrument, a long wooden device—as much as 6 m long according to Vitruvius[27]—which having been first of all levelled, either by means of its water trough or using plumb lines, was then aimed at graduated poles to provide back-and fore-sights in the customary manner. On the face of it, the long bed of the *chorobates* does suggest its usefulness for aqueduct surveying on the grounds of accuracy alone, as Vitruvius says.[28] On the other hand the inescapable objection to the device's general use is its being excessively unwieldy and on any number of aqueduct lines I have visited I can believe it must have been so; indeed, on all sorts of steep, bumpy or rough terrain it is hard to believe the instrument was any use at all, nor where sharp curvatures arose.[29] Also, just recently, Jean-Pierre Adam in France has thrown some interesting light on the *chorobates* as a working instrument by making a replica and testing it.[30] His main conclusion is interesting because he achieved an accuracy (against a modern theodolite as standard) of 4 cm in 344 cm, that is 1.16 percent, over a distance of 51.3 m. For aqueduct surveying over long distances and gradients of, say, 1 in 1000, that is not impressive even allowing for the replica's relatively short length of 1.5 m.

Myself, I am inclined to think that more consideration should be given to the possibility that the Romans used the A-frame level for water channels, setting the instrument with its apex uppermost so that the plumb-line intersected a mark of scale on the bar. The A-frame level, or *libella*, is usually associated with carpentry and building contruction, a small-scale instrument in other words, and yet it has its place in the setting out of canals and aqueducts over many centuries. Islamic writers describe plumb-bob levels for canal work and Ibn-al-Awam specifies a slope of 1 in 200 for irrigation ditches.[31] In Juanelo Turriano's "Twenty One Books" the *libella* is shown[32] in use on canals and there are numerous references from his time and later including what appears to be a graduated instrument, in effect a clinometer. To what extent the *libella* continued to be used in engineering surveying is difficult to say but it is very significant that in 1915 Ernest McCullough was writing, "The early miners in California set grade pegs on hundreds of miles of ditches and roads with this primitive instrument".[33] There is such a thing as a traditional method and the A-frame level is not to be ruled out as a Roman aqueduct instrument whose compactness would have been an advantage but whose inherent limitations are still hard to reconcile with the gentle gradients achieved in practice.

When all is said and done the Nîmes aqueduct, with its minimal gradients, must have been the most challenging task for its surveyors and sure enough there is more than a hint that the job went wrong. For example, at the place where the excavation of the railway cutting exposed the aqueduct, there was revealed not one tunnel but two, side-by-side, and virtually the same piece of construction (Fig. 5). The explanation which is occasionally encountered is that one of these tunnels, the one showing no signs of incrustation, was once used, and once only, to drain a marsh. In reality is it not conceivable that it was intended to be part of the aqueduct, was inaccurately levelled, and the reason why it shows no evidence of being used is because it never was? It was substituted by a new tunnel which corrected the levelling error and it is that tunnel which shows every sign of sustained use.

I believe it is possible that there is other evidence of surveying mistakes, and more dramatic. We must now consider the Pont du Gard itself.

THE PONT DU GARD

The bridge has been frequently and comprehensively described with regard to every detail of its size, shape and composition.[34] Suffice to say here, it is 48.77 m high, 275 m long and contructed of large blocks of a softish, golden-coloured limestone[35] laid without mortared joints. Conceivably the spandrels contain a proportion of rubble concrete. Features of the means of construction are evident in the big projecting stones and the recessed ledges which supported scaffolding and centering; the separate voussoir rings which probably used common centering

Fig. 5. The "cross-section" exposed by railway construction to the south-west of Sernhac. The tunnel to the right is heavily incrusted; the one to the left is clean.

THE PONT DU GARD AND THE AQUEDUCT OF NÎMES

Fig. 6. Constructional numbers on one of the voussoir rings of an arch in the second tier of the Pont du Gard. The sequence reads: FR.S.II; FR.S.III; FR.S.IIII.

positioned in sequence across the width of an arch (four rings in the bottom tier, three in the middle); and the unusually interesting numbers (Fig. 6) on a few of the voussoirs which delineated their fit and contructional position.[36] The largest arch is the one which crosses the river, a clear span of 24.52 m and 29.32 m between pier centres. To the south of this biggest arch are three of 19.2 m (23.76 m between the piers) and there are smaller arches at the extremities. The pier centres of the middle tier match those below and the arch spans are therefore the same as their low-level counterparts. Because the spans vary somewhat the imposts on certain piers are at different levels. The piers at water level have upstream cut-waters but there do not appear to have been any on the downstream side, the usual Roman arrangement.

One other feature of the bridge never fails to be emphasised but it is surely not original, rather a sign of old age. Along the crest it is impossible not to notice a slight upstream curvature; (Fig. 7) the offset may be a metre or two. That it is not recent in origin is evident from old photographs. Nor can it be deliberate and the opinion that here is a structural device designed to give the Pont du Gard resistance to the pressures of wind and water is fanciful in the extreme. For one thing the convexity is not visible at river level. Also, because it only occurs in the upper part of the bridge, it gives the structure a pronounced overhang which adds anything but extra stability. Moreover, no such characteristic is claimed for any other Roman aqueduct bridge to my knowledge, not even the more slender specimens such as Segovia or Mérida.[37] No, the explanation must be that in the course of time the Pont du Gard has twisted. This might have been due to settlement, or perhaps was induced by cutting the middle piers for the one-time roadway, or conceivably Pitot's road bridge has leaned too hard from the opposite side. There is a worrying aspect to all this because the belief that what, in reality, is a defect should be applauded as clever design shows complacency over the bridge's future security as well as ignorance about its contruction.

THE PONT DU GARD AND THE AQUEDUCT OF NÎMES

Fig. 7. The 'curve' of the Pont du Gard is best appreciated on the down-stream side. This view is from the north. Also visible are projecting stones to support scaffolding.

Just recently the Pont du Gard and its design have figured centrally in the writings of Professor George F. W. Hauck[38] and it is not inappropriate to deal here and now with his approach and conclusions. Hauck's position, essentially, is that retrospective calculation will reveal aspects of the Pont du Gard's design and its structural behaviour and that thereby light is thrown on the ability and knowledge of the Roman civil engineer. Hauck has conducted stress analyses of the side wall of the conduit, proved that the no-tension condition is very nearly met, and concluded that therefore the design is good. More impressive though is a thorough stress and stability analysis of the whole bridge when subjected to side winds as well as the effects of the Gard in full spate, a worst condition brought about when wind and water roar down the valley simultaneously. By applying the middle-third rule to two levels of the bridge it transpires that in a hurricane force wind whose ground velocity is 215 kph the bridge is initially at risk at the base of the middle tier, while at the lower wind velocity of 150 kph a factor of safety of 2 against being blown over is achieved at the same level. In addition, when the factor of safety of weight against wind is 2 in the middle level it is 2.23 at the base and this reserve, so Hauck argues, is just enough to cover the extra pressure on the bottom piers when the Gard is in the condition witnessed by Mérimée. Professor Hauck evidently regards all of this as evidence of good, even precise, design. I wonder. A rigorous mathematical demonstration that the Pont du Gard is still standing because it has not been hit by a wind of 215 kph and a massive flood of the Gard at the same moment in the last 2000 years is a somewhat lavish proof of the obvious. Moreover, while there emerges an evaluation of the bridge's behaviour in modern terms, it is difficult to see how in any sense we are being introduced to a Roman view of the structure and its performance, nor indeed that the Roman engineer had a view of these things, at all.

Calculations of the above type as a means of satisfying the analytical curiosity of present-day engineers are clearly harmless up to a point but when they begin to assume the role of historical analysis the position is not so simple.[39] Engineers' questions are not necessarily historical questions and any claim that analysis in the Hauck style can be revealing of the ideas in the mind of the Pont du Gard's designer or the structural concepts he understood is potentially very misleading. Personally I am profoundly sceptical that we have, or can have, any grasp of how a Roman civil engineer pictured the behaviour of a structure or comprehended the loadings and forces at work. In other words the notion that we can work back to the thoughts, and especially design thoughts, in men's minds from the objects they made, without the crucial contribution of the written word, is, I think, extremely doubtful. And I suspect this to be true in other cases where retrospective analysis—mathematics, models or even full-scale replicas—has been resorted to. In the structural field Robert Mark's work on the Pantheon[40] and Gothic cathedrals[41] is a case in point. The real issue here of course is that of the essential difference between craftsmanship and engineering science. Know-how is not knowledge and it cannot be made so because modern analysis can demonstrate, *post facto*, that ancient practitioners unwittingly and fortuitously had got it right.[42] Ancient swordsmiths were not metallurgists because modern metallographic analysis proves how superlatively they could forge a blade.

Let us return to more straightforward matters. Another of George Hauck's observations concerns the novelty of the Pont du Gard's size. He writes, "Yet the Pont du Gard, with its extraordinary height and soaring arches, was a dramatic departure from anything the Romans had done before".[43] Of course it is easy to see what is meant here; the Pont du Gard *is* a massive and dramatic piece of work (the more so to modern eyes because it is so complete) but within the sequence of Roman bridge building Hauck's comment needs some qualification.

By the end of the first century B.C. Roman engineers had assembled a good deal of bridge building experience and they were acquiring more during the Augustan period. Some impressively large spans had been constructed:[44] the Ponte Milvio, for example, featured spans of up to 18 m as early as 109 B.C.; the Pons Fabricius in Rome has two spans of 25 m dating from 62 B.C.; the Pont St. Martin clears 37 m and the Ponte della Badia 35 m. Two Augustan bridges at Ascoli Piceno have spans of 22 m (the Ponte di Porta Cappuccina) and 15 m (the Ponte di Cecco) and the most impressive of all is the great Ponte di Augusto at Narni, 32 m in its biggest arch and 160 m long. The Hauck view is that the Pont du Gard is itself Augustan in which case its pedigree is obvious and the spans of its arches are not exceptional. What is unexpected is their application to water supply.

Lastly we must consider the Pont du Gard's upper tier, the range of small arches carrying the water channel itself. The spans of these thirty-five arches vary subtly because the engineer cleverly adapted them to match the spacing of the big arches below. Indeed, a number of writers[45] have been greatly impressed by the aesthetic fashion in which the third tier was integrated into the design without considering that the whole exercise was, very possibly, an unwanted and unexpected nuisance. In short, perhaps the third level was an afterthought, so to speak, a remedial modification to overcome what we were considering just now, a surveying error.[46]

THE PONT DU GARD AND THE AQUEDUCT OF NÎMES

Three factors conspire to suggest this possibility. The first is that the Pont du Gard is unique amongst aqueduct bridges, ancient ones that is, in being three-tiered at all, never mind in combining two giant tiers with one dwarf.[47] Everywhere else it was the two-stage bridge (or of course one) that was used and gradually developed into the final form of the Roman aqueduct bridge as at Mérida or Metz. Secondly, it is odd that if a bridge of the height we see was always intended then quite modest increases—no more than 2 m—to each of the lower tiers would have sufficed. Structurally this does not represent more than a trivial increase in scale well within the capabilities of bridge and builders alike (Fig. 8). Thus the inference could be that the bridge was up to its second stage *before* the need for greater elevation, about 4 m, was appreciated. And thirdly, there is the revealing profile of the aqueduct's gradient. It has often been noticed that the conduit descends relatively steeply from Uzès to the Gard (a gradient of 1 in 1450) and then is seriously deficient in slope from there on. Of the 17 m of fall available in 50 km, 11 m is used up in the first 16 km, leaving only 6 m over 34 km to get the water from the Gard to Nîmes. Moreover, this is the case *with* the third tier so that without it the consequences of the surveying error, if such there was, would have been very serious indeed.

Of course we shall never know but one other point is relevant to the curious surveying of the Nîmes aqueduct and this takes us downstream to the last topic. Within Nîmes, the *castellum divisorium* appears to have been situated more than generously above the Roman levels to which it delivered water. The minimum difference of level at delivery is of the order of 10 m and at a maximum nearer to 15 m. Now this is odd because here is an aqueduct which falls a mere 17 m in 50 km but then commands an elevation of 10–15 m above the city whose furthest customer was no more than about 1550 m away. In other words a further indication that overall the surveying of the Nîmes aqueduct was faulty is that self-evidently the *castellum* was too high. Put it 3 or 4 m down the hill and the fall of the aqueduct is materially improved—around 20%—without impairing, one would have thought, water flows around the city.

There is of course more to the *castellum* than its level relative to Uzès upstream and Nîmes below. It was fundamentally an hydraulic device and must be considered accordingly.

THE *CASTELLUM DIVISORIUM*

The Nîmes *castellum* (Fig. 9), half-way up the Rue de la Lampèze, is one of two which are well-preserved, the other being at Pompeii.[48] Basically it is a low, circular basin nearly 6 m across and 1.4 m deep. As we see it today the *castellum* is completely open but in its heyday it

Fig. 8. The Pont du Gard as built is shown on the left; the conjectural two-tier bridge is to the right.

Fig. 9. Two views of the *castellum* in Nîmes.

would have been enclosed by a circular wall and tiled roof (or perhaps a dome?), decorated with frescoes showing dolphins and other aquatic motifs and fitted with a bronze balustrade around the edge of its promenade. In short it was a public building to be admired.

The hydraulic features of the *castellum* which survive are readily described; they are not by any means so easily explained. The aqueduct, 1.2 m wide and 1.1 m high,[49] arrives from the east and at its point of entry to the *castellum* it was equipped with some form of vertical sluice-gate whose grooves are visible in the side-walls of the conduit. Some sort of mechanism for raising

THE PONT DU GARD AND THE AQUEDUCT OF NÎMES

and lowering the sluice was connected to it by rods or wires passing through holes in the pavement. How the castellum worked has exercised many minds. Their explanations range from the unconvincing to the excessively complex and sometimes both. Two who have written on the problem recently are Professor A. Trevor Hodge followed by the same Professor Hauck who tackled the Pont du Gard. In 1984 Hodge[50] developed an ingenious argument to answer the long-standing question of how Frontinus measured the *quinaria*. Frontinus' *The Aqueducts of Rome* contains a lengthy discussion of the flow in the city's aqueducts, the need to gauge them and the problems arising as a consequence. But how it was done is a puzzle. Hodge's proposal is that a sluice gate was lowered into the flowing channel causing the level to back up above the sluice with a consequently increased rate of flow under the sluice and the formation of a hydraulic jump down-stream where the former flow conditions re-established themselves (Fig. 10). Two

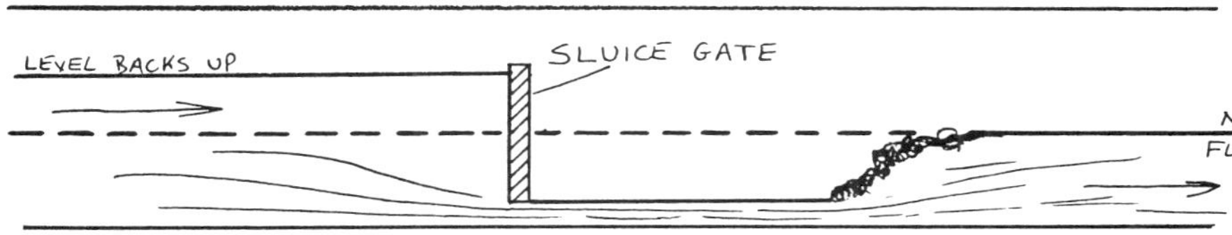

Fig. 10. The essence of the Hodge proposal for Frontinus' method of gauging an aqueduct. In the case shown the lowered sluice-gate is causing free discharge; with submerged discharge the level to the right of the gate would be re-established very much earlier.

measurements—the gap under the sluice and the depth of water above—will determine the flow and hence by this means, and it is sound both theoretically and practically, an aqueduct might be gauged. Overall it is quite a persuasive explanation the main question being, had Roman hydraulics advanced to a point where such refined concepts had been grasped? Hodge himself writes, "It was certainly within the bounds of what was practically feasible in the ancient world; no less certainly it raises the possibility that Frontinus, or at least the engineers working under him, may have been a good deal more sophisticated in their knowledge of hydraulics than some modern commentators have suspected."[51] True, I would not have suspected it myself and furthermore there are practical aspects of what is proposed which remain problematical.[52]

The Hodge theory is taken further by Hauck[53] who bases his extention of the idea on a conviction that flow measurement and checking discharge was a preoccupation of Roman water-supply engineers. But why should it have been? Is it not more likely that capturing the source and conveying it to its destination was the crucial issue and providing the supply was sufficient all was well; if there was too much water it would be run off while a deficient flow could not in any case be rectified by gauging it. It has to be remembered that Frontinus' problem was different in so far as he was investigating leaks, theft and contradictions. At any rate, the Hauck proposal for the Nîmes *castellum* sluice is not one gate but two, running in separate grooves over the full height of the channel and thus able to overlap completely. In normal flow, and thus virtually all day long, both gates were fully raised and all the water streamed underneath. Once a day, so the argument runs, both gates would have been judiciously adjusted to allow a measurement of the flow in the manner, to all intents and purposes, of the Hodge proposal for Frontinus' single gate. If anything Hauck is willing to ascribe an even greater sophistication of hydraulic concepts to Roman engineering than Hodge. By any reckoning a two-gate system requires more insight into open-channel flow than a one-gate system. At one point Hauck claims[54] that "Our findings suggest that the Roman engineers grasped a principle generally thought to have been discovered by 19th century hydraulic engineers". In fact it is eminently likely that the principle *was* discovered in the nineteenth century and when one examines the complexities of solving open-channel flow problems in the eighteenth and nineteenth centuries the chances that Roman engineers made the break-through mearly two thousand years before seem exceptionally remote and fundamentally at odds with what can be ascribed to intuition in the absence of calulation.

In reality Hauck's elaborate analysis of the *castellum* sluice is unacceptable for two reasons: it is actually at odds with the archaeological evidence and there is in any case a much simpler explanation. His two sluice system would have required suitable and separate guides within the walls of the aqueduct. Running two sluice gates up and down against a flow of, say, 350 l/sec. at a velocity of about 0.5 m/sec. is not that easy. Moreover, all one can detect in the walls of the channel is a single, narrow groove. Of course, the passage of time will have degraded the evidence but all the same the provision for twin sluices is simply not to be seen. Much more revealing are the deposits of calcium carbonate on the walls and roof of the conduit. The conduit is very heavily incrusted—it is true for the whole length of the Nîmes aqueduct—and the deposits are thickest along the roof and in the upper reaches of the side walls. In short Hauck is wrong; the flow was not normally along the floor of the channel at the entrance to the *castellum*. In fact it must have run full, or nearly so, for a good deal of the time, and this suggests that the discharge was *over* a two-part sluice whose bottom plate was a sill fixed in position and whose top half only moved up and down. A fixed lower sill is confirmed by the original archaeological findings of A. Pelet who discovered the *castellum* in 1844. He reported traces of lead along the floor of the channel so it is pretty clear that the lower fixed plate was sealed by lead run into a groove. There is more. The incrustations on the roof of the aqueduct extend beyond the sluice-gate groove, in fact right to the edge of the pavement slab. Therefore Hauck's contention that two gates resided against the roof of the channel in normal flow appears to be flatly contradicted by the evidence. (Fig. 11)

Fig. 11. Deposits whose thickness increases up the sides of the aqueduct and which are even thicker across the roof suggest that at the point where it entered the *castellum* the channel ran full over a low-level sluice.

THE PONT DU GARD AND THE AQUEDUCT OF NÎMES

Hodge and Hauck and others have missed the most obvious and much simpler explanation. The *castellum* sluice gates never were intended to measure discharge or regulate channel flow at all. They were, quite simply, designed to control the level. Upstream, at one or more locations, there were other outlets, probably with their own sluices, meeting the needs of further water-supply or perhaps irrigation or waterpower. Such alternative uses might, conceivably, have been developed at dates after the aqueduct's first construction. In general support of this interpretation of the sluice system we can notice that (a) generations of millers and irrigators have resorted to the judicious adjustment of sets of gates to distribute water here and there according to demand and circumstances, (b) there is evidence from elsewhere that aqueduct waters were re-assigned with the passage of time, Rome itself being on example and Barbegal another[55] and (c) Philippe Leveau[56] has recently reported the extent to which other sections of the Nîmes aqueduct show independent evidence of changes of use and the re-designation of water away from the city to mills or farms.

Finally what of the three outlets in the bottom of the *castellum*? Clearly they were not simply kept open all the time or they would have taken much too large a proportion of the flow. More likely each was fitted with a vertical pipe (the slots into which something locked are still visible) so that excess water, over and above the flow required for the other ten outlets, was poured into the sewers to clean them.[57] It is entirely possible that the overflow levels of these three stand-pipes, by changing the pipes themselves, could be varied. In other words the variation in sluice-gate levels envisaged above is paralleled in a similar fashion by the three stand-pipes within the *castellum*. Thus we are looking at a consistent design concept and an elementary practical technique: the distribution and division of the available flow is being regulated and adjusted by judicious, and possibly quite subtle, variations of discharge levels in a network of overflowing sills. (Fig. 12) As I remarked just now, generations of millers and irrigators have used a similar routine. In effect it is the open channel equivalent of valves in a pipe system.

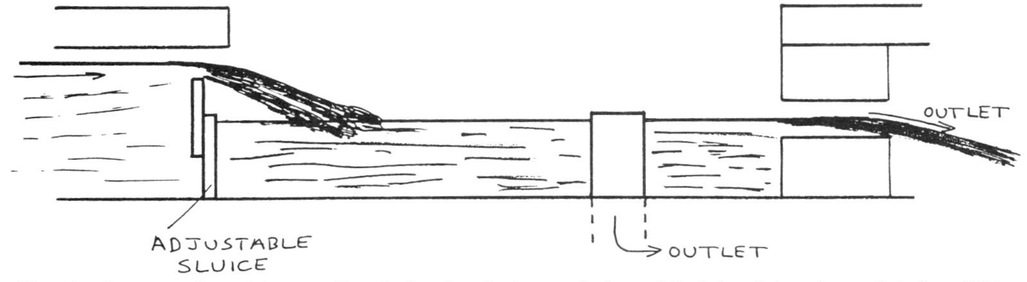

Fig. 12. Cross-section of the *castellum* indicating the inter-relation of the inlet sluice, the outlet pipes (10 in all) and the over-flow pipes (3 in number) used in combination to "balance" the *castellum's* flow.

Possibly the three lower outlets served one other purpose from time to time. By removing the stand-pipes completely the *castellum* could be drained. In this case, of course, the entrance sluice would have to be closed and the whole of the aqueduct's flow diverted, somewhere upstream, to other uses or simply wasted. The main objection to this procedure, as given already as a fault in the Hauck thesis, is that the incrustations on the roof of the channel, extending as they do beyond the sluice gate grooves, indicate that the sluice was never closed completely, or at most it only was in the early days. On the other hand it is possible that to carry out cleaning and maintenance of the *castellum*, certainly in later years, all the flow had to be diverted at some point, or points, well upstream. Nor should we rule out an even more comprehensive arrangement, namely closing the aqueduct at Uzès and overhauling the whole system, *castellum* and all, in one go.

THE PONT DU GARD AND THE AQUEDUCT OF NÎMES

The *castellum divisorium* was the end of the aqueduct and it brings to an end this survey of a few aspects of the engineering features of the Nîmes' water-supply. It remains now to discuss one or two points and try to draw some conclusions.

DISCUSSION AND CONCLUSIONS

One question yet to be considered is how much water the aqueduct could deliver. It is in fact a very difficult question to answer but since the matter is so often of interest it ought to be dealt with.

Oddly enough the calculation is both easy and complex at the same time.[58] A *theoretical* capacity of the conduit can be calculated from its slope, cross-section and roughness. The slope is the one parameter we can be sure of. Roughness has to be estimated by comparison with modern materials while determining the cross-sectional area of flow is subject to two uncertainties: one is that Roman aqueducts were bigger than required merely for the flow of water because they had also to allow room for workmen to carry out repairs and cleaning; and secondly the accumulation of deposits would gradually modify the cross-section. At least the marks of incrustation can reveal the nominal depth of flow and this is particularly true of the Nîmes aqueduct which has very thick deposits indeed. Over the years estimates of the channel's capacity have thus varied widely, from as little as 100 l/sec to as much as 2300 l/sec (if it is assumed that the conduit is running full). But in addition it is just as important to know the capacity of the source. Unfortunately estimates vary considerably. Therefore the not very satisfactory conclusion has to be that the source at Uzès was evidently quite capable of supplying Nîmes with a reasonable quantity of water along an aqueduct that was certainly adequate when new and still serviceable after many years of incrustation had reduced its cross-section. However, for the sake of argument and in order to make a general observation about this question of quantities, let us assume a relatively modest flow of 250 l/sec (say nearly 5 million gallons/day). What do we do with this number? what does it mean? The population of Nîmes is uncertain; it could not have been as low as 5000 nor as high at 500,000. A figure of 50,000 is frequently quoted as a plausible estimate. Even so we cannot derive a realistic per capita consumption from it. The *hoi polloi* were not on the mains, so to speak, so that their access to the supply and their use of it is highly problematical. The aristocracy's consumption is equally difficult to comprehend and beyond that is the intractable problem of relating population, assuming that is known, to the quantities run through sewers, baths and public fountains, without doubt the biggest consumers of all. It is important to bear in mind the extent to which Roman water-supply was not merely a practical proposition meeting a public need. Achieving an ostentatious display of civic ability and pride was important as well, and one which did not exclude public health and domestic comfort as a bonus. Two thousand years later the Victorians were seized by the same sense of civic satisfaction; regrettably it seems to have deserted us in the twentieth century.

Examination of the Pont du Gard and the Nîmes aqueduct leaves unexplained the fundamental surveying problem; how was a high degree of levelling accuracy—notwithstanding the errors discussed above—achieved with inherently inaccurate instruments? For *very* shallow gradients one possible explanation is that the constructed channel was used as its own level. In other words as it was built the channel was filled, or partly filled, with water. If the slope required was less than 1 in 5000, say, then the fall per 100 m would have been only 2 cm (or less). As construction proceeded it is imaginable that this drop could be provided by slightly sloping the channel's water-proof lining along each length of the built aqueduct. However an accurate initial survey was still crucial and how that was done is very puzzling. It is plausible—although contrary to what is usually assumed—that different surveying instruments and techniques were used for the successive stages of an aqueduct survey. Without a doubt a *chorobates*, with as long a bed as possible and located securely at carefully chosen stations from which there were good sight lines, was essential for the initial levelling which above all had to be accurate. This survey would not have to follow the aqueduct line although it would need to be near enough to it to allow the setting up of bench marks through which the construction would be laid. Probably the technique for setting the gradient of the aqueduct depended on the slope required, an A-frame level or similar device being suited to steepish slopes and water-channel levelling, as suggested above, being only really practical when the gradient was exceptionally gentle.

Next there is the question of design. Unfortunately it is all too easy for ancient technology to find itself more at the mercy of hindsight than in its debt. So often the reconstruction of how things were done is flawed by a tendency to over-elaborate, to attribute more than is necessary for explanation or warranted by the evidence. Another defect, and in the context of the Nîmes aqueduct the views of Professor Hauck typify the risk, is to assert that ancient design was a rational process capable of precise and correct results. Statements which claim that because the circular pipes of the *castellum* ran half-full (which is the optimum) they represent "an example of first-rate hydraulic design" or then, in relation to wind loading on the bridge, "The conclusion of this approximate stress analysis is that in the design, the wind loads, difficult to predict as they are, were reasonably taken into account," convey a misleading impression of the Roman engineer's command of concepts and expectations of design. And in any case, how was it done? How were structures and aqueducts supposed to have been designed to a standard which satisfies modern criteria?, a question which is made the more baffling given that the Hauck position does, here and there, admit to the profound limitations of the Romans' command of hydraulic and structural concepts. Nor incidentally is the sample very large. What might be impressive would be a demonstration that fifty, or even ten, bridges or aqueducts had all been designed to exactly the same limits. That the Hauck approach can reveal conclusions rather than coincidences is not easy to justify unless the sample is suitably large and unless some opinion is proferred as to how Roman design was conducted. Ancient design has to be judged on its own terms; attacking it with lots of modern analysis to see if it comes up to par is a contradiction in terms.

Is it possible to have any idea of the terms within which ancient design was conducted? In so far as Roman bridges and aqueducts had to be somehow proportioned in order to be built at all then a design process of some sort was a necessity. It is ridiculous to imagine that one day a gang of workmen with masonry turned up on the banks of the Gard and got busy. Exactly how the size and shape of the Pont du Gard were determined is an intriguing question and two sources who might, however obliquely, throw a little light on the problem are Philo of Byzantium and Vitruvius. They both discuss a procedure for calibrating catapults which is more controlled and objective than mere trial-and-error. The relevant parameters are identified and assembled into a quantitative relationship whose solution is intricate to the point of requiring the determination of cube roots. It is inconceivable that ancient structural design could have been based on objective testing, and difficult to imagine it for aqueducts, but what we might reasonably entertain is the idea that design formulae, or 'proportioning' formulae, were evolved which in some way, however rudimentary, related the parameters relevant to the problem. For bridges such a formula might embrace pier width, arch span, arch thickness and depth at the crown of the arch while for aqueduct channels the relevant parameters could be slope, width, depth of water and volume of flow. By the time the Nîmes aqueduct and the Pont du Gard were built the accumulation of experience in both fields was considerable for the Roman period alone, not to mention the Greeks before that, and the proposition is realistic in principle. However, I am bound to admit that for bridges the reconstruction, for the ancient world, of the type of design rule which was after all to persist into the nineteenth century has eluded my calculations. It is an interesting question therefore as to whether a design rule can be resurrected by an examination of the artefacts designated by it even when the sample is quite large.[59]

THE PONT DU GARD AND THE AQUEDUCT OF NÎMES

All the same I believe that the design of the Pont du Gard was essentially straightforward. Only one decision was crucial and that was the choice of an arch of 24.5 m to cross the river in its normal state in one span. From there the use of some empirical rule of the type referred to above, and applied in conjunction with the width of the valley, would readily determine the lay-out of both the large tiers. The questionable third tier could be rather easily adjusted, at whatever stage in the proceedings the need for it became apparent, to match the arches below. The width of the top of the bridge was that of the water channel itself while the widths of the lower tiers must have been arrived at merely by providing sufficient space for workmen to carry out construction and maintenance.

Finally back to the Pont du Gard's date. When this problem was raised at the beginning it was hinted that an appeal to engineering technique and structural style might be helpful, so here is what can be done. It is simply a fact that the reign of Augustus was rich in civil engineering in Rome, in Italy, and beyond, especially in Gaul and Iberia. As the emperor's right-hand man Marcus Agrippa was very influential as well and, as the engineer in charge, more intimately involved. He was responsible for aqueducts in Rome, for overhauling the city's sewers and the Tiber's banks and there is a strong suggestion that a Pons Agrippae once stood close to the position of the Pons Aurelius. Augustus re-built the via Flaminia which included the great bridge at Narni and the Ponte Augusto at Rimini. As mentioned earlier two big bridges at Ascoli Piceno are Augustan and even more revealing, stylistically and technically, is the Pont St Martin. One of its features is independent arch rings, four of them, separated by concrete infill. The technique of repeatedly using the same centering across the width of a bridge is exactly what is found in the Pont du Gard.[60] The Pont du Gard shares some constructional similarities with the Ponte di Augusto at Narni as well. Both bridges utilised projecting stones and recesses to form ledges so that scaffolding and centering could be supported during construction and for subsequent maintenance. So far as this evidence goes it is not unreasonable to think of the Pont du Gard as being part of the Augustan/Agrippan tradition and close to it. Somehow one senses that the builder of the Pont du Gard knew about road bridges and to produce an aqueduct crossing of 50 m elevation he piled them three high.

Henry James observed that the Pont du Gard is a massive solution to the problem of channelling a modicum of water to a small town. And the truth is that Roman engineers soon realised that a pile of road bridges, however well tried and understood were the basic techniques, was a laborious and generally over-elaborate approach to the problem; perhaps it was judged to be too expensive as well. During the first century, and into the second, aqueduct bridge building evolved markedly. Bridges such as those at Tarragona, Segovia, Mérida, Cherchel and Metz illustrate a transition in both structural form and materials used.[61] The bridge-stacking arrangements of the Pont du Gard give way to a style in which the piers are more continuous from bottom to top and the increasingly vestigial arches of the lower tiers become mere bracing between the tall piers. At Metz they appear to have been dispensed with altogether. The evolution of Roman bridge building is too big a subject to elaborate here; suffice to say the changes did occur and another manifestation is the move towards a more regular and rational use of concrete. The great aqueduct bridges at Mérida[62]—Los Milagros and San Lazaro—which date from the second century exemplify these developments at an advanced stage for here the piers are continuous, rubble concrete cores are encased in bossed masonry, flat arches of brick brace the structure longitudinally while lateral bracing is provided by deep counterforts running the full height of the piers. The Pont du Gard is not like this and I believe it is older.[63]

One other factor suggests an early date. Had the Pont du Gard been built later by 50 or 100 years it might be conjectured that a siphon would have been used. Siphons were used at Lyon as a matter of routine although admittedly the crossings being attempted were, with one exception, deep and long. The exception is interesting. On the Mont d'Or aqueduct the siphon across the

THE PONT DU GARD AND THE AQUEDUCT OF NÎMES

Ruisseau de Limonest was 400 m long and reached a depth of 30 m to its siphon bridge.[64] Now these dimensions are comparable to the Gard crossing so that it is pertinent to observe that in at least one case a siphon was preferred to a bridge of the size of the Pont du Gard. Another revealing parallel is to be found at Toledo. There a deep gorge above a violent river—the Tagus is probably more menacing than the Gard—was crossed by a bridge of three 28-metre spans and 50 m high. The siphon it carried was 50 m deep so while the Toledan example is to a bigger scale it *is* an entirely comparable situation.[65] So it is arguable that had the Pont du Gard been built well into the first century, or later, rather than at the beginning, a siphon might have been a very practical alternative.

Toledo provides a final and symptomatic comment on our subject's pervading influence because even there the Pont du Gard very nearly had the last say. In the 1930s, when the Spanish engineer Alphonso Rey Pastor[66] proposed an exceedingly splendid reconstruction of the bridge he imagined had once spanned the Tagus, yes, of course, it was a Pont du Gard look-a-like he chose.

NOTES AND REFERENCES

1. O. Brogan, *Roman Gaul* (G. Bell, London, 1953), p. 81.
2. This translation is taken from D. B. Steinman and S. R. Watson, *Bridges and Their Builders* (Dover, New York, 1957), p. 51.
3. An edition of Faraday's journal describing his travels with Humphry Davy in Europe in the years 1813–1815 has been prepared by Brian Bowers and Lenore Symons and it will be published in 1991 under the title *Curiosity Perfectly Satisfyed.*
4. From Merimée's *Notes d'un voyage dans le Midi de la France*, 1835 and quoted in V. Lassalle, *Le Pont du Gard et l'Aqueduc Romain de Nîmes* (Editions de la Source, Montpellier, 1980), p. 4. (my translation).
5. Following his escape from the Chateau d'If Edmond Dantès tracks down Gaspard Caderousse in an inn called the *Pont du Gard.*
6. Henry James, *A Little Tour in France* (Sidgwick and Jackson, London, 1987), p. 170.
7. Op. cit. (6), p. 169.
8. Op. cit. (6), p. 169.
9. Op. cit. (6), p. 170.
10. Op. cit. (6), p. 170.
11. Several old views are reproduced in Emile Espérandieu, *Le Pont du Gard et L'Aqueduc de Nîmes* (Laurens, Paris, 1926); see also Henry James, op. cit. (6), pp. 168, 171 and V. Lassalle, op. cit. (4), p. 6.
12. Espérandieu, op. cit. (11), p. 57.
13. The essence of the restorations is set out in Espérandieu, op. cit. (11).
14. Illustrated in R. S. Kirby and P. G. Laurson, *The Early Years of Modern Civil Engineering* (Yale University Press, New Haven 1932), p. 211.
15. A very full contemporary description is George Rennie, "Description of the Bridge-Aqueduct of Roquefavour, on the line of the Canal of Marseilles", *Proceedings of the Institution of Civil Engineers*, Vol. xiv, (1854–5), pp. 190–238. See also the interesting remarks by Philippe Leveau, "Research on Roman Aqueducts in the Past Ten Years", *Future Currents in Aqueduct Studies* (Cairns, Liverpool, forthcoming).
16. It should be noted that the fact is not all that remarkable; large numbers, perhaps the majority, of Roman engineering relics—bridges, dams, harbours, mills, mines—have not been dated accurately.
17. Details of Agrippa's life are to be found in Meyer Reinhold, *Marcus Agrippa: A Biography* (Humphrey Press, New York, 1933).

THE PONT DU GARD AND THE AQUEDUCT OF NÎMES

18. Agrippa's water-supply work in Rome is discussed in Sir T. Ashby, *The Aqueducts of Ancient Rome* (Clarendon, Oxford, 1935); see also Harry B. Evans, "Agrippa's Water Plan", *American Journal of Archaeology*, Vol. 86 (1982), pp. 401-411.

19. A. F. L. Rivet, *Gallia Narbonensis: Southern Gaul in Roman Times* (Batsford, London, 1988), p. 167.

20. Nicolayev's contention is rejected by Lassalle when he writes "M. Nicolaiev has recently maintained that the work must be somewhat later, the Pont du Gard being, according to him, the outcome of an evolution in aqueduct bridges towards more slender construction. But it is difficult to place the structure in an evolving sequence because, as we have seen already, its exceptional qualities seem to be less a matter of time than place": V. Lassalle, "Le Pont du Gard et l'aqueduc de Nîmes", *Dossiers de l'Archéologie*, No. 36, Oct-Nov, 1979, pp. 60-61.

21. Good descriptions of the aqueduct's route and the surviving components can be found in E. Espérandieu, op. cit. (11); V. Lassalle, op. cit (4); G. Hauck, *The Aqueduct of Nemausus* (MacFarland, London, 1988), Chap. 1; and A. Grenier, *Manuel d'Archéologie Gallo-Romaine* (Picard, Paris, 1960), Tome IV, Pt. 1, pp. 89-97.

22. O. A. W. Dilke, *The Roman Land Surveyors* (David and Charles, Newton Abbot, 1971).

23. William E. Kreisle, "History of Engineering Surveying", *Journal of Surveying Engineering*, Vol. 114, No. 3, August 1988, p. 102.

24. For some discussion of these wide-ranging recommendations see Hugh Plommer, *Vitruvius and Later Roman Building Manuals* (Cambridge, 1973), pp. 26-299.

25. Jean Burdy, "Les Aqueducs Romains de Lyon", *L'Araire*, No. 66, 1986, pp. 48-50. Evidently the Brévenne aqueduct had *chutes* also.

26. But see Note 29.

27. Vitruvius, *De Architectura*, Book viii, Chapter 5.

28. The *chorobates* is described and discussed in Dilke, op. cit. (22) pp. 74-78 and also in E. R. Kiely, *Surveying Instruments* (Columbia University, New York, 1947) pp. 20-24.

29. On the other hand merely to establish the difference in level between two points it would, presumably, have been possible to select stations for the instrument which minimised the problems of setting up and sighting accurately. Levelling as such is not the same operation as setting a grade.

30. Jean-Pierre Adam, "Groma et Chorobate" *Mélanges de l'Ecole Française de Rome Antiquité* Vol. 94, 1982/3, pp. 1003-29.

31. Also, Islamic surveyors appear to have worked at night using lights as targets. In order to avoid errors due to distortion near ground level in the heat of the day it is an obvious ploy. We do not know if Roman surveyors were as shrewd but in setting an aqueduct such as the one which supplied Carthage over 132 kms, no less, it would probably have been an advantageous technique.

32. Pseudo-Juanelo Turriano, *The Twenty-One Books of Devices and of Machines* (Colegio de Ingenieros, Madrid, 1984), Vol. 1, p. 219 (and pp. 109-110).

33. Quoted in E. N. Stone, *Roman Surveying Instruments* (University of Washington, Seattle, 1928), p. 236.

34. See, for example, Espérandieu, op. cit. (11); Lassalle, op. cit. (4) and (20); Hauck, op. cit. (21); Grenier, op. cit. (21).

35. The stone was quarried north of the village of Vers and about 4 kms from the site. The quarries are still worked.

36. For example one sequence is clearly marked FR.S.II; FR.S.III; FR.S.IIII; FR.S.V meaning *frons sinistra* (front left) 2,3,4,5.

37. The lateral bracing of aqueduct bridges was sometimes achieved by buttressing the piers; Mérida is one example and Fréjus another.

38. See for example George F. W. Hauck, "Structural Design of the Pont du Gard", *Journal of Structural Engineering*, Vol. 112. No. 1, Jan 1986, pp. 105-120; George F. W. Hauck. "The Roman Aqueduct of Nîmes", *Scientific American*, Vol. 260. No. 3, March 1989, pp. 78-84; George Hauck, *The Aqueduct of Nemausus*, (McFarland & Co, London, 1988).

THE PONT DU GARD AND THE AQUEDUCT OF NÎMES

39. For some further elaboration of my reservations about these matters see my contribution to A. Trevor Hodge, *Future Currents in Aqueduct Studies*, (Francis Cairns, Liverpool, forthcoming).
40. Robert Mark and Paul Hutchinson, "On the Structure of the Roman Pantheon", *Art Bulletin*, No. 48, March 1986, pp. 24-34.
41. Robert Mark, *Experiments in Gothic Structure* (MIT Press, Cambridge, 1982).
42. Important reflections on this point can be found in A. Rupert Hall, "On Knowing, and Knowing how to . . .", *History of Technology*, Vol. 3 (1978), pp. 91-103.
43. George F. W. Hauck, "The Roman Aqueduct of Nîmes", *Scientific American*, Vol. 260, No. 3, March 1989, p. 78.
44. These statistics are taken mostly from Piero Gazzola, *Ponti Romani* (Olschki, Firenze, 1963).
45. For example A. Trevor Hodge, "Aqueducts" in I. M. Barton, *Roman Public Buildings*, (Exeter Studies in History, No. 20, 1989), pp. 139-140; and also the lyrical appraisal by G. Baldwin-Brown, "Roman Engineering Works and Their Aesthetic Character: the Pont du Gard", *Journal of Roman Studies*, Vol. 22 (1932), pp. 47-54.
46. So far as I know the only other consideration of this idea is Carlos Fernandez Casado, *Ingenieria Hidraulica Romana* (Colegio de Ingenieros, Madrid, 1983), pp. 413-415, 455.
47. Strictly speaking the Cherchel aqueduct of Oued Ileluin is three-tiered but here the lowest level is dictated as much by topography as structural form and so the design is not altogether comparable with the Pont du Gard.
48. The *castellum* is described by André Nadal, *Le Castellum* (Chastanier, Nîmes, 1974) and also by A. Grenier, op. cit. (21), pp. 97-101.
49. A. Grenier, op. cit. (21), p. 97 is wrong when he refers to "un ouverture haute de 1 m.80, large del 1m.20"; the height is 1.1 m.
50. A. Trevor Hodge, "How did Frontinus Measure the Quinaria?", *American Journal of Archaeology*, Vol. 88 (1984), pp. 205-216.
51. A. Trevor Hodge, op. cit., (21), pp. 213/214.
52. At least two operational problems arise. Frontinus' measurement of flow was based on the use of adjutages of varying cross-section—so many *quinariae*—set at a standard depth measured to the *centre* of the pipe or orifice involved. This standard depth is taken by Hodge, and others, to be 12 cms, just enough to accommodate the largest adjutage quoted by Frontinus, the *centenum-vicenum* of an assumed 11.5 cms radius. However, in an aqueduct of even modest flow it is obvious that the depth between the upstream surface and the *centre* of the sluice opening, once the gate had been lowered, would be much more than 12 cms. If by contrast—and I emphasise the contrast because it is not trivial—the 12 cms is taken to the *top* of the sluice opening, which is the technique Hodge (op. cit.(50), p. 213) appears to be advocating, then the basis of aqueduct gauging is *not* derived from the concept of the *quinaria* as such but rather it is a more than elementary modification of it, the derivation, to me at least, being less than obvious. Secondly there is the unconsidered question of what the gaugers would have done if the flow under the sluice, instead of streaming down the channel to form a hydraulic jump, that is to say free discharge, had merely re-filled the channel immediately below the sluice, in other words submerged discharge. What now, one wonders, would Frontinus have imagined was the standard depth? Did he realise that in this case the flow is proportional to the difference in level? How Frontinus measured the *quinaria in an aqueduct* remains a good question.
53. George F. W. Hauck and Richard A. Novak, "Water Flow in the Castellum at Nîmes", *American Journal of Archaeology*, Vol. 92 (1988), pp. 393-407; and for a shorter account see Hauck, op. cit. (43).
54. Hauck, op. cit. (43), p. 82.
55. In both places the water from long existing aqueducts was diverted to milling establishments, on the Janiculum Hills in Rome's case and to a purpose-built flour mill at Barbegal near Arles.
56. Philippe Leveau, "Research on Roman Aqueducts in the Past Ten Years", in A. Trevor Hodge, op. cit., (39).

THE PONT DU GARD AND THE AQUEDUCT OF NÎMES

57. Rather as suggested by Lassalle, op. cit., (4), p. 22.
58. For a recent and lavish example of the calculation see George F. W. Hauck and Richard A. Novak, "Interaction of Flow and Incrustation in the Roman Aqueduct of Nîmes", *Journal of Hydraulic Engineering*, ASCE, Vol. 113, No. 2, Feb. 1987, pp. 141–157.
59. J. B. Ward-Perkins claims for the Pont du Gard that its design is based on the proportions: "4 units for the central arches, 3 for the lateral arches, 1 for the arches of the uppermost tier, and 6 for the height of the whole structure". It is an interesting and frequently quoted 'design' which is unconvincing because in fact the proportions suggested are only approximate when measured against the Pont du Gard itself. For such proportionality to be *really* persuasive it needs to be exact. See J. B. Ward-Perkins, *Roman Imperial Architecture* (Penguin Books, Harmondsworth, 1987), p. 224.
60. However it must be recorded that the Pont St Martin—or Pont sul Lys to be strictly accurate—has been subject to re-construction and how far such work preserved the original Roman features is uncertain. See Piero Gazzola, op. cit. (44), p. 97.
61. For Tarragona, Segovia, Mérida and Cherchel see Carlos Fernandez Casado, op. cit. (48); Metz is in Grenier, op. cit (21), pp. 206–208.
62. Most fully described in Carlos Fernandez Casado, *Acueductos Romanos en España*, (Instituto Eduardo Torroja, Madrid, 1972), no page numbers, 3rd chapter.
63. Two qualifying observations are necessary here. The Gard is a river prone to heavy flooding. If the Pont Du Gard's engineers were aware of this and were anxious therefore to ensure the structure's security by constructing as few piers as possible then large spans were imperative *whichever* century the bridge was built in. Comparison with the style of the later aqueduct bridges which do not, with the exception of Metz, cross big rivers is then less useful as evidence for the developmental sequence. Also, there is the fact that the form of the Pont du Gard provides two 'platforms' from which to work upwards, to erect the tiers above. This might have been an important consideration in itself. In the history of structures generally too little consideration tends to be given to *constructional* procedures as against *structural* ambitions.
64. Jean Burdy, op. cit. (25), p. 62.
65. Norman A. F. Smith, "Attitudes to Roman Engineering and the Question of the Inverted Siphon", *History of Technology*, Vol. 1 (1976), pp. 66–67.
66. José A. Garcia-Diego, "Restoration of Technological Monuments in Spain", *Technology and Culture*, Vol. 13, No. 3, July 1972, pp. 426–429.

4

Vitruvius and the elevated aqueducts

F.D. Prager

Vitruvius's *Architecture* is a classic work packed with technical information and full of historical problems. This essay concerns the work's position on the technology and history of elevated aqueducts, which is beset by questions about the aqueducts themselves as well as the aqueduct text.

The ancient aqueducts were magnificent. Their channels had been elevated on boldly constructed arcades, which extended high above the plains, to conduct water flowing by gravity from distant hills into the cities and towns. Even the ruins are still impressive (Fig. 1).[1] In some cities, mainly in Spain, Roman aqueducts still operate. In Rome itself, the Porta Maggiore is still intact and it is possible to visualize how the

Figure 1. Elevated aqueducts: ruins of the Aqua Claudia, with remnants of the Anio Novus on top, a few kilometres south-east of Rome. While many arches were destroyed or used as quarries for building material, some arches (seen here in the middle of the picture) were repaired or supplemented by lower arches.

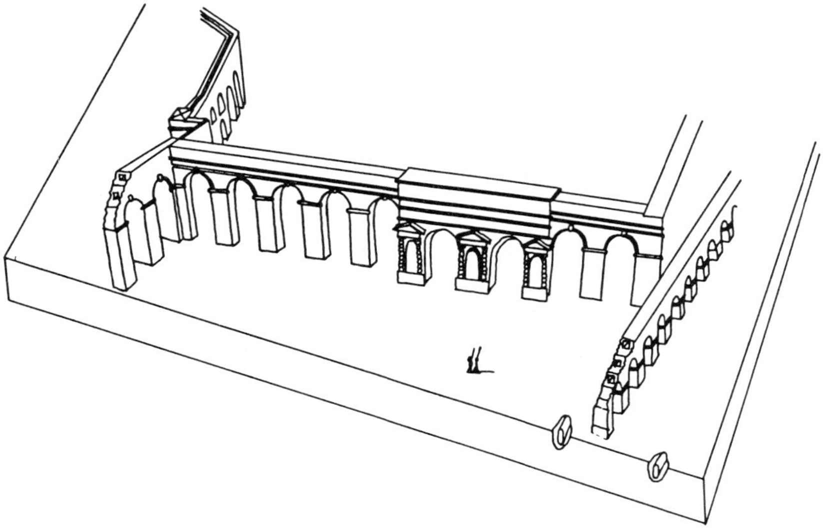

Figure 2. The major aqueducts entering ancient Rome. At right, the pre-Vitruvian ducts: The Anio Vetus below ground, and near it, somewhat above ground, the Aqua Marcia, carrying the Aqua Tepula and then the Aqua Julia. More highly elevated, and extending over the Porta Maggiore, the post-Vitruvian ducts: the Aqua Claudia and the Anio Novus on top.

aqueducts built at different times converged and extended over or near or under its arches (Fig. 2).

It has also been possible over the centuries to envisage in principle the entire course of the ducts (Fig. 3). However, accurate mapping was not easy. Some kilometres outside Rome, the elevated ducts disappear into the ground; conjecture or difficult searching is needed to establish the further course of each duct towards the source-head. Only in the present century was it possible to establish clear facts about this vast underground network (Figs. 4 and 5). This task was promoted mainly by a team of distinguished workers: Ashby, an English expert on Roman architectural history; Reina, an Italian expert on surveying optics; and Van Deman, an American archaeologist of Rome.[2] The results of their teamwork are still in need of study.

Other problems are encountered in the Vitruvian treatise, particularly in the notes on elevated aqueducts.[3] If these notes were originally accompanied by drawings or maps, which is unlikely, the illustrations have vanished long ago.[4] The text was criticized in late antiquity as being too wordy for practitioners: about A.D. 30 the abbreviator Cetius Faventinus called it 'extensive' as well as 'learned' and pointed out the need for a popular abridgement, which he produced — beginning with the aqueduct part — in order 'that more common workers be not prevented by the long and complex verbiage from studying the contents'.[5]

Today, by contrast, the old descriptions seem rather too short and

Frank D. Prager 107

Ductus autem aquæ fiunt generibus tribus, riuis per canales ſtructiles, aut fiſtulis plumbeis, ſeu tubulis fictilibus, quorum eæ rationes ſunt. Si canali/bus, vt ſtructura fiat q̄ ſolidiſſima, ſolumq; riui libramenta habeat faſtiga/ta ne minus in centenos pedes ſemipede, eæq; ſtructuræ confornicentur, ut minime ſol aquam tangat.

Figure 3. Aqueduct text and picture from *M. Vitruvius per Jocundum solito castigatior factus* . . ., Venice, 1511, folio 81r. At *c. fornix*, the covered channel construction.

dry for historical evaluation. According to a recent commentary by Callebat, Vitruvius 'disregards the different data affecting the arcaded aqueducts, which concern their economics . . . aesthetics . . . and technology . . .'; and also shows 'an irritating inability to distinguish the essential from the details'.[6]

We will investigate these difficulties. How does the Vitruvian text relate to the actual elevated aqueducts? How verbose or laconic is it? Why did it confuse the different readers? How did its author try to handle the matters which were essential at his time?

Vitruvius's qualifications as an authority on aqueducts

One of the problems about Vitruvius until recently was his dates. Because of Latin peculiarities encountered in the text, the *Architecture* was sometimes called a medieval production, and its dedication to the Emperor Augustus a forgery. However, reliable evidence shows that the first draft originated about 25 B.C., allowing for a variation of five to ten

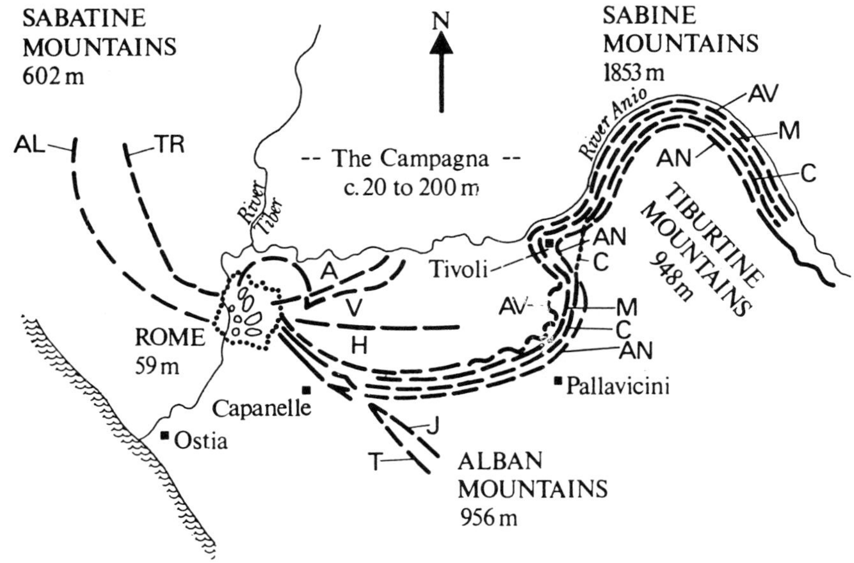

Figure 4. General plan of Rome's aqueducts

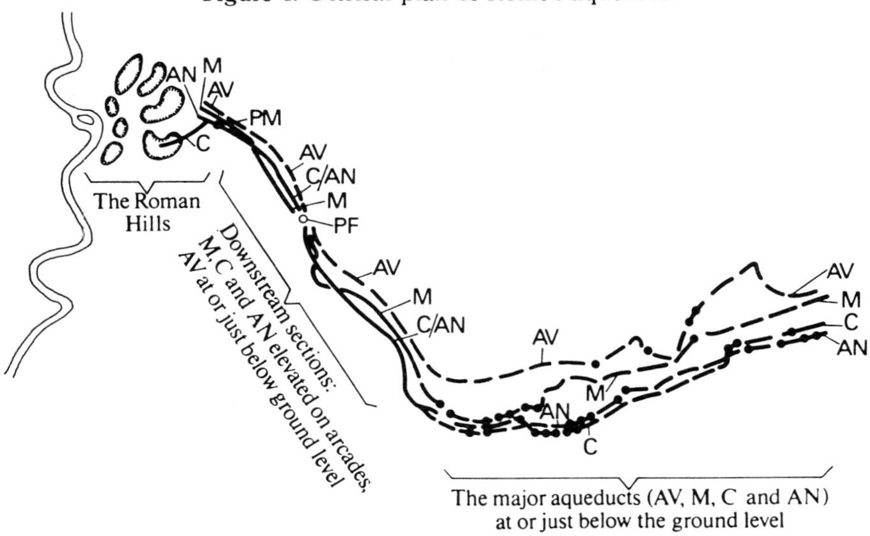

Figure 5. Detailed plan of the major aqueducts near Rome

KEY

A Aqua Appia (4th century B.C.)
AV Anio Vetus (3rd century B.C.)
M Aqua Marcia (2nd century B.C.)
T Aqua Tepula (2nd century B.C.)
J Aqua Julia (1st century B.C.)
V Aqua Virgo (1st century B.C.)
AL Aqua Alsietina (1st century A.D.)
C Aqua Claudia (1st century A.D.)
AN Anio Novus (1st century A.D.)
TR Aqua Trajana (2nd century A.D.)
H Aqua Hadriana (Alexandrina) (2nd/3rd centuries A.D.)
PM Porta Maggiore on Via Praenestina
PF Porta Furba on Via Tuscolana
● *putei* and *cippi* located by T. Ashby
■ towns
——— Elevated on continuous arcades
— — — At or just below ground level
- - - - - - - Deeply below ground level

years either way for preparation and completion. The work is therefore definitely of the Augustan age.[7]

Other problems relating to the technical notes in the work were introduced by subsequent writers who paraphrased, abridged or copied the work and in some cases misunderstood and corrupted the engineering expressions, which were converted into meaningless phrases. In modern times some of these errors were emended by such writers as Fra Giocondo, a Renaissance engineer, and Rose, a philologist of the nineteenth century.[8]

There is no firm evidence that Vitruvius had the qualifications of a water engineer or for teaching such work. He himself makes the pertinent assertions that he introduced a new system of urban water distribution and that he had hydrological experience and book-learning.[9] As an old man he writes nostalgically about the traditions of the 'ancients' in fields of architecture other than aqueduct building.[10] He does not claim that he invented anything concerned with elevated aqueducts. If he planned or supervised water systems, whether in Italy or the major provinces, the work is likely to have taken place under Agrippa, who was Augustus's son-in-law, counsellor, general and waterworks administrator.[11]

There are some items of information relating to remote places in the provinces. Vitruvius says he discussed water supply with a nobleman of Numidia;[12] and Algeria has furnished archaeological information which might indicate his presence there. At Thibilis (now Announa), parts of public baths and aqueducts belonging to the aquae Thibilitanae were excavated and inscriptions found.[13] Two of the inscriptions contain the name Vitruvius as well as the given name Marcus, which is usually assigned to the architect; whereas one includes the name Mamurra, which has been asserted and denied to be the cognomen of Vitruvius.[14]

The Vitruvian treatise mentions building traditions in Africa and other regions. It has also been asserted — and denied — that the author came from Numidia.[15] At this point it is pertinent to note two inscriptions, the final one in the paragraph above and another which has hardly been studied in connection with aqueduct arcades.

This Mamurra inscription, which is on a marble plate, is described by the excavator only as having been found in front of 'an edifice with arches' and reads:

M * VITRVVIVS
MAMVRRA
ARCVS
S * P * F

MARCUS VITRUVIUS
MAMURRA
BUILT THESE ARCHES
AT HIS OWN EXPENSE[16]

The other plate, which seems to come from the tomb of a woman related to Marcus Vitruvius, is a fragment which reads:

VITRVVI POT . . . .
TVLA * V * A XX . . . . .
M * VITRV . . . .

VITRUVIUS' . . . .
. . . . LIVED . . . YEARS
MARCUS VITRUV . . . .[17]

Together, the two inscriptions show that a man or two men called Marcus Vitruvius, who might not be our architect but might have been related to him, lived in Thibilis; that he, or one of them, financed the building of arches and perhaps technically directed the project; and that this building occurred near the 'waters of Thibilis', at an unknown date in antiquity, conceivably, as part of an aqueduct. Unfortunately, the treatise is silent on these matters.

In post-Vitruvian antiquity our architect or his treatise was mentioned occasionally.[18] We find a positive although not an enthusiastic view in Pliny the Elder's *Natural History*, an encyclopedia dedicated to the emperor Titus in A.D. 77. It mentioned Vitruvius as a scientific and technical authority on 'stones' and is silent about the treatise, but closely paraphrases the aqueduct chapters.[19] We find a somewhat similar view a few decades later in Frontinus's work on aqueducts where 'Vitruvius the architect' is mentioned in connection with the distribution system.[20] Pliny and Frontinus were naturally interested first of all in their own achievements, but both treated Vitruvius as a recognized builder and teacher in the indicated fields.

As already mentioned, a more critical attitude was adopted about A.D. 300 by Faventinus, but the treatise was considered authoritative. Certain sections, especially those on foundations and piers, remained very influential in the Middle Ages, while other sections, principally those on the classic orders, were completely forgotten in the Romanesque and Gothic periods; they found a revival of interest in the Italian Renaissance, but were promptly forgotten again in the Baroque period.[21] Among humanists and for some time among architects in the 1500s, Vitruvius gained enormous fame, and such writers as Cardano admired him as a unique genius of both science and the arts; they blamed the 'obscurities' of his treatise on his copyists.[22]

In the late 1700s, renewed classical studies made it clear that Vitruvius had sometimes been overestimated, and that his significance for Graeco-Roman art and science must be evaluated realistically. Some modern writers may then have over-reacted, declaring his work and most or all of Roman science, to be negligible.[23]

Some writers of the 1800s even charged him with plagiarism. This charge was surprisingly made by Rose, one of those who purified the text of errors introduced by copyists. He noted remarks by Vitruvius which had close parallels in the work of the earlier writer Varro — a source acknowledged in the treatise — and promptly concluded that Vitruvius 'copied everything' from Varro. Another antiquarian, Oder, found this theory exaggerated, but still believed that Vitruvius made false pretences about his sources and that he was 'the typical half-educated proletarian'.[24] Such disdainful views, as well as the former hero-worshipping, should be rejected.

Vitruvius's notes on elevated aqueducts

In his introduction on architectural harmonies, Vitruvius discusses matters to be studied by the architect. He states in Book I chapter i paragraph 7 and 8 that those matters include 'physics . . . of conducting water' as taught by Ctesibius or Archimedes and also 'canons of music'. Shortly thereafter he points out that there is a 'common bond' between these matters.[25] The statement was later paraphrased as suggesting that in aqueducts utility and beauty 'should be combined'.[26]

Perhaps the statement can be paraphrased more clearly in modern terms: there is aesthetic satisfaction in functional perfection which is achieved by such structures as aqueducts. Other paraphrases of course are also possible. It would lead us too far to enter more deeply into this reference to early aqueduct physics and into the suggested relation of these physics to the architectural harmonies discussed in the introduction; but one cannot simply say that Vitruvius disregards the aesthetics of the aqueducts.

In Book II he considers the selection and preparation of timber, stone and concrete. He discusses various mixtures of mortar, including those which 'neither the waves nor the force of the water can dissolve'. In Books V and VI he considers the construction of building foundations and frameworks; he points out in detail that in substructures arches should 'discharge the load' of overlying structures and channel this load into piers and thereby into bases.[27] In the aqueduct field these discussions are inherently applicable. The author also expressly says they should be 'committed to memory.'[28] The modern commentary to the aqueduct rules is silent about these discussions and simply asserts that Vitruvius disregards their problems.

In Book VIII on water supply, chapters i-iii contain hydrological observations, chapter iv includes tests for good water, and chapter v begins with a discussion of instruments for levelling an aqueduct grade line. This leads in chapters v and vi to a discussion of aqueduct structures and routings which are mixed with remarks about related topics.[29] The chapter and paragraph numbers are incidentally not authentic and are only retained here to facilitate cross-reference to other discussions.[30] Using the well-known English translation by Morgan, the text reads as follows:

v.3. . . . If there is to be a considerable fall, the conducting of the water will be relatively easy. But if the course is broken by depressions, we must have recourse to substructures.

vi. 1. There are three methods of conducting water, in channels through masonry conduits, or in lead pipes, or in pipes of baked clay. If in conduits, let the masonry be as solid as possible, and let the bed of the channel have a gradient of not less than a quarter of an inch for every hundred feet,[31] and let the masonry structure be arched over, so that the sun may not strike the water at all.[32]

vi. 3. If, however, there are hills between the city and the source of supply, subterranean channels must be dug . . .

vi. 4. But if the water is to be conducted in lead pipes . . .

vi. 5. . . . [and] if there is a regular fall from the source to the city . . . with depressions in it, then we must build substrctures to bring it up to the level[33] as in the case of channels and conduits. If the distance round such depressions is not great, the water may be carried round circuitously; but if the valleys are extensive, the course will be directed down their slope.[34]

It is interesting to compare this text with the paraphrase written by the abbreviator Faventinus who found Vitruvius too wordy:[35] the abbreviation omits the entire subject-matter of chapter vi, paragraph 5, while it slightly shortens the other chapters and rearranges their sequence. In effect, it totally alters the meaning of the text. It is also interesting to compare the Vitruvian text with the notes of Callebat.[36] These notes, when dealing with chapter vi, paragraphs 1-5, disregard the Vitruvian discussions on aqueducts and related technology in Books I, II, V, VI, and VIII (chapter v, paragraph 3) and then charge that Vitruvius disregarded important data.

The text and the attempts at restatement and evaluation present undeniable problems. The ancient abbreviator had an impoverished view of what a practitioner needed, while the modern commentator has a more richly developed view of the data essential for the treatise than he himself can keep in mind. We will now take a further look at these data.

Vitruvius and the earlier aqueduct traditions

The earliest elevated aqueducts may be found in Mesopotamia.[37] Groundwater was collected in long, wide infiltration galleries, now known as 'qanaats', and conducted to fields and cities by wide, shallow channels, which at some points were elevated on bridges to carry them over ravines. In northern Iraq a ruin of such a bridge has been found with cuneiform inscriptions dating it to about 700 B.C. It used corbelled arches, which were twenty-five to thirty feet high.[38]

Ancient travellers, such as Herodotus, saw some of the eastern irrigation works,[39] and it is conceivable that the Romans learnt of such bridges, although the surviving contemporary literature does not describe this detail. In any event, however, this was only a small beginning, which was technically obsolete long before Vitruvius, as we will see.

Other types of aqueduct were developed by the Greeks. Vitruvius mentions Herodotus among other writers;[40] for example, he may have known from them or from restatements of their reports about the tunnel constructed by Eupalinus on Samos and its clay-pipe aqueduct.[41] He describes similar aqueducts (VIII, vi, 8-11), and, as already mentioned cites Greek aqueduct 'physics'. The exact content of these hydraulic writings are unknown, since most of those cited by Vitruvius are lost. Most of the actual Greek pipe-line structures are also lost, although some locations are reported.[42] No Greek elevated aqueducts are known.

Nor is there any elevated structure among the many pre-Roman water disposal channels found so far in Italy.[43] However, the Roman constructions of pre-Vitruvian times,[44] provide one major source of information about aqueducts, including elevated aqueducts. We must briefly compare their basic forms with Vitruvius's text and then reconsider the position of this text in the historic evolution of the Roman aqueduct arcades.

The basic construction of the Roman ducts as shown by their ruins confirms the reconstituted Vitruvian text,[45] for example as to aqueduct gradients. As specified in Book VIII, chapter vi, paragraph 1, a very flat, minimum slope was used, which was much flatter than was customary in Greek practice.[46] The ruins similarly confirm the reconstituted text about access shafts of tunnels, which was another detail Roman practice had refined.[47] The practice also confirms the Vitruvian teachings in such areas as mortar cement mixtures.[48] The text specifies the standard dimensions for lead-pipes to be found among Roman remains.[49] In these respects the practice at least reasonably supports the Vitruvian specifications.

According to the artefacts, other portions of the text are inaccurate or indefinite. This applies for example to the uses of masonry in some types of tunnel, the placement of aqueduct reservoirs,[50] and the construction and operation of inverted siphons.[51] Clearly, the text is far from being a perfect description of the practice.

We may also inquire how the Vitruvian text relates to earlier aqueduct writings, where we can find such writings. Aristotle had mentioned aqueduct projects three centuries before Vitruvius, but only to assert that tyrants order such projects to keep their people occupied and poor. When he described his ideal city, he considered wells and cisterns rather than aqueducts.[52] The Ctesibian books cited by Vitruvius are lost and it is only known from Vitruvius and others that they described demonstrations and uses of pneumatic and hydraulic forces.[53] Again, we possess Archimedes's work on floating bodies,[54] but nothing of his on the flow of water. It is conceivable that he wrote a study of the pipe-line aqueducts which existed in this home town of Syracuse when he defended it against the Romans.[55] Vitruvius may have known such a study, since as already mentioned, he writes of both Ctesibius and Archimedes as teachers of aqueduct physics. However, knowledge of elevated aqueducts could hardly come from such a study because the Syracuse pipe-lines only extended through tunnels. Another Greek author mentioned by Vitruvius is Philo of Byzantium, who was a pupil of Ctesibius. According to an eastern tradition Philo's work on pneumatics dealt with the conduction of water; however, the surviving fragments are totally silent about aqueducts, whether elevated or not.[56]

Evidently Vitruvius fully acknowledged earlier aqueduct writings: he mentions more than are known from the surviving literature. If the existing and relatively pertinent remarks of Herodotus, Aristotle, Archimedes and Philo are taken as typical examples, it appears that there had been little of any relevance to aqueducts in general and nothing about elevated aqueducts in the earlier writings. Vitruvius thus appears to have been the first who tried to formulate rules for this field, or at least to outline their essentials.

Vitruvius and the elevated aqueducts of the Romans

The evolution of Roman aqueducts began about 312 B.C. with the first duct, Aqua Appia, which ran entirely underground like the known pre-Roman ducts.[57] The subsequent major ducts gradually developed elevated structures. The layout or course of these major ducts was established about 270 B.C. by Anio Vetus, which extended into the mountainous source area farther east of Rome. The duct described a peculiar S-curve (Figs. 4 and 5).[58] For the first twenty kilometres, it followed the semicircular course of the valley of the Anio River down to Tivoli. Then, substantially maintaining its level, it meandered between the rises and depressions of the Campagna while generally remaining close to the ground surface in a cut-and-cover construction. It emerged only where it spanned small ravines on low bridges. The arches of the bridges used voussoirs with radial joints (as later described in Vitruvius VI. viii, 3); evidently the older corbelled arches were already outdated. Due to its winding and zigzagging or meandering course, the Anio Vetus was much longer than the direct distance from its source to Rome. Therefore it lost much hydraulic head and discharged itself only a few metres above sea-level.[59] Although it was embedded in the ground it must also have collected much heat, since the ground itself becomes hot from the sunshine.[60]

A much more effective design began to be evolved between the middle of the second century B.C. and Vitruvius's time, but the names of the men who conceived and completed it are not recorded. The principles of the new design may be found incorporated for the first time in Aqua Marcia built about 140 B.C. This duct cuts across many detours of Anio Vetus south of Tivoli and further downstream, where it uses tunnels following more direct lines. As a result this duct remains in higher ground, closer to the Alban Mountains. Finally, near Rome Marcia was slightly elevated above the ground on continuous arcades (Figs. 2, 4 and 5). This duct was famous not only for the good taste of the water provided by its spring but also for its coolness when delivered in Rome; Vitruvius specifically mentions the quality of its water.[61]

Thereafter, several smaller aqueducts were built, which either joined Marcia and were superimposed on it, or used underground courses elsewhere (Fig. 4). One of these was called Tepula (lukewarm); another Julia (after Julius Caesar); a third was Virgo which followed a very winding course through deep-cut tunnels near and across the old Appia; and a fourth, Alsietina, brought water from the west. The last three were built in Vitruvius's time, but it is not known whether he took any part in their construction.[62]

The next Roman aqueduct was the major addition of Aqua Claudia, whose construction was completed in A.D. 47 under the Emperor Claudius.[63] Its course was close to the old S-curve of Anio Vetus and Marcia, but its middle section from upstream of Tivoli to Capanelle was significantly different (Figs. 4 and 5). This section began with deep-cut tunnels of several kilometre's length, which extended straight across wide curves of the earlier aqueducts, and continued with surface-cut tunnels closer to the Alban Mountains and correspondingly higher above the River Tiber than the previous ducts. As a result, the downstream section of Claudia from Capanelle to Rome could be and was elevated on high arcades (Figs. 1 and 2). The entire new duct traversed the irregular ground with unprecedented directness. By comparison both Anio Vetus and the famous Marcia have 'timid contours'.[64] In the smoothened S-curve of Claudia, the ancient Roman aqueducts 'reached at last their full development' in the expression coined by a leading archaeologist, that confirmed the opinion of a leading historian.[65] For the first time, spring water from the calcareous Tiburtine Mountains reached all levels of Rome and was delivered in a cool condition.[66]

The downstream arcades of Claudia begin close to those of Marcia where the tunnels of both ducts end. They extend to the entrance of Augustan Rome, the Porta Maggiore, which is now in the middle of the city not far from the main railway terminal. These arcades are about nine kilometres long and generally parallel to one another, but the Claudia still uses a more direct line, while the Marcia crosses it repeatedly at a lower level. Some five years after the Claudia, another major aqueduct, the Anio Novus, was finished. Novus, which began close to the Claudia, followed its bold new course and was carried downstream by the same arcades (Figs. 1, 2, 4 and 5).

The new design, which was built almost straight across the rolling countryside, has been interpreted as expressing an idea of power.[67] If we review its development once more with particular reference to Vitruvius, we can see its beginnings in the Marcia which traversed the irregular ground with a much greater degree of directness. Although this duct became famous, the principle of its construction was disregarded by subsequent builders and not developed any further for more than a century. Then Vitruvius formulated the concept suggested in the Marcia. He wrote that 'the conducting of water' should take place above the plains on raised substructures (VIII. v. 3) and through the hills in tunnels (vi. 3) by means of a channel which slopes positively but minutely, is built to avoid solar heating of the water (vi. 1), and is free of unduly long detours (vi. 5). Within a few decades, the new principle as formulated and developed by Vitruvius reached a state of full development as already described.

Some centuries after Vitruvius, the Aqua Hadriana (Alexandrina) even began to eliminate the basic S-form of the former major ducts and to run east in a substantially straight line, but it was not extended into the mountains and was built on a relatively low level. Some higher aqueducts built in the provinces used long, substantially straight arcades, which adhered to the principle of directness foreshadowed by the Marcia and epitomized by Vitruvius.[68]

Of course, we must not visualize the builders of these ducts as reading Vitruvius's *Architecture* at every moment of their planning and constructing activity. They kept this text in mind, and at hand. As already mentioned, they were provided with an abridgement about A.D. 300 and another abridgement followed about a hundred years later.[69]

The aqueducts were built by contractors and their engineers.[70] These men must have had experience and interest regarding the choice of elevated arcades, surface channels, near-surface or deep-cut tunnels, or possibly inverted siphons. Nevertheless, as mentioned, a considerable degree of uniformity was maintained throughout the imperial period. There had been no such uniformity in the times of the Tepula, Julia, Virgo and Alsietina. We may assume that the change was caused in part by the Vitruvian formulation. It is another question whether the ultimate uniformity was a technical success, or a cause of technical stagnation, or a mixture of both.

According to Frontinus, it was a success.[71] He describes it in words similar to Vitruvius, although he writes of a completed achievement, whereas his predecessor had used the future tense. Vitruvius had summarized some of his aqueduct rules by saying that builders will provide for the conducting of water 'most successfully' when they follow those rules; while Frontinus, referring to the Claudia, called it a work 'most magnificently completed'.[72] Frontinus also noted the 'ancients had brought the water only to rather low levels', which was true of the earliest Roman ducts, especially the Anio Vetus and even of the constructions directly preceding the Claudia. As Frontinus continues, technical progress was achieved by 'subtle exploration of the levelling art' and its use became safe as the empire became free of 'interference by enemies'.[73] This political safety was established by Caesar and Augustus and lasted for several centuries. Vitruvius who lived at the time, was of course not responsible for these political changes and he did not in any way influence 'the levelling art', but, as mentioned, his text did contribute to its new utilization.

The remarks of Frontinus were at one time interpreted as belittling Vitruvius for his supposedly 'ancient' views, particularly on the proper aqueduct gradients. However, these views were presented by confused copyists, not by Vitruvius.[74] Further in his book on *Aqueducts*, Frontius did not charge Vitruvius with such views, since the book is quite in keeping with the proposition that the Vitruvian formulation contributed to the new utilization of improved techniques.

Nevertheless, the imperial aqueducts were not an unqualified success. The masonry channels were made 'as solid as possible', but they could not be made solid enough to operate without continual repairs. The alarming loss of water Frontinus put down to 'theft', but the actual major cause was leakage, which even armies of repairmen could not effectively control.[75] Therefore the Roman aqueducts decayed, largely due to their own shortcomings, although their decay was of course connected in many respects with the general decline of the Roman civilization and empire. After the few centuries of the imperial age, no further aqueducts were built for a millennium.[76]

Vitruvius was still read during and after those times. His reference to Aristotle and others may even have contributed, in some measure, to the gradual renewal of learning. This in turn brought intensified Vitruvian studies during the Italian Renaissance and new attempts to supply water through elevated aqueducts modelled on the Vitruvian principle, particularly near Rome and Paris.[77] Builders then began gradually to recognize that more solid ducts were needed and ducts of welded steel and reinforced concrete as used today developed. As a result, the Vitruvian aqueduct rules have become a matter of historical interest, where study remains difficult.[78]

Notes

The author wishes to thank Professor A. Rupert Hall and Dr. Norman A.F. Smith for their careful review of an earlier draft and for their valued suggestions.

1. Figures 1 and 2 were drawn for this study by Dr. Mark Podwal and were based respectively on a photograph by Fratelli Alinari and on a model in the Museo della civiltà Romana, Rome. Figure 3 has been reproduced by courtesy of the Astor, Lenox and Tilden Foundations at the New York Public Library, Rare Book Division. Figures 4 and 5 were drawn by the author. Figure 4 was developed from C. Singer and others, *A history of technology*, Oxford, 1954-8, vol. 2, p. 669 for general orientation in accordance with Figure 5, which was traced from reduced photographic copies of the large, detailed maps (1:25000) appended to V. Reina and others, *Livellazione degli antichi acquedotti romani*, Rome, 1917. For brilliant illustrations of aqueducts throughout the Roman world, including aerial views, close-ups and all types of architectural drawings, see C. Fernández Casado, *Acueductos en España*, [Madrid], 1972.

2. T. Ashby, *The aqueducts of ancient Rome*, Oxford, 1935. V. Reina and others, *Livellazione degli antichi acquedotti romani*, Rome, 1917, pp. 3f., *passim*. Esther B. Van Deman, *The building of the Roman aqueducts*, Washington, 1934 (reprinted 1973), pp. iiif., *passim*. About the work of Ashby, Reina and Van Deman, see G. Lugli, *La tecnica edilizia romana*, Rome, 1957 (Reprinted London and New York, 1968), vol. 2, pp. 332f. For the continuation of Van Deman's work, see Marion E. Blake, *Roman construction in Italy from Tiberius through the Flavians*, Washington, 1959 and *Roman construction in Italy from Nerva through the Antonines*, Philadelphia, 1973.

3. Vitruvius, *De architectura*, Book VIII, chapter v, paragraph 3 to chapter vi, paragraph 5. (For such citations, this study uses the form Vitruvius viii. v. 3-vi. 5.) The Vitruvian statements in English are quoted from trans. M.H. Morgan, *Vitruvius: the ten books on architecture*, Cambridge, Mass., 1914 (reprinted New York, 1960), mainly pp. 243-5.

4. G. Tabarroni, 'Vitruvio nella storia della scienza e della tecnica', *Atti della Accademia delle Scienze dell' Istituto di Bologna*, classe di scienze morali, memorie, 66 (1976), 18f. Summarizes the aqueduct notes.

5. Cetius Faventinus, *Liber artis architectonicae* in ed. V. Rose, *Vitruvii de architectura*, Leipzig, 1899, p. 285 and on aqueducts, pp. 287-91. A newer edition is H. Plommer, *Vitruvius and later Roman building manuals*, Cambridge, 1973.

6. L. Callebat, *Vitruve, de l'architecture, livre* VIII, Paris, 1973, pp. 144, xlviii.

7. M.H. Morgan, *Addresses and Essays*, New York, 1910, pp. 224-72. P. Thielscher, 'Vitruvius' in Pauly and Wissowa, *Real-Encyclopaedie der classischen Altertumswissenschaft*, 2nd ser., Stuttgart, 1961, vol. 17, col. 458f. Tabarroni, 'Vitruvio', 8-11.

8. Vitruvius V. preface. 1-5. See the editions *M. Vitruvius per Jocundum solito castigatior factus*, Venice, 1511; ed. V. Rose and H. Mullr-Strubng, *Vitruvii de architectura*, Leipzig, 1867; and trans. Morgan, *Vitruvius*. See also Morgan, *Addresses and Essays* and Tabarroni, 'Vitruvio', 11-13.

9. Vitruvius VIII. iii. 27 and vi. 2.

10. Vitruvius II. preface. 4; II. viii. 8-10, 16-18; IV. ii. 2, 3; V. ix. 8; VII. v. 5-8.

11. Van Deman, *Roman aqueducts*, pp. 9-11. Callebat, *Vitruve*, p. x.

12. Vitruvius VIII. iii. 25.

13. R. Bernelle and others, *Académie d'Hippone: comptes-rendus des réunions*, Bône, 1890-1, *Passim*. Also in *Corpus inscriptionum latinarum*, vol. 8, supplement 2, Berlin, 1894, nos. 18858-19110. Also see Pauly and Wissowa, *Real-Encylcopaedie*, vol. 2, col. 307 on Aquae Thibilitanae.

14. Thielscher, 'Vitruvius', col. 419-58. P. Ruffel and J. Soubiran, 'Vitruve ou Mamurra?' *Pallas*, ii (1962), 149f. Tabarroni, 'Vitruvio', 2-5.

15. Vitruvius II. iii. 2; IV. v. 2; VII. vii. 2. Tabarroni, 'Vitruvio', 5f. L. Sontheimer, *Vitruv und seine Zeit*, Tübingen, 1908, pp. 14-16.

16. Bernelle, *Académie d'Hippone*, 1891, pp. xlviif. *Corpus inscriptionum latinarum*, no. 18913.

17. *Corpus inscriptionum latinarum*, no. 19006.

18. Tabarroni, 'Vitruvio', 30f.

19. Pliny the Elder, *Historia naturalis*, book I, comments about book XXXV; book XXXI, xxxi; trans. H. Rackham, *Natural history*, London and Cambridge, Mass., 1947-62, vol. 1, pp. 156f and vol. 8, pp. 412 and 4.

20. Sextus Julius Frontinus, *De aquaeductibus urbis Romae*, trans. C. Herschel, *The two books on the water supply of the city of Rome of Suxtus Julius Frontinus*, Boston, 1899 (2nd ed., London and New York, 1913), p. 18, containing the Latin text of Frontinus's chapter 16 with translation facing it.

21. E. Jüngst and P. Thielscher, 'Vitruv über . . . Grundbau', *Mitteilungen des deutschen archaeologischen Instituts, römische Abteilung*, 51 (1936), 144-80.

22. L.B. Alberti, *De re aedificatoria*, Florence, 1485; trans. G. Orlandi, L'architettura, Milan, 1966, pp. 441f., *passim*. G. Cardano, *De subtilitate*, Paris, 1550; ed. Lyons, 1580, pp. 568-71. A. Palladio, *Architettura*, Venice, 1570; facsimile reprint, Milan, 1968, Book 4, p. 4, *passim*.

23. For this trend in art, see F.W. Schlikker, *Hellenistische Vorstellungen von der Schönheit des Bauwerks nach Vitruv*, Berlin, 1940, p. 6 in science, W.H. Stahl, *Roman science*, Madison, 1962, pp. 92-5.

24. V. Rose, *Anecdota graeca et graeco-latina*, Berlin, 1864-70, vol. 1, p. 9. Against Rose and Vitruvius, see E. Oder, 'Ein angebliches Bruchstück Demokrits', *Philologus*, Supplementband, vol. 7, 1899, pp. 343f. and pp. 338, 340, 351. Against Oder on Vitruvius, see Sontheimer, *Vitruv*, pp. 55f, 124 and Schlikker, *Hellenistische Vorstellungen*, p. 7.

25. Vitruvius I. I. i. 7 also reflects more specific aqueduct problems of VIII. vi. 6. It is overlooked by Callebat commenting on Vitruvius's aesthetics and sources in his *Vitruve*, pp. 144, xxvi-xxxix.

26. Alberti (trans. Orlandi), pp. 99, 101.

27. Vitruvius II. iv-vi; V. xii. 2-7; VI. viii. 3-7. Jüngst and Thielscher, 'Vitruv', 156-78.

28. Vitruvius V. preface. 2.

29. Including urban water distribution basins in Vitruvius VIII. vi. 1-2 and lead pipe standards in VIII. vi. 4-5.

30. Ed. and trans. F. Granger, *Vitruvius: on architecture*, London and New York, 1931-4, vol. 1. pp. xxixf.

31. 'Ne minus in centenos pedes sicilico'. The manuscript text 'ne minus ... semipede' was based on confusion of an abbreviation symbol (s or inverted s); see ed. Rose and Müller-Strübing, *Vitruvii*, p. 207.

32. Such arching over is shown by 'c. fornix' in Fig. 3. Sometimes it is replaced by roofing over; see John H. Parker, *The aqueducts of ancient Rome*, Oxford and London, 1876, plate 21.

33. 'ad libramenta', 'to the level [of the substructure]' — not to a level as high as the higher discharge levels in the city. A 'low' level of the substructure for the pipeline siphon is specified in Vitruvius VIII. vi. 5 after the text quoted here.

34. About the latter variant, see N.A.F. Smith, 'Attitudes to Roman engineering and the question of the inverted siphon', in A. Rupert Hall and Norman Smith (eds.), *History of technology: first annual volume, 1976*, London, 1976, pp. 45-71.

35. Cetius Faventinus, *Liber artis architectonicae*.

36. Callebat, *Vitruve*.

37. M.S. Drower, 'Water-Supply', in C. Singer and others, *A history of technology*, Oxford, 1954-8, vol. 1, pp. 531-5.

38. T. Jacobsen and S. Lloyd, *Sennacherib's aqueduct*, Chicago, 1935, pp. 6-18. T. Safar, 'Sennacherib's project for supplying Erbil with water', *Sumer*, 3 (1949), 23-8 and (Arabic, with illustrations) pp. 71-86.

39. I. 185, 193; II. 99; III. 117; ed. F.R.B. Godolphin, *The Greek historians* (including the complete and unabridged historical works of Herodotus translated by G. Rawlinson), New York, 1942, pp. 78f., 82, 128f. 211f.

40. Vitruvius VIII. iii. 27.

41. Herodotus III 60 (ed. Godolphin, p. 190). A. Burns, 'Ancient Greek water supply', *Technology and Culture*, 15 (1974), 389-412 and 'The tunnel of Eupalinus', *Isis*, 62 (1971), 176-85.

42. Smith, 'Attitudes to Roman engineering', p. 52 mentions a number of Greek installations and cites pertinent literature.

43. W.J. Anderson and R.F. Spiers, *The architecture of Greece and Rome*, vol. 2 revised and rewritten by T. Ashby, London, 1927, pp. 30, 123-5.

44. Frontinus (trans. Herschel), pp. 16-22 (chapters 13-20).

45. See note 31.

46. Burns, 'Water supply', 392-4. Reina, *Livellazione*, pp. 31, 75f., *passim*, C. Germain de Montauzan, *Les aqueducs de Lyon*, Paris, 1909, pp. 167-72.

47. Vitruvius VIII. vi. 3 (after the part quoted above, reconstituted in *M. Vitruvius per Jocundum*, folio 81v). About the Greek and Roman traditions, see Burns, 'Water supply', p. 407 and Montauzan, *Les aqueducs*, pp. 285f.

48. Frontinus (trans. Herschel), pp. 155f. Anderson and Spiers, *Architecture of Greece and Rome*, pp. 26f., 32, 41. Jüngst and Thielscher, 'Vitruv'. M.S. Briggs, 'Building-Construction', in Singer, *A history of technology*, vol. 2, pp. 397-418; Lugli, *La tecnica*, *passim*.

49. Callebat, *Vitruve*, pp. 161-3.

50. Callebat, *Vitruve*, pp. 158-60, 170.

51. Smith, 'Attitudes to Roman engineering', pp. 54-9, 68.

52. Aristotle, *Aristotelis opera*, Berlin, 1831-70, pp. 1313 b 20-5 and 1330 b 1-17 (*Politica* v. 11 and VII. 11). Vitruvius refers to Aristotle in other respects in VII. preface. 2 and IX. preface. 2.

53. A.G. Drachmann, *Ktesibios, Philon and Heron*, Copenhagen, 1948, pp. 1-41.

54. Archimedes, *Archimedis opera omnia*, ed. J.L. Heiberg, Leipzig, 1880-1, vol. 2, pp. 345-426.

55. Plutarch, *The lives of the noble Grecians and Romans*, Life of Marcellus, Modern Library ed., New York, n.d., pp. 376-80. Burns, 'Water supply', pp. 389-97.

56. Vitruvius VII. preface. 14. Philo of Byzantium, *Pneumatica*, ed. F.D. Prager, Wiesbaden, 1974, pp. 47-53, 132, 230f.

57. Frontinus (trans. Herschel), pp. 143f.

58. See note 1 for derivation of Figs. 4 and 5.

59. Van Deman, *Roman aqueducts*, pp. 29-66. See also Fig. 4.

60. Figs. 2, 4 and 5.

61. Vitruvius VIII. iii. 1. Frontinus (trans. Herschel), pp. 150-3, 163. Van Deman, *Roman aqueducts*, pp. 67-146. For illustrations, see Lugli, *La tecnica*, plates 51, 1; 64, 4; 171, 4.

62. Frontinus (trans. Herschel), pp. 163-74. Van Deman, *Roman aqueducts*, pp. 23-8, 147-86.

63. Van Deman, *Roman aqueducts*, pp. 187-270. Blake, *Roman construction in Italy from Tiberius through the Flavians*, pp. 26f., 79. Frontinus (trans. Herschel), pp. 175-83.

64. Frontinus (trans. Herschel), p. 154.

65. Van Deman, *Roman aqueducts*, p. 13. Similarly Frontinus (trans. Herschel), p. 176.

66. Frontinus (trans. Herschel), p. 16 (chapter 13); pp. 163, 183 (but note that on p. 165 the names *Claudia* and *Anio Novus* are erroneously printed on a photograph). About the maximum height of Roman aqueducts, see Smith, 'Attitudes to Roman engineering', pp. 65-7.

67. F. Klemm, *A history of western technology*, London, 1959, p. 48.

68. Casado, *Acueductos en España*, *passim*, mainly on Spanish and African aqueducts. Blake, *Roman construction in Italy from Nerva through the Antonines*, pp. 273-80. Also see K.O. Dalman, *Der Valens-Aquaedukt in Konstantinopel*, Bamberg, pp. 4-18, plates 2-20.

69. Ratilius Taurus Aemilianus Palladius, *De re rustica*; IX. xi; ed. J.G. Schneider, *Scriptorum rei rusticae*, Leipzig, 1795, p. 200. Cetius Faventinus, *Liber artis architectonicae*.

70. Frontinus (trans. Herschel), pp. 177f, 193. Montauzan, *Les aqueducs*, pp. 357-419.

71. Frontinus (trans. Herschel), pp. 186-200; see also note 72 below.

72. Vitruvius VIII. vi. 6. Frontinus (trans. Herschel), p. 18 (chapter 20).

73. 'Veteres humiliori derectura perduxerunt, sive nondum ad suptile explorata arte librandi, seu quia . . . infra terram mergebant ne facile ab hostibus interciperentur', Frontinus (trans. Herschel), p. 18.

74. Frontinus (trans. Herschel), p. 194. Ed. Rose and Müller-Strübing, *Vitruvii*, p. 207.

75. Frontinus (trans. Herschel), pp. 256-60.

76. G. Panimolle, *Gli acquedotti di Roma antica*, Rome, 1972, pp. 34f.

77. P. Mastrigli, *Acque, acquedotti e fontane di Roma*, Rome, 1923, pp. 121-67. R.J. Forbes, 'Hydraulic engineering', in Singer, *A history of technology*, vol. 2, p. 691.

78. See notes 31 and 33 above. Difficulties may also be found in the phrase 'eaeque structurae confornicentur ut minime sol aquam tangat' (Vitruvius VIII. vi. 1).

a. The words 'eaeque structurae' may refer back to the several substructures just mentioned in VIII. v. 3 not just to the solid channel structure of VIII. vi. 1. For such referring back one might more strictly expect the term 'illae substructiones; however, the Vitruvian language is flexible in such respects for he also writes 'signa . . . inveniuntur' and 'eaeque inventiones' in VIII. i. 3-4.

b. The word 'confornicentur' is capable of different interpretations. In addition to letting the structure be 'arched over', it may mean letting it be 'arched under' by pile-connecting arches. The supporting arches of aqueducts were called 'fornices' as in Frontinus and in an African inscription dated 153 B.C. (*Thesaurus linguae latinae*, VI, 1, Leipzig, 1926, pp. 1125f). According to Cetius Faventinus, *Liber artis architectonicae* (chapter vi), a levelled duct could be provided by tunnels, ground-level channels or by 'structura . . . arcuatili'. The latter term, clearly referring to supporting arches, may perhaps be taken as showing the abbreviator's understanding of the Vitruvian 'confornicentur'.

c. The comma before 'ut', inserted in many editions of the Latin text, is unauthentic. (Ed. and trans. Granger), Vitruvius, vol. 1, pp. xxixf.

d. 'Minime' may mean 'minimally' as well as 'not at all'. (Lewis and Short, *Latin dictionary*, under 'parvus-minor-minimus-minime').

e. In view of points a. to d. it is conceivable that the phrase means 'Let these structures be arch-constructed in such a way that the sun can only minimally strike the water'.

f. This in turn could mean 'Let them, for this purpose, be arched over to provide shade for the water, *and* arch-supported as to minimize the length of the duct'.
g. However, the modified translation e. and interpretation f. would not substantially change the Vitruvian rule interpreted by Morgan; it would just provide a specific reason for the subsequent rule in VIII. vi. 5 that circuitous detours should be used only 'if the distance round such depressions is not great'. On the other hand the modified interpretation f. would present a problem, as it would give a double meaning to the word 'confornicentur', which in other Vitruvian Books (for example in v. v. 2) has only one of the two meanings. Therefore the possibility of the modified translation is disregarded in this study.

5

Resistance to technological innovation: the history of the pile driver through the 18th century

M.N. Boyer

Writing a history of the pile driver, we are hampered by the lack of interest displayed in the subject by our predecessors. Only in the 14th century A.D. do we find substantial evidence about the operation of that device, and only in the 15th do we begin to see pictures of pile drivers. Even then, most of the sketches were imaginative, drawn to promote this or that improved pile driver; these machines seem to have encountered a resistance that prevented their adoption. It is the purpose of this paper to determine what pile drivers were actually in use in Europe, especially France and the Netherlands, between the 14th and 18th centuries and to propose an explanation for the clear preference by builders for the simpler rather than the more mechanically sophisticated versions.

The scarcity of source materials for the history of the pile driver is illustrated in a quick survey. The beginnings are obscure; the pile driver was far too humble to appear in the royal annals of Assyrian conquests or in the Greek epics. Yet archaeology tells us that there were piles under houses built by the Lake Dwellers in prehistoric times. One might suppose that the Greeks would have described the pile driver in their treatises on machines, or the Romans in their discussions of construction, but no. Vitruvius does indeed recommend driving piles under temple foundations in marshy areas,[1] but he does not explain how to do this. It was not worth his while to discuss common machines,

DR. BOYER is professor emerita at York College of the City University of New York. A version of this paper was read before the Committee for the History and Philosophy of Science, Technology, and Medicine at Brooklyn College on October 27, 1983. For criticisms and suggestions, the author is indebted to T. H. Boyer, Norman A. F. Smith, Bert S. Hall, Thomas B. Settle, and Gustina Scaglia.

[1] Vitruvius, *De Architectura*, ed. from Harleian manuscript 2767, trans. Frank Granger (London and New York, 1931), 3.4.2.

an attitude illustrated in the following remark: "Besides, there are innumerable mechanical devices about which it does not seem useful to enlarge because they are to hand in our daily use."[2]

During much of the Middle Ages the situation with regard to the history of the pile driver was not much better than in antiquity: references are mostly incidental and uninstructive. About 1200 Alexander Neckham does mention pile drivers, but only as follows: "Now the solidity of the foundation is tested [*exploratur*] with piles driven into the bowels of the earth."[3] Villard de Honnecourt in the 13th century gives a sketchy drawing of a machine for sawing off the tops of piles underwater, but we are supposed to know how to get them there before cutting them off.[4] There are brief references to pile drivers in chronicles and more information in accounts. Pile drivers are mentioned in 1238 at the cloister of Bloemkamp in the Netherlands; in 1289, 1329, 1330, and later in the accounts of London Bridge and the Tower of London; at Rochester beginning in the late 14th century; at Poitiers in 1386; at Albi in 1408–10; and at Lyons in 1416 and later.[5] Finally, in the years 1387–89 we find copious information on pile drivers in the accounts of the Orléans bridge, enabling us to determine the details of their construction and use.[6]

If written information on the early history of the pile driver is scarce, the situation as regards pictorial evidence is even worse. The earliest pile driver illustrating a literary text is in a 1519 manuscript of Caesar's *Gallic Wars*, where the Roman commander is shown supervising the building of the bridge across the Rhine.[7] The lack of interest on the part of early illustrators does not mean they were averse to depicting construction techniques as such. From the 13th century on, many are the pictures of the founder inspecting the building of a church or the

[2]Ibid., 10.1.

[3]L. F. Salzman, *Building in England down to 1540: A Documentary History* (Oxford, 1952), pp. 87–88.

[4]Theodore Bowie, ed., *The Sketchbook of Villard de Honnecourt* (Bloomington, Ind., and London, 1959), plate 60.

[5]For Bloemkamp, see Franz Maria Feldhaus, *Die Maschinen im Leben der Völker* (Basel and Stuttgart, 1954), pp. 197–98; *Atlas de l'ordre cistercien*, s.v. "Bloemkamp, Netherlands, province of Frisia, commune Wouserdael, diocese Utrecht." For London, see Salzman, pp. 88, 328. For Rochester, see Salzman, p. 86. For Poitiers, Archives municipales de Poitiers, Cahier 25. J 7, 9, 14. For Albi, Archives communales d'Albi CC 168, fol. 72r. For Lyons, see Marie Claude Guigue, ed., *Registres consulaires de la ville de Lyon ou Recueil des délibérations du conseil de la commune de 1416 à 1423* (Lyon, 1882), pp. 7, 288–89.

[6]Archives départementales du Loiret CC 920; Marjorie Boyer, "A Fourteenth-Century Pile Driver: The *Engin* of the Bridge at Orleans," *History of Technology*, vol. 9 (1984).

[7]Marjorie Boyer, *Medieval French Bridges: A History* (Cambridge, Mass., 1976), fig. 1.

construction of the Tower of Babel. These illustrations frequently show hoists but not pile drivers. Perhaps one reason is simply that in real life pile drivers were comparatively rare. They were used only during construction in marshy areas or during building in water, as of bridges, jetties, or piers. The former situation is one that builders preferred to avoid when they could, and the latter does not occur in the texts usually illustrated by artists.

By the 15th century, the problem of driving piles effectively had caught the attention of engineers, including those preoccupied with building military machines. Drawings in technical treatises include those of the *Anonymous of the Hussite Wars, Das Mittelalterliche Hausbuch,* and the notebooks of Leonardo da Vinci.[8] In the 16th century there were more of these elaborate pile drivers, some of them with several windlasses and numerous pulleys, designed to secure mechanical advantages and to save manpower. Designs were proposed in which the ram, reaching a certain height, dropped of its own weight, or in which a catch automatically released the ram at the desired height, and the ram was automatically rehooked after hitting the pile. Such a pile driver, sketched by an anonymous artist, is shown in figure 1.

The technological sophistication of the improved pile drivers provokes our admiration, but there is little evidence that contemporaries shared our attitude. They continued to employ the medieval pile drivers with a blatant disregard for what 19th-century historians were fond of calling progress. Historians of technology well know that the invention of a mechanically superior device does not ensure its adoption. Indeed, the construction practices of 14th- and 15th-century France were still being followed in the 18th century, as I have shown elsewhere in the case of bridge building.[9] The present inquiry demonstrates that this conservatism was equally true with regard to the pile driver. Those employed in the 17th and 18th centuries were still essentially the simple ones of the 14th and 15th, as shown in the construction manuals of Cornelis Meijer (1696), of Hubert Gautier (1716), and of Jean Rodolphe Perronet (1782–83).[10]

[8]Bert S. Hall, *The Technological Illustrations of the So-called "Anonymous of the Hussite Wars" in Codex Latinus Monacensis* (Wiesbaden, 1979), fol. 8v, 37v,; Helmut and Willy Storck, eds., *Das Mittelalterliche Hausbuch nach dem originale des Fürsten von Walburg-Wolfegg-Waldsee im Auftrage des Deutschen Vereins für Kunstwissenschaft* (Leipzig, 1912), Tafel 44; *Il codico atlantico di Leonardo da Vinci nella Biblioteca Ambrosiana di Milano,* reprodotte e pubblicato della Regia accademia dei Lincei (Milan, 1894), fol. 289r; Feldhaus (n. 5 above), pp. 238, 239.

[9]Boyer, *Medieval French Bridges,* p. 144.

[10]Cornelis Meijer, *Traité des moyens de rendre les rivières navigable, avec plusieurs desseins de jetée* (Amsterdam, 1696); Hubert Gautier, *Traité des ponts* (4th ed., Paris, 1765); and Jean Rodophe Perronet, *Description des projets de la construction des ponts de Neuilly, de Mante, d'Orléans et autres,* 2 vols. (Paris, 1782–83).

The pile drivers demonstrably in use from the 15th through the 18th centuries were the hand ram, the *hye*, and the *engin*. Each of the last two types had what contemporary documents call a tower and modern oilmen a rig. From the top hung a pulley over which ran a rope with the ram on one end, but the *hye* had a windlass on the other end, while the *engin* was powered directly by a crew of men pulling on individual cords running into the great rope. The meaning of *hye* is unambiguous: whether in the 14th, 15th, or 18th century, it always indicates a pile driver with a winch (windlass). *Engin*, on the other hand, was a comprehensive term. In 14th- and 15th-century France there were

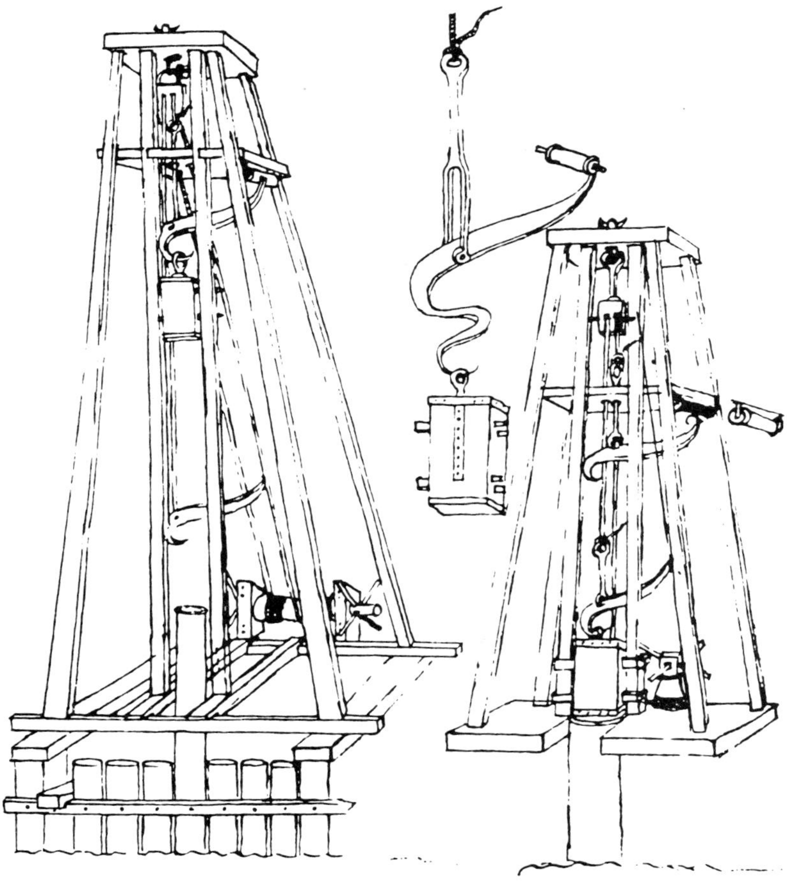

FIG. 1.—In this design, the ram is automatically released at the desired height and rehooked upon striking the pile. MS Lat. 2941, Biblioteca Nazionale Marciana, Venice, fol. 44r. The artist is unknown, but this is mainly a copy from Taccola's *De Machinis*. (Photo courtesy of the Biblioteca Nazionale Marciana and Columbia University.)

engins de guerre, engins to raise the portcullis, and *engins* to drive piles. Each of these machines had a pulley and a rope—although frequently not much else in common. Yet at this time, in the case of pile drivers, *engin* seems always to have meant the simple pile driver with men pulling on the rope.[11] Later there seems to have been no single word indicating the pile driver without the winch. For the sake of convenience in the present discussion, the word *engin* has been used to designate the simple pile driver in which the ram was raised by manpower, the one without the winch.

The hand ram was the least used of the three types of pile driver. No documents known to me attest its use, but it appears in three pictures, one a view of a city, one a construction scene, and one a builder's manual. The hand ram occurs in the Catena view of the city of Florence ca. 1470 (fig. 2). The irony in this picture of Florence is striking. It was in 15th-century Italy that engineers proposed some of the most elegant solutions to the problem of driving piles. Yet at Florence in the Catena view the builders are using a hand ram which could equally well have been employed by the prehistoric Lake Dwellers. The apparatus is of the simplest. It consists in four uprights fastened together by timbers to form a rectangle supporting two boards on which stand two men facing each other, one on either side of the pile to be driven. The workers are lifting a length of tree trunk and dropping it on the pile.

It seems worthwhile to emphasize that, from the technological point of view, the mode of pile driving in the Catena picture could have been employed by men a few thousand years earlier. Here may indeed be the solution to a problem which has long plagued historians of technology: how the Lake Dwellers put the piles under their houses. It has been conjectured that prehistoric men used gigantic hammers. That there was such a technique, even in the 17th century, is shown by Cornelis Meijer (fig. 3, *1*). Yet Meijer considers the usefulness of such a mallet very limited, applicable only to cases in which the soil is very light and little force is required to make the piles enter the ground.[12] Feldhaus rejected the suggestion of mighty hammers as unreasonable.[13] It is much more likely that the Lake Dwellers used a hand ram.

The hand ram was still being employed in the 17th century. Meijer recommends it in cases in which the ground is very difficult to pene-

[11]At Orléans in the bridge accounts the *engin* is always the pile driver with 14 to 16 men pulling on the ram. Archives du Loiret CC 920, fol. 20v. "L'engin a batre les pelz aux orgaux du dit pont" [i.e., The engin to drive the piles at the starlings of the said bridge].

[12]Meijer, pp. 13–14.

[13]Franz Maria Feldhaus, *Technik des Antike und des Mittelalters* (Potsdam, 1931), p. 19.

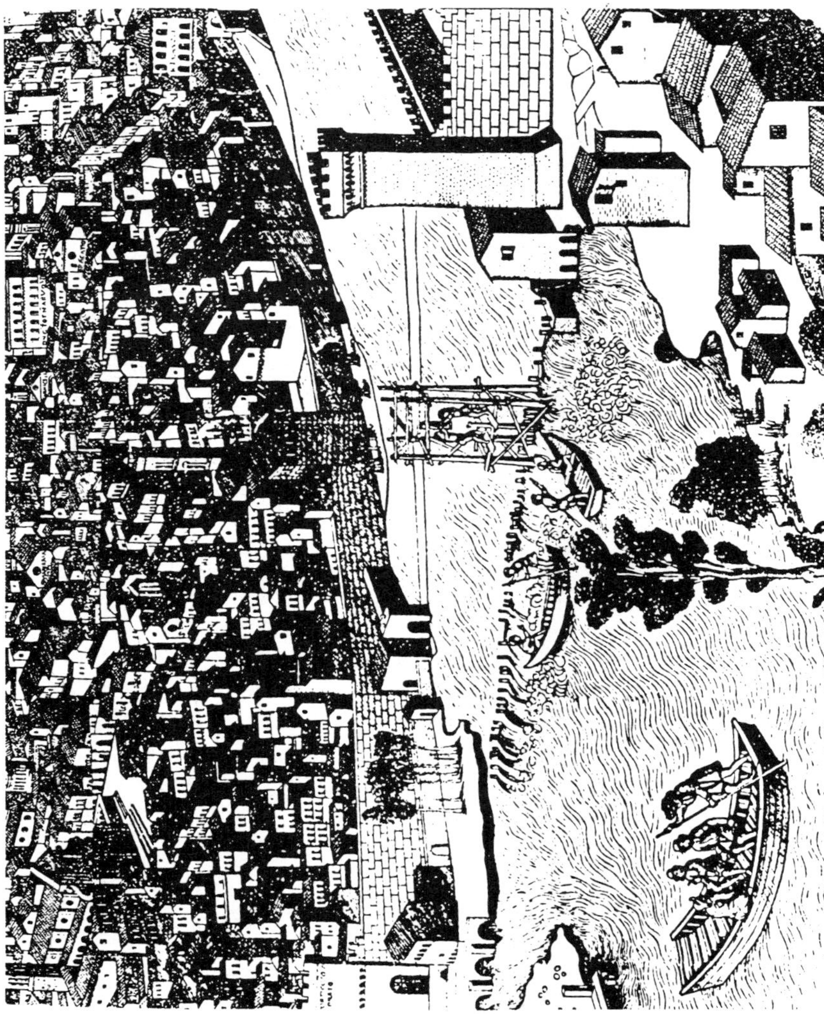

FIG. 2.—Hand ram in operation, as shown in the Catena view of Florence, ca. 1470. (From Attilo Mori and G. Boffito, *Piante e Vercute di Firenze studio storico, topografico, cartografico* [Florence, 1926].)

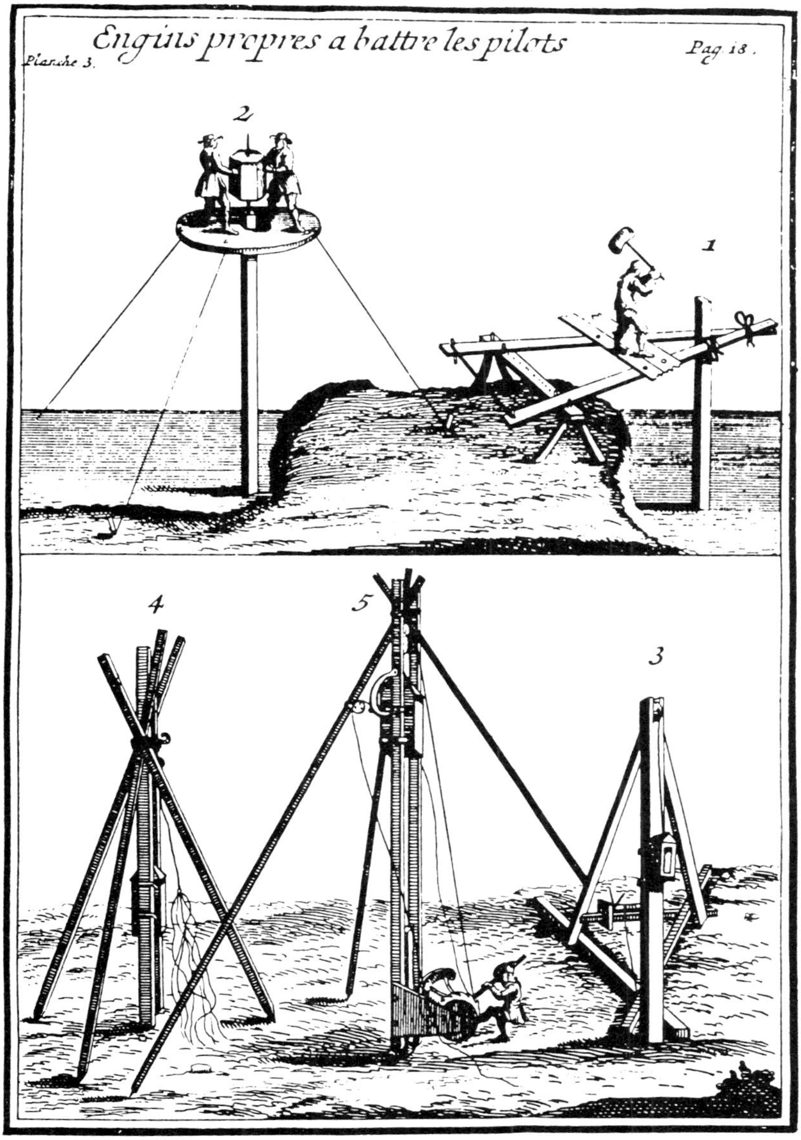

FIG. 3.—At upper right a pile is being driven with a large mallet, at upper left with a hand ram; below are examples of the *engin* and *hye* with running rams. The center *hye* has both a windlass and a catch release. (From Cornelis Meijer, *Traité des moyens de rendre les rivières navigables avec plusieurs desseins de jetée* [Amsterdam, 1696], pl. 3, p. 18.)

trate and the piles longer and thicker than those appropriate for a mallet. His hand ram is more technically sophisticated than the Florentine one. He describes it as a heavier and larger mallet, pierced vertically by an iron rod, one end of which is fixed in the pile to be driven. The ram is provided with protruding horizontal handles, and on either side a man stands on a platform to raise the weight and allow it to fall along the iron rod (fig. 3, *2*). Meijer's hand ram seems to suggest imagination rather than utility, and it immediately raises doubts that it was used in just that form.

There is another picture of a hand ram in use a little later than Meijer's, that is, ca. 1700.[14] The operation is on the same technological level as that in the Catena view of Florence more than 200 years earlier, but here the two men lifting the hand ram are working in the Netherlands. They occupy a spot in the upper right-hand corner of a picture showing the building of a shipping lock. Most of the illustration is taken up by two *engins*, each in its own boat, the two boats being fastened together by boards. Such a machine must have been unwieldly. Perhaps, if the work involved were a minor matter, it was more convenient to use a hand ram than to maneuver two *engins* into place. The hand ram was certainly cheaper in terms of manpower, for each *engin* required up to thirty men to pull on the ram.

The most popular of the three types of pile driver was the *engin*. *Engins* can be subdivided according to whether the ram was a free-falling or a running ram. A free-falling ram is shown in the *Anonymous of the Hussite Wars* on folio 37 verso, where it is suspended from the point where the uprights meet. This was not the only type of free-falling ram; there was also the gibbet ram in which the driving head was suspended from a beam perpendicular to a solitary upright, an arrangement especially suited to pile driving from a boat, as was the case at Orléans. This was the type shown in the above-mentioned French manuscript of Caesar's *Gallic Wars*. In contradistinction to the gibbet ram was the running ram, which moved up and down between or along uprights.[15] (See fig. 3, *3, 4, 5*.) Whether the *engins* mentioned in the 14th and 15th centuries were running rams is frequently unclear. However, later writers of construction manuals—Meijer, Gautier, and Perronet—do not even consider the free-falling ram. They prefer the running ram to all others, and they assume it will be provided with a release catch.

[14]Johann van Veen, *Dredge, Drain, Reclaim: The Art of a Nation* (5th ed., The Hague, 1962), p. 66.

[15]Salzman (n. 3 above), p. 328. Medieval London accounts refer to the gibbet ram (the small one) and the running ram (the great one).

In the 14th and 15th centuries, almost all mentions of pile drivers specify the *engin*. However, the *hye* (the pile driver with the windlass) was employed in 1413 at the ceremonial opening of construction of the timber bridge, the Pont-Nôtre-Dame at Paris,[16] and there was an *hye* briefly in use at the bridge at Orléans in 1389. There the pile driver with the windlass was used only twice, once in March for one day and once in July for two.[17] During those four months the builder, a wheelwright, was able to reduce the number of men needed for operation from six to two, but that is the last we hear of the *hye* at Orléans. The Orléanais continued to employ the *engin*, but not the *hye*, at least into the 16th century. Authors of the 17th and 18th centuries describe the *hye* as equipped with both a windlass and a catch with a rope for releasing the ram.[18] (See fig. 3, 5.)

The resistance to the adoption of the *hye* is the more striking because of the proliferation in the 16th and early 17th centuries of drawings of pile drivers with numerous features to produce mechanical advantages and save labor. In this period, for example, two inventors—the Italian engineer Buonaiuto Lorini and the Amsterdam carpenter Lamprecht Gerritsz—advocated the adoption of their improved pile drivers on the grounds that they spared expense, conserved manpower, and were easier on the men who did the work. Lorini states that the old pile drivers required twenty-five to thirty men who quickly became tired, whereas his needed only three, two to turn the crank and one to manage the two ropes controlling the direction of the ram.[19] The chief features of Lorini's machine were two pulleys, a rope for each, an unusually long windlass crank whose parts he calls levers, and a flywheel (presumably to make the process of winding the rope smoother). Despite the drawing and the text, many points about operation remain obscure. To build such a pile driver might have been extraordinarily difficult, especially since the only dimension given was the diameter of the flywheel.

If one suspects that Lorini was more interested in dazzling the reader with ingenuity than in providing him with useful information, an attempt to construct a pile driver from the description furnished by Gerritsz can only be called hopeless. The details on his pile driver constituted the bare minimum necessary to obtain a ten-year patent in 1595 from the States General of the Low Countries without letting his

[16]A. Tuetey, ed., *Journal d'un bourgeois de Paris* (Paris, 1881), pp. 30–31.

[17]Archives du Loiret CC 920, fol. 30r and 32v.

[18]Meijer illustrates the catch. See also Gautier (n. 10 above), p. 159.

[19]Buonaiuto Lorini, *Le fortificatione di Buonaiuto Lorini corrette & ampliate di tutto quello che mancaua per lor compita perfettione con l'aggiunta del sesto libre* (Venice, 1609), pp. 233–34. See also Frieda van Tyghem, *Op en om de Middeleeuse Bouwwerf de Gereedeschappen en Toestellen Gebruikt bij het bouwen van de Vroege Middeleeuwen tot omstreeks 1600. Studie Gesteund op beelende, gescheven archaeologische Bronnen* (Brussels, 1966), p. 240.

competitors know anything useful. Gerritsz claims that his advanced pile driver needed only six or seven men rather than twenty-eight to thirty for operation. He says that after each volley of twenty to twenty-two blows with the old type, the men needed a rest pause, six or seven being required during the driving of each pile.[20] Because Gerritsz's pile driver was not as fatiguing, it eliminated the need for rest pauses. Rest pauses, however, persisted. In the 18th century Perronet mentions them,[21] and Gerritsz's machine does not seem to have come into general use. A hundred years later, ca. 1700, pile drivers in the Netherlands were using some thirty workers. Nor does Gerritsz's claim of superior speed for a pile driver with mechanical advantages seem to have been admitted by others. Later writers on construction techniques consider the *hye* to be slower than the pile driver without the winch.[22]

The two claims to superiority of the *hye* rested on its facility in delivering heavier blows to drive larger piles and on its saving of manpower compared with the pile driver where men pulled on the ram. Yet a series of lighter blows seems to be more efficient in driving piles than fewer, heavier blows. In medieval Orléans, where a great deal of information on pile driving exists, there rarely were large piles, and probably the Orléanais preferred a larger number of men. There the pile-driving season coincided with the period of low water which forced boatmen into unemployment. During their slack season a number of them worked on the bridge at Orléans.

* * *

The whole question of the lack of adoption of sophisticated machines has to do with the separation, reaching all the way back to antiquity, of the learned and the artisan traditions. The well-known dichotomy between the theoretician and the practitioner, the architect and the laborer, persisted throughout the Middle Ages, as learned treatises were written and artisan traditions developed in two discrete streams with little intercommunication. Theoreticians did not expect to know the exact details of construction, and workmen expected neither to learn from treatises nor to write down what they knew. Hence the cursory nature of the sketches of Villard de Honnecourt, which we today find so uninformative. This attitude toward theory and practice was displayed in 1437 by a notary drawing up a contract for the rebuilding of two piers and adjoining arches of the bridge at

[20]G. Doorman, *Octrooien voor uitwindigen in de Nederlanden* (*de 16e de tot 18de eeuw*) (The Hague, 1940), p. 93, 1–9–1595. fol. 119, G 21. A patent for a pile driver using only four men was granted in 1636, and it was equally brief in its description, 3–9–1636. fol. 85, G 378.

[21]Perronet (n. 10 above), 1:97.

[22]Meijer, p. 17.

Orléans. His only real concern is to make sure the piers are built in such a manner as to provide foundations for the houses to be constructed on the bridge in order to bring in rent to maintain the structure. After specifying many details, he turns the matter over to the workmen: "As those who do the job will see, while working on it, what has to be done to support the beams for the buildings which will be made in time to come."[23] Somewhat the same sentiment is expressed about the same time by Taccola in his *De Machinis*: "Not everything can be said clearly, but persons who know their work will see in the drawing what needs to be done."[24] In the early 18th century, Hubert Gautier still accepts the dichotomy between the theoretical and the practical. His recommendation to the architect requiring scaffolds is to take the advice of carpenters who know how to make them.[25]

The dichotomy between theoreticians and practitioners of technology tended to leave craftsmen without access to the newest ideas and inventions. In medieval France there was not much consultation among experts in construction except over relatively short distances, and in bridge building experienced men were reluctant to undertake a job far from their base. Ideas do not seem to have spread widely, and bridges were built by local residents.[26] Pile drivers were built by men on the spot. Such artisans commonly had neither the training nor the incentive to learn more about complicated machines. The Orléans wheelwright who abandoned the attempt to produce an *hye* was obviously having mechanical problems and may have become convinced that it was hopeless to attempt to make a machine capable of competing with the *engin*. The level of technical education prevalent in the Middle Ages seems to have lasted into the 18th century and militated against the adoption of the elaborate pile drivers of a Lorini or a Taccola or a Leonardo da Vinci. To construct these required a theoretical education not available to the usual builder. Hubert Gautier, in his treatise on bridges, explains that in 1715 he sent to the *Journal des savants* five questions concerning bridge construction to be resolved. The mathematician La Hire sent in an answer, but Gautier was much disappointed. La Hire's response was useless: he had used algebra,

[23]Archives du Loiret III al fol. 130v. Published in my "Moving Ahead with the Fifteenth Century: New Ideas in Bridge Construction at Orleans," *History of Technology* 6 (1981): 18–19.

[24]Mariano di Jacopo, called Il Taccola, *De Machinis*, ed., annotated, and trans. Gustina Scaglia (Wiesbaden, 1971), p. 103.

[25]Gautier (n. 10 above), p. 72.

[26]See my "Moving Ahead with the Fifteenth Century" (n. 23 above), pp. 12–13, and my "Working at the Bridge Site in Late Medieval France," forthcoming in the *Proceedings* of the Colloque, Artistes, Artisans et Production Artistique au Moyen-Age, Université de Haute-Bretagne, Rennes II, May 2–6, 1983.

something no builder knew.[27] If the obverse of the coin was that craftsmen had an inferior technical education, the reverse was that the machines kept breaking down. Even the simple pile driver powered by men pulling directly on the ram needed frequent repairs during the two-month pile-driving season at Orléans in 1387–89. How much more exasperating must have been the case of the complicated pile drivers!

Finally, in considering the reluctance to adopt innovative designs in pile drivers, one must take into account the question of the relative reliability of machines, the constraints their use imposes on operators, and the availability of a large supply of inexpensive manpower. No hint of a labor shortage is contained in the accounts of the bridge and hospital of Orléans for 1387–89, and in the 18th century Gautier tells us that men are never lacking at a construction site. Perronet had the services of two regiments of soldiers in building the new bridge at Orléans. Gautier's and Perronet's fundings came from the French government, and it is not surprising that they were unmoved by considerations of cost which had seemed so important to Lamprecht Gerritsz in Amsterdam in 1595.

Gautier, although head of the Ponts et Chaussées under Louis XV, distrusted machines. On the subject of cofferdams he mentions a number of contrivances to exhaust water—Archimedes' screws, waterwheels, pumps, and other machines—but he complains that some of them do not raise water high enough without incredible difficulties; that waterwheels take up too much space; and that, anyway, machines break down. For draining water from a cofferdam, he writes: "I do not find anything more natural than the arms of men with a tub [*baquet*] with two handles, who without interruption day and night ceaselessly dip out the water at different heights."[28] If the men changed off every hour or two, there were no problems of fatigue, and if anyone should become indisposed, there was always someone to replace him. In Gautier's view, men were so much more reliable and satisfactory than machines!

It is obvious that, in France and the Netherlands, from the 14th to the 18th century, people preferred the simple pile driver in which men raised the ram by pulling on a rope to the more sophisticated pile driver with the windlass. They did so for a variety of reasons. Probably there was a preference for tradition, an unwillingness to adjust to new ideas and to learn and observe the rules for operating unfamiliar machines. These very human reactions are still with us and account in

[27]Gautier, p. 21.
[28]Ibid., p. 81.

part for the present-day reluctance, on the part of some, to adopt the computer. There was probably also an appreciation of the superior speed of the simple pile driver and a lack of interest in the chief point in which the *hye* outshone its rival—its ability to save on manpower. Before the industrial age, in contradistinction to our own age, labor was cheap and materials expensive. Besides, people probably preferred full employment to efficiency. Finally, technical expertise was in short supply among builders, so that a pile driver with a foolproof construction, or at least one where the repairs were easy, was bound to be popular. The history of the pile driver demonstrates once again that to concentrate on "firsts" in ideas or inventions is to distort the picture of how people lived in the past.

6
Rebuilding the bridge at Albi, 1408–1410

M.N. Boyer

> Expenses and charges made to repair and finish the cutwater at the base of the pier of the bridge over the Tarn at Albi which is under the iron cross and to build and repair the said base of the said pier.[1]

This statement introduces fourteen folios of MS CC 167 of the communal Archives of Albi, detailing the wages paid workmen, the materials purchased, and other disbursements by the town in repairing the bridge in August and September of 1408. Also, the Albi town accounts for 1409–10 (MS CC 168) relate the expenditures for reconstruction of two arches of the bridge. From these two manuscripts much can be learned about construction methods and the use of work crews in medieval bridge building—subjects on which we have very little information.

An eighth-century manuscript of the *Mappae claviculae*[2] includes a formulary, based on ancient practices, for laying in water the foundations of buildings and, of course, bridges. Apart from this, the interest of medieval writers seems to have been pretty much limited to passing references to materials used in bridge construction; or, in the fifteenth century, to a discussion of machines, as hoists, pile drivers, and pumps, useful in various types of construction including bridges.[3] In the sixteenth century Ramelli[4] included among his diverse machines a sketch

1 Archives communales of Albi, CC 167, fol. Ir.

2 Victor Mortet, "Un Formulaire du VIIIe siècle pour les fondations d'édifices et de ponts d'après des sources d'origine antique. Nouvelle édition critique," *Bulletin Monumentale*, LXXI (1907), 442–65. See also Sir Thomas Phillipps, "Letter . . . Communicating a Transcript of a MS Treatise on the Preparation of Pigments, and on Various Processes of the Decorative Arts Practised During the Middle Ages, Written in the Twelfth Century, and Entitled 'Mappae clavicula,'" *Archaeologia*, XXXII (1847), 183–244.

3 F. M. Feldhaus, *Technik der Antike und des Mittelalters* (Potsdam, 1931).

4 Capitano Agostino Ramelli, *Le Diverse e Artificiose Macchine* (Paris, 1588), p. 171.

of a cofferdam; but for a full-length treatise devoted to bridge building the world had to wait until the eighteenth century. In 1716 Hubert Gautier published his *Traité des ponts,* in which he deplored the lack of books on bridge construction.[5]

The accounts of Albi are therefore especially valuable for their record of what actually occurred at the construction site. There are other French medieval bridge accounts, as for example those of Romans and of Orleans, but they are not as informative or as detailed as those of Albi, nor do they cover as long a period.[6] Thus at Orleans, after the bombardment of 1940, the bridge accounts survive only from 1386–89, and 1428–31, and 1435–36. At Albi, on the other hand, thirteen annual communal accounts written in Provençal have survived from the fourteenth century, and eight dating from between 1402 and 1410. Publication of the fourteenth-century accounts was provided by A. Vidal in his *Comptes consulaires d'Albi* (1900) and his *Douze Comptes consulaires d'Albi du XIVe siècle* (1906–11).[7] On the other hand, the town accounts from the early fifteenth century remain in manuscript only.

There is a further advantage in studying the medieval bridge at Albi rather than at Romans or at Orleans. At the latter two places the bridges were washed out in the eighteenth century and rebuilt in a different style, but it is still possible to visit the old town of Albi and see the pointed arches of the Pont de Tarn. Documentary evidence places the erection of the bridge at Albi at least as early as about 1037, at which time the viscount, bishop, nobles, and people of the town combined to erect the bridge on land donated by the chapter of St. Salvi.[8] The Pont de Tarn is mentioned again in a chronicle of the twelfth century, and a thirteenth-century toll list for the bridge has survived.[9] In 1269 the bishop, lord of the town, found that the income from the bridge failed to compensate him for its maintenance, and he turned over the tolls and the responsibility for the structure to the municipality. This is why the

[5] Hubert Gautier, *Traité des ponts où il est parlé de ceux des Romains & de ceux des modernes* (4th ed.; Paris, 1765).

[6] Ulysse Chevalier, "Notice historique sur le Pont de Romans," *Bulletin de la Société départementale d'archéologie et de statistique de la Drôme,* II (1867), 308–32. Archives communales d'Orléans, CC 920, 930, 967.

[7] The first title was published in the *Bibliothèque méridionale,* Series I, Vol. V, at Paris; the second in the *Archives historiques de l'Albigeois,* Vols. I and II, at Toulouse and Albi.

[8] *Gallia christiana* Vol. I (Paris, 1715), ad instrumenta, p. 4.

[9] A. Vidal, "Costumas del Pont de Tarn d'Albi," *Revue des Langues Romanes,* XLIV (1901), 481–513.

records of the Pont de Tarn are to be found in the communal archives of Albi.

* * *

In 1359, when the extant accounts of work on the bridge began, it was already more than three hundred years old. The recurrent repairs to the structure raise the question: how medieval is a bridge that has survived into the twentieth century only by virtue of repeated reconstruction? Certainly the bridge at Albi has undergone significant changes. About 1400 the pier nearest the suburb of Cap del Pont supported a gate, the Porta Nostra Dona, protected by a drawbridge; and after 1368 by a *guachil*, a wooden fort covered with brick and surmounted by a watchtower with a pointed roof. Besides the Porta Nostra Dona and the *guachil*, the house of P. Soelh was also on this pier. Three other piers are identified in the accounts, one as being under the iron cross, one under the toll house, and one under Pos Galaup's house, while the bridge ended in the Porta Nova in the town fortifications. Nor were these buildings the only excrescences obscuring the severe outlines of the bridge, for in 1397 the town had installed gargoyles to carry off the rainwater from the roadbed.[10] However, nothing of the superimposed structures survives today except the iron cross. In 1888 E. Degrand, in *Ponts en Maçonnerie*, condemned the bridge as without ornament and described it as a barbarous work, offering no interest save for the ogival arches. He was critical of the irregularity of the bridge, for "the openings of the seven arches composing it vary from 9.75 meters to 16," the piers, badly aligned, are too wide compared to the arch openings, and the tympans are almost all of different heights.[11] (Here Degrand ignores the eighth and smallest arch—only 4.1 meters.) Upstream the cutwaters are too pointed, Degrand continued, and downstream they are square and insignificant. The accuracy of Degrand's observations may be tested by consulting the accompanying illustrations, especially the plan of the Pont de Tarn in Figure 1.

William Emerson and Georges Gromort, in their book *Old Bridges of France*, were more generous in their judgment. They wrote that the brick work "which forms such a pleasing contrast with the original stone structure on the upstream side" and entirely covers the downstream elevation was added in the nineteenth century.[12] Emerson and Gromort were correct in pointing out that in the early nineteenth century there were important alterations to the bridge, including

[10] Vidal, *Douze Comptes*, II, 231.

[11] E. Degrand and Jean Résal, *Ponts en Maçonnerie*, II (Paris, 1888), 64–65.

[12] (New York, 1925), p. 33.

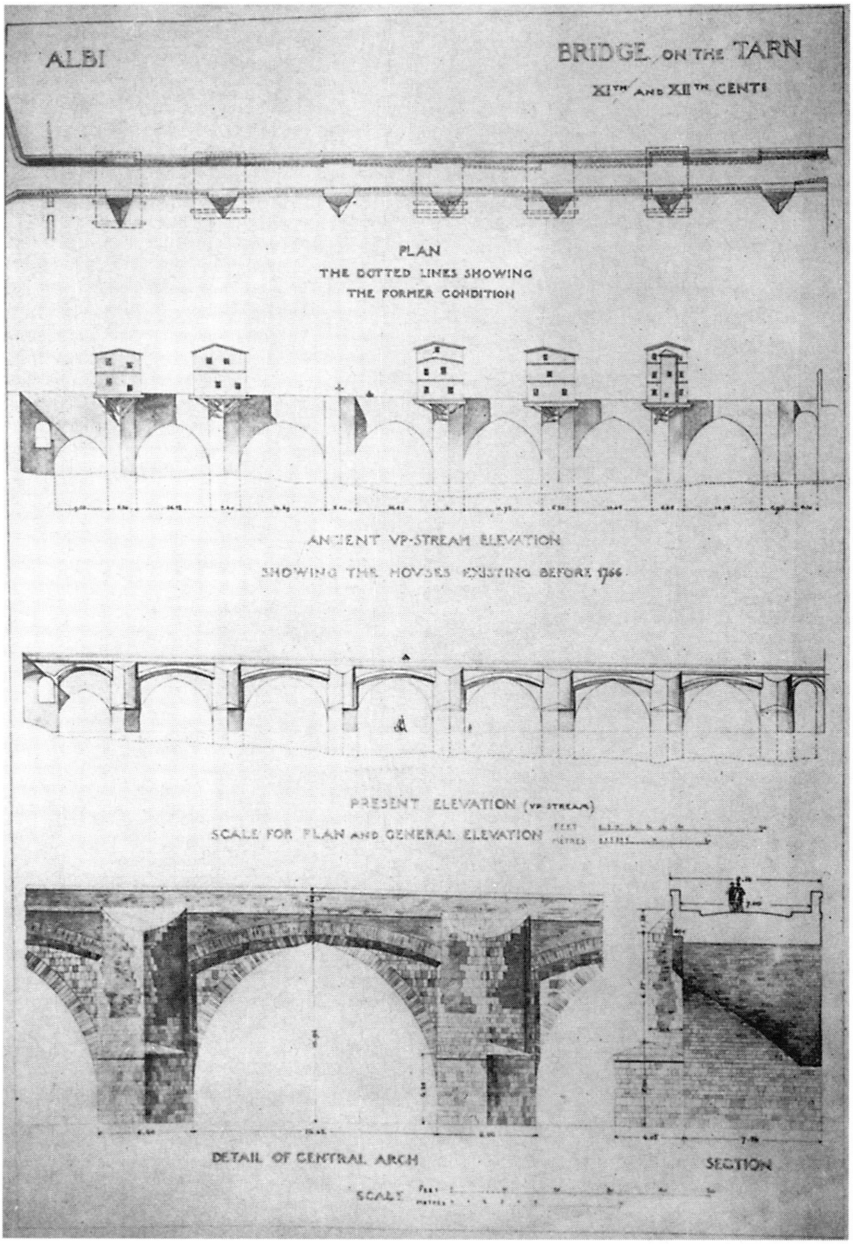

FIG. 1. — Plan of the bridge over the Tarn at Albi. (Reproduced from Emerson and Gromort, *Old Bridges of France*, with the permission of the American Institute of Architects.)

FIG. 2. — View from upstream of the bridge, showing the pointed cutwaters. (Reproduced from Emerson and Gromort, *Old Bridges of France*, with permission of the American Institute of Architects.)

FIG. 3. — View of Bishop's Palace and the River Tarn from the Cathedral Tower of Albi. (From negative by Lindsley F. Hall. Picture supplied through the courtesy of the School of Architecture of the University of Oregon.)

widening the structure. However, before speaking of the pristine character of the bridge, one should remember that documentary evidence shows extensive repairs to the Pont de Tarn in every century from the fourteenth to the nineteenth.

The maintenance of the bridge over the Tarn was taken seriously by the town of Albi, and in those years of the period 1359–1410 covered by the accounts, there do not seem to have been periods longer than about three months in which the bridge was impassable. The shortness of these periods is in contrast to the years at the Pont-de-la-Guillotière at Lyons, the Pont-Saint-Bénézet, and the bridge at Romans.[13] If the consuls at Albi did not move sufficiently quickly to prevent disruption of service, they at least instituted repairs shortly after any disaster. In the fifty years prior to 1410 the Pont de Tarn was a frequent source of concern to the town. Especially the drawbridge was in frequent need of repairs. It was rebuilt in 1359–60, in the days of the English occupation of the countryside.[14] One year the drawbridge would require a new chain or rope; another year, hinges and planks. In 1381–82 a railing was installed to make it safe for animals, and in 1391 the drawbridge itself fell into the river. From time to time the bridge buildings needed attention. For example, in 1368–69 the stone tower on the drawbridge pier collapsed into the Tarn. When the fortifications had been repaired, work was needed on the toll house. For most of the forty years following 1359 the drawbridge and superstructure had been the chief concern, but beginning in 1397–98 the pier foundations required attention. In that period workmen were rebuilding the pier under P. Soelh's house, an operation which lasted from the middle of July to the end of September. In 1403–4 it was necessary to reconstruct an arch, and the repairs of 1408–10 constitute the subject of this paper.

* * *

As a record of bridge construction in 1408–10, the manuscripts offer a wealth of details. Master Peyre Ayral, notary, treasurer of Albi, was a conscientious administrator, generous with details on the materials purchased and the names of the workmen. In many cases he gave the lengths of the pieces of lumber, the type of wood and the uses to which it was put, the number of bricks, the cartloads of stone, and the tally

[13] M. C. Guigue, "Notre-Dame de Lyon; recherches sur l'origine du pont de la Guillotière," *Mémoire de la Société littéraire . . . de Lyon* (1874–75), pp. 227–365. Claude Devic and J. Vaissette, *Histoire générale de Languedoc,* ed. Ernest Roschach (Toulouse, 1872–1905), new ed. VI, p. 610. For Romans, see n. 6 above.

[14] Vidal, *Douze Comptes,* II, 231.

of each particular type of nail. However, after meticulously listing and describing individual beams, he suddenly noted, without itemizing, the purchase of twenty cartloads of timber. Furthermore, precise accuracy as to the materials used in any one construction job is illusory, for frequently it is impossible to determine whether the materials were used in the Ronel tower in the town wall or the Pont de Tarn, and if in the latter, in what part of it. It is constantly borne in upon the reader of Albi CC 167 and CC 168 that the purpose of the manuscripts was to explain and justify the expenditures of the town of Albi, and information on building is clearly subordinate to this aim. Hence the surprising lacunae, so that after learning of the felling of the trees and the trimming of the timbers, one suddenly finds a scaffolding in place. Nevertheless, it is possible to obtain a fairly clear picture of work practices and building methods current at Albi.

The treasurer, Master Peyre Ayral, was interested in people. There is a small-town air about the Albi accounts, for the names of almost all the workmen, and frequently their relationships, are given. The town found it necessary to treat its employees well, and again and again we find the justification for raising the pay or providing extra benefits for the workers on a particular job: "Because otherwise they did not wish to be there."[15] By providing additional incentives, the town was able to obtain an adequate supply of local unskilled labor for construction on the bridge in 1408–10.

It was, however, another matter with the experts. Skilled labor at the actual construction site consisted of masons and *fustiers.* Of some nine *fustiers* who worked on the reconstruction of the third arch at Albi in 1409, three were from out of town.[16] Perhaps "carpenter" is the best translation for *fustier*, for these men cut down trees, trimmed the logs, sawed them into boards, and nailed or tied them together for scaffolds and centering. They acted as contractors, agreeing to provide scaffolds and centering, employing and supervising laborers, renting out hoists, and driving piles.

In the Middle Ages in the building trade there was much more employment for carpenters than for masons, even in the case of a masonry bridge. In the reconstruction of the Pont de Tarn at Albi more *fustiers* were employed for longer periods than masons. This was so not only in the case of the rebuilding of two arches but also when the break-

[15] See, for example, Albi, CC 167, fol. VIIr.

[16] One *fustier* was from Villefranche de Rouergue (Aveyron); one from Réalmont, arrondissement Albi (Tarn); and one from Campagnac, canton Castelnau-de-Montmiral, arrondissement Gaillac (Tarn).

water at the base of a pier was reconstructed. In the latter case, between August 16 and September 5, 1408, the *fustiers* worked on fourteen days, but the masons on eight. In total days worked the differences are even more impressive, for four *fustiers* worked forty-four man-days, the three masons only fifteen. Their spheres of work overlapped, for masons helped to put up scaffolds, and *fustiers* took pick or iron bar in hand to aid in demolishing the old arches. Although the masons and *fustiers* co-operated on the job, it was a different matter when the town provided food for the workmen during the reconstruction of the cutwater. Then three masons insisted on a separate table, "because they did not wish to be with the above-named," the *fustiers* and their helpers.[17]

In rebuilding the bridge all operations unrelated to masonry and expert carpentry were performed by laborers, or at least by unskilled workers. At Albi, the town felt the disadvantages of the system under which men who mixed mortar were merely laborers. More than once there are complaints that the mortar was "massa grossa," literally "too fat,"[18] and the solution to the problem in each case was to add more sand. Also, each time the same man was called in to mix the mortar correctly, but he was paid no more than other laborers. During the reconstruction of the cutwater, in 1408, the masons and *fustiers* were paid 5 sous daily, but everybody else, whether master or man, father or son, was paid 2 s. 1 d. per day. Later, during the rebuilding of the arches there was more differentiation in payments to workers. Frequently, workers were paid 2 s. 6 d., and on rare occasions masons and *fustiers* received more or less than 5 s. daily. Although women had worked on the bridge in 1368–69, "serving" masons with mortar and water,[19] and although in 1408–9 they were employed on occasion at the Ronel tower, they do not seem to have worked on the bridge in 1408–10.

The town of Albi was not especially fortunate in its workmen. In the reconstruction of arches for the bridge, the town consuls let two contracts, one to a *fustier* for the centering and one to a mason for the brick arch. The "master of the arches and centering" for the two arches rebuilt in 1409–10 was Arnaut Albert, *fustier*. He seems to have been competent in his line, but he was so dilatory in answering the summons of the town to start work that in August 1409 the consuls obtained a court order (delivered by a royal sergeant) to force him to appear and work on the scaffolding and centering. After competition of the third

[17] Albi, CC 167, fol. Xr.

[18] Albi, CC 168, fol. LIIIv. and LXIIv.

[19] Albi, CC 151, fol. LXXIIIIr. ff. (Vidal, *Douze Comptes*, I, 77).

arch in brick, he was so slow in removing the centering that the town hired another *fustier* to do it. Nevertheless, the consuls continued their agreement by which Master Arnaut Albert supervised the building of the scaffolding and centering for the next arch.

The consults were more dissatisfied with Master Anthoni Mauri, the mason who rebuilt the third arch, and did not contract with him to reconstruct the next one. He was unscrupulous, and the town found itself obliged to pay his debts. For example, the consuls had arranged with the hospital of St. James to lend a bed to Master Anthoni. The latter sold it, and when the hospital officials insisted on its return, the municipality agreed to replace it. Furthermore, the consuls had to pay the wages of the workmen, sums which ordinarily would have been paid by the contractor, because otherwise the men refused the job, "doubting that they would be paid by the said Master Anthoni."[20]

In a further unfortunate incident, the town found itself paying for having piles for one of the centerings driven a second time. A *fustier* insisted that the piles had not been driven deeply enough, but when his pile driver was brought into action, "it was found that the piles refused to enter farther."[21] The treasurer's conclusion was that the whole incident represented a successful attempt by the *fustier* to force the town to hire his *corda*.

* * *

The first operation in rebuilding the cutwater was to clear off the driftwood and the next was to put in a cofferdam around the place to be repaired. Thursday, August 16, 1408, eight men climbed on the *arca* to demolish it and to dig down in the river bed for the foundations. The piles measured 4.96 meters long, and behind them were driven thinner pieces to form two inclosures, one inside the other, the timbers being fastened at the top by cross pieces. Work on the cofferdam continued on Friday and Saturday, and Monday the men threw *bart* (a clay mortar)[22] between the rows of piling to make them watertight.

Tuesday, August 21, the process of exhausting the water inside the cofferdam began. The equipment consisted of buckets made of thin planks by a carpenter (*fustier*), bound with iron rings by a blacksmith, and pulled by six ropes. This contrivance vaguely suggests Agricola's pump for emptying water in mines—an endless chain of ironbound buckets mounted on a rope—but his machine[23] implies a sophistication

20 Albi, CC 168, fol. LIXv.

21 Albi, CC 168, fol. LXXIIr.

22 Vidal (*Douze Comptes*, Vol. II, Glossary) defines *bart* as a mortar of clay and hay (foin), but usually the accounts mention only the clay as an ingredient.

23 Georgius Agricola, *De Re Metallica*, trans. from the first Latin ed. of 1556 by Herbert C. Hoover and Lou H. Hoover (New York, 1950), pp. 173, 174, 175.

foreign to the processes of construction at Albi. Apparently the men drained the cofferdam merely by the use of pails. This method was used at Orleans in 1435–36, was depicted by Ramelli in 1588, and recommended above all others by Hubert Gautier (1765).[24] Gautier especially valued it because, unlike pumps, "baquetage" never broke down from mechanical failure, and if the men changed off every two hours, there was no excessive fatigue.

At Albi the process of exhausting the water inside the cofferdam went on around the clock from Tuesday, August 21, until Saturday, August 25; and from Sunday, August 26, to the following Saturday, September 1. During these two weeks the town provided food for the men to enable them to stay on the job. Men received 2 s. 1 d. per day and 1 s. 3 d. per night. Perhaps the differential meant that the day was considered to be five-eighths and the night three-eighths of the twenty-four hours, but the work of draining the water cannot have been too arduous, as is proved by the very long hours during which the men were employed. During the first week six men stayed on the job around the clock for four days, and two did so for five. The second week four men were at work each in a continuous period of ninety-six hours. At Albi between August 21 and September 1 at night, the work crew varied from five to sixteen men, and in addition to the draining of water, it is probable that piles were driven under the foundation of the cutwater. Certainly some laying of masonry took place. After the end of the process of exhausting the water, the masons worked another three days. At least part of the time the masons and their helpers operated from boats.

The total labor that went into repairing the *arca* and the pier included, for the *fustiers* 44 man-days of work, for the masons 15, and for others 226 man-days, not counting more than 23 days of work transporting mortar, bricks, timber, and *terra de Piulet* (clay from the banks of Piulet brook). There were 87 man-nights of work. With the exception of some additional mortar applied toward the end of September, all the work was done between Tuesday, August 14, when laborers cleared driftwood off the cutwater, and September 5, when the masons completed their repairs to the pier. Because of the necessity of "keeping water from the *arca*,"[25] the operation required a concentrated effort unnecessary in most other construction projects.

* * *

Rebuilding the two arches was a more lengthy operation than the repairs to the cutwater at the base of the pier. Some of the wood left

[24] Ramelli, *op. cit.*, p. 171; Gautier, *op. cit.*, I, 87. [25] Albi, CC 167, fol. XIv.

over from the latter had been carried back to the town hall to be used again, but it was necessary to cut (in the forest) most of the lumber needed for scaffoldings and centerings of the arch reconstructed in 1409 (called, in the records, "the third arch counting from the town.")[26] On April 12, 1409, Master Arnaut Albert, master of the arches and centering of the arches of the Pont de Tarn, came to Albi to supervise the selection and felling of the trees for the job. With his assistants he prepared thirty cartloads of timber which were transported to Albi. This was by no means the total amount of wood used in the reconstruction of the third arch, and other timbers were purchased during the operations.

The immense quantities of wood necessary to build a masonry bridge were a cause of surprise and dismay in fourteenth- and fifteenth-century France. In 1389–90, the town of Romans found the purchase, felling, and transportation of thirty-three fir trees to the site of the bridge, for the centering of an arch, extremely expensive.[27] At Albi, again and again the timber assembled before beginning work on some part of the bridge was found to be inadequate, and more had to be procured.

To rebuild the third arch of the bridge over the Tarn, sufficient supplies had been assembled at Albi to justify the start of work on Tuesday, September 17, 1409. On that day, two masons and six other men were at work dismantling the old third arch, and the project had reached the stage where the scaffolding was in need of planks, "so that the men could demolish and take apart safely the old *arcvout*."[28] Also, Master Arnaut Albert was sent for to make the centering. The scaffold served, too, as the "*sobre pon* [literally 'upper bridge'] over the water by which people go"[29]–in other words, a temporary bridge on which to cross the river while the arch was impassable. The *sobre pon* was chiefly composed of poplars, but also included wood from nut trees. The scaffolding (or *sobre pon*) was partly tied and partly nailed.

Construction of the centering involved no little skill and judgment. The scaffolding seems to have been largely put together in the absence of Master Arnaut Albert, but not so the centering. The thirty carloads of timber mentioned earlier were insufficient for both the scaffolding and the centering, and one of the first things that Master Arnaut Albert did upon reaching Albi was to make (with the help of three assistants) four great beams for the centering. Then another two cartloads of timbers were needed, because the original centering was not high enough. As the men were nailing on the plans for the intrados, there occurred an

[26] Albi, CC 168, fol. XLIIIr.

[27] Chevalier, *op. cit.*, p. 316.

[28] Albi, CC 168, fol. XLIIIv.

[29] Albi, CC 168, fol. LXXIIIr.

untoward incident that set them back in their work. The centering was put together from materials assembled on the dry gravel bed of the Tarn under the arch, and then it was hoisted into place. The *gravier del Tarn* was submerged during high water, and more than once the town was inconvenienced because of its practice of storing timbers there. For example, on October 8 the centering was in danger from high water, so it was tied with a long rope; but, nevertheless, it was felt necessary for eight men to watch all night to protect it.[30] When the waters had subsided and the driftwood had been cleared away, the workmen used two great pins to repair a break in the centering, several days being required to make good the damage.

In the meantime, Master Anthoni Mauri, the mason who had contracted to rebuild the third arch, had ordered four great stones to be placed on either side of the arch, and on October 21, to celebrate the laying of the first stone, he was given a present by the consuls.

After three days of adjusting the centering because it was now too large, the men began driving piles. At least some of the latter were 4.46 meters long. They were shod with iron, thirty-seven pounds being assigned to a blacksmith for this purpose. After the piles had been driven, beams were laid across them, and the centering hoisted into place. After the final touches (in the shape of struts and the planks on which the voussoirs were built), the brick arch was laid. On Tuesday, November 19, the consuls sent to Campagnac to request Master Arnaut Albert to come and remove the centering. He duly came, but he announced that it was too early to think of taking away the support for fear the brick arch would fall. Three weeks later, the consuls, losing patience, hired another *fustier* with his helpers to saw off the piles, December 11–14, and take away the centering. This critical operation was accomplished without harm to the brick arch.

* * *

The final construction of the period 1408–10, as far as the bridge was concerned, was the rebuilding of another arch. This arch is described as "lo arc segon" (the next arch), that is, contiguous to the third arch counting from the town. It seems preferable to assume that "lo arc segon" was the second rather than the fourth arch, because its centering was assembled on the gravel bed of the Tarn, and lifting it into place was not a major operation.

Arnaut Albert was also master of the arch and centering for this project. His duties were described as to change the *sobre pon* to the

[30] Albi, CC 168, fol. XLVIIr. and XLVIIv.

next arch and to alter and repair the centering and lift it into the arch to be repaired.[31] The mason who had contracted to make the brickwork was Master Thomas Ubert. The town used as much as it could of timber previously employed in the third arch. When, in December 1409, the centering had been removed from that arch, it was necessary partly to dismantle it to preserve the brickwork. It was also essential to rebuild the centering, for in 1410, as today, the second and third arches differed in width.

On Wednesday, July 9, 1410, work began on taking apart the old centering, there being present two *fustiers* and five other men in addition to three valets of Master Thomas Ubert. Master Arnaut Albert arrived about a week later, at which time the old masonry work was being dismantled. Next, the *sobre pon* was moved to its new position. On Monday, July 21, Master Arnaut Albert and ten other men tried to raise the centering into the arch, only to find it was too large. After a certain amount of sawing and trimming of timbers, three days were spent raising and setting the centering in place. Next the centering was nailed, lashed, and reinforced with more struts, and the rest of the facing to support the voussoirs was added. Then the men were obliged to drive wedges under the centering to raise it before it was finally ready for the masonry. Master Arnaut Albert's method of working was evidently one of continual improvisation. Despite the fact that the town had purchased ropes to enable him to measure the arch, he evidently relied very little on them.

In MS Albi, CC 168, information on the masonry work of this *arcvout* is almost entirely lacking, presumably because Master Thomas Ubert, who contracted to make the brick arch, seems to have performed his obligations to the satisfaction of the town. We know that on September 9 he received a present for laying the first stone, and there are a certain number of notations on laborers making mortar or carrying bricks; but otherwise the details are missing of what must have been a routine and successful job.

* * *

Materials for the reconstruction of the bridge at Albi were all available locally. The timber came from nearby forests and was used shortly after it was cut. The manuscripts speak of stone being removed from the old arches, and apparently it was re-used in the bridge, but there was a shortage of stone for construction at Albi, and the attempts of the town to find new quarries in 1408–9 were unavailing. Accordingly, the commune relied very heavily on brick, for the making of which there

[31] Albi, CC 168, fol. LXIIr.

were two brickyards at Albi. Timber and stones were delivered by ox-drawn carts, bricks, lime, and sand by pack animal. In earlier manuscripts it was noted that *terra de Piulet* (clay from Piulet brook) was frequently delivered by boat, but in 1408–10 it was carried on pack horses. Sand was obtained from the bed of the Tarn. Frequent journeys were necessary to request *fustiers* and masons from other towns to work at Albi, to insure a supply of lime from the operators of the kilns, to purchase ropes for the hoists, and otherwise insure the smooth operation of the construction.

In 1408, in rebuilding the cutwater, the builders used two ideas new to Albi. One of these concerned materials, the other the method of founding a pier. In 1397–98, in rebuilding the breakwater at the base of the pier under P. Soelh's house, the workmen employed brick, sand, and *bart* (a mortar of clay and hay), but the core of the cutwater was *terra de Piulet* stamped down by the feet of the workmen.[32] In 1408, in reconstructing the breakwater under the pier with the iron cross, the materials consisted of ten cartloads of stone plus an unstated number of bricks, 5,996 kilograms of lime, 216 hectoliters of sand, 754 hectoliters of *terra de Piulet,* plus the amount carried from the brook to the bridge in eleven days. In addition, the builders finished the *arca* off with 10 hectoliters of flagstones, and they used a different type of mortar or *betum*[33] where the masonry came in contact with water. The *betum* contained, in addition to some of the sand and lime mentioned above, 605 liters of vinegar and 800 eggs. In the Middle Ages the use of vinegar to increase the resistance of mortar to water was common, according to Bertrand Gille,[34] but the use of eggs seems to have been rare.[35] John T. Metcalf informs me that the purpose in using eggs was to introduce air into mortar, thus increasing not only its weather resistance but its workability.

The other innovation at Albi in 1408 was the use of a cofferdam in rebuilding the cutwater, which made it possible to drive piles under the foundation and to lay the masonry where it was dry. The cofferdam was not a new idea, as it was used in antiquity and recommended in the eighth-century formula of the *Mappae clavicula,* apparently based on

[32] Albi, CC 160, fol. LXXXVIIIv.; Vidal, *Douze Comptes,* II, 239. F. de Verneilh, "Architecture civile du moyen-âge. Construction des ponts," *Annales archéologiques,* XX (1860), 99, mentions masonry piers having a core of earth.

[33] Albi, CC 167, fol. XIIr.

[34] Maurice Daumas (ed.), *Histoire des techniques,* I (Paris, 1962), 580.

[35] There is a solitary case reported in L. F. Salzman, *Building in England Down to 1450: A Documentary History* (Oxford, 1952), p. 153.

Vitruvius. However, the Albi accounts for the fourteenth century do not indicate the use of this but rather of another method of founding piers. This system, which also dates from antiquity, consisted in driving piles to form an inclosure inside which rubble was thrown to form a platform above water level. On this the masonry was laid. This method was employed at Old London Bridge[36] and at Orlęans in 1386–89, and had the immense advantage of avoiding not only technological difficulties and expense but the necessity for a large supply of manpower at one time.

* * *

Despite the unusually full sources on the bridge over the Tarn at Albi, much of the history of the structure remains obscure. It is uncertain of what material the original bridge was built. Jolibois[37] and others have assumed that at once time the entire bridge was of stone, and this view would seem to be supported by the fact that the ruins of the tower which fell in 1368 were composed of stone and wood. However, brick was a material much used at Albi in the fourteenth and fifteenth centuries. The cathedral of Saint Cecilia, built 1282 to *ca.* 1390 is of brick. As for the bridge, between 1368 and 1410 the accounts show the purchase of more bricks than stone for this structure. In the sixteenth century a mason employed by the town to inspect the bridge reported that the piers had been made of small stones intermixed with brick at places and that, because the stones were small, it had been necessary to clamp them together with iron.[38] His description tallies with that given in the accounts. The report mentions the poverty of materials available when the bridge was built.

From the seventeenth and eighteenth centuries, pictures of the bridge have come down to us.[39] In 1642 a view of Albi shows a substantial portion of the bridge, which has lost the old entrance gate on the left bank, but now has a house on every pier except one. Perhaps not much can be said for the accuracy in detail of this picture, which confuses the bishop's palace with the cathedral and depicts the arches of the bridge as semicircular. Nevertheless, it is known that there were houses on the bridge and that they were removed after the flood of 1766 had heavily damaged the Pont de Tarn.

When a bridge has been repaired as often over the centuries as the

[36] Gordon Home, *Old London Bridge* (London, 1931), p. 35.

[37] E. Jolibois, "Le Vieux Pont d'Albi," *Revue historique, scientifique et littéraire du département du Tarn*, II (1878–79), 197.

[38] *Ibid.*, pp. 200–201.

[39] *Ibid.*, p. 196.

Pont de Tarn at Albi, it is not to be supposed that the dimensions or the superstructure will have remained identical from one period to another. Yet it would appear that the outline of the bridge has not materially altered. According to the documents, the medieval bridge was narrow, as is proved both by the thirteenth-century toll list, which did not allow for objects transported in carts, and by the fourteenth-century town accounts, in which materials carried by packhorses were taken directly onto the bridge, but timber carried by oxcart was unloaded in the streets of the town and transported by manpower to the construction site on the bridge. An inspection of the present-day Pont de Tarn shows that it has been widened. It is extremely probable that already in 1400 the Pont de Tarn had pointed cutwaters upstream only, for in the sixteenth century a mason complained of this on the ground that the water, swirling together suddenly on the downstream side, tended to undermine the piers. Nothing was done, however, and today, four hundred years after this criticism, there are still pointed cutwaters on the upstream side only (Fig. 2).

Today there are seven piers in the bridge over the Tarn at Albi (Fig. 3). The picture of 1642, which gives an incomplete view of the bridge, shows six piers, while the town accounts of about 1400 give distinguishing features for four. This does not, however, mean that there were not seven at that time. Indeed, the third arch from the town seems to have been at approximately the same place in 1400 and now, for at low water it was and is over a dry gravel bed not far from the left bank of the river. Furthermore, it is clear that the silhouette of the bridge must have been determined between the late twelfth century and about 1500, because the shape of the arches is pointed.

One of the factors making for the preservation of the main features of the Pont de Tarn was the matter of finances. Maintenance was very expensive, and the revenues of the bridge failed to cover it. Albi was constantly short of funds, especially during the reconstruction of the fortifications required by the Hundred Years War. Like other medieval proprietors of bridges, Albi was not inclined to make preventive repairs but to wait for an emergency, such as the fall of the drawbridge or the collapse of the stone tower. Even in such cases, only essential portions were rebuilt. Since reconstruction was always piecemeal, the bridge continued to preserve its general character. It is therefore proper to speak of the "medieval bridge" over the Tarn at Albi.

7

Le Pont du Rialto: un chantier public à Venise à la fin du XVIe siècle

D. Calabi and P. Morachiello

I. — Les procédures

A Venise, entre 1588 et 1594 : le grand chantier public qui s'ouvre au Rialto, le centre de la ville, au croisement des principales voies d'eau et de terre, fait participer un certain nombre d'institutions d'État et de maîtres d'œuvres à la réalisation d'un ouvrage extraordinaire tant par ses dimensions que par l'ampleur des moyens techniques utilisés.

Depuis presque un siècle, le nouveau pont de pierre sur le Grand Canal est un sujet de discussion, mais on continue de procéder à la restauration partielle ou à la remise en état complète du vieil édifice de bois. En 1551, le Sénat, bien décidé à clore une longue histoire de précarité et d'alertes successives avait réagi en constituant une juridiction « spéciale ». Déjà, en 1514, quand le grand incendie avait frappé l'ensemble de l'île du marché, le Sénat avait chargé d'abord sept Sages (*Savi*) puis trois Commissaires aux constructions du Rialto (*Provveditori sopra le Fabbriche di Rialto*) de faire faire le relevé de ce qui existait et de rassembler des projets pour la réalisation de bureaux, de boutiques et d'entrepôts publics ; de la même façon, en 1551, il désigne des Commissaires à la construction du pont du Rialto (*Provveditori sopra la Fabbrica del Ponte di Rialto*)[1]. Le palais ducal terminé, et la paix revenue, il était temps de « restaurer »[2] (*ornar*) aussi l'aire du marché ; aux magistrats nouvellement élus était confiée la tâche spécifique et temporaire de recueillir des informations, des avis et des maquettes, à Venise et ailleurs, pour l'édification d'un pont en pierre, destiné à remplacer l'ancienne construction en bois qui menaçait ruine.

LES CHANTIERS URBAINS

FIG. 1. — Bureau de l'adjoint à l'Urbanisme, Commune de Venise, vue aérienne du pont du Rialto, dans *Venizia : Forma Urbis,* Venise, Marsillo, 1986. (Documents reproduits grâce à l'aimable autorisation de l'Archivio di Stato de Venise)

Ces magistrats auraient donc dû superviser les travaux du Rialto, en assumant « per tempora » les charges qui incombaient aux Sages chargés d'embellir la ville (*Savi all'ornar*). Mais dans l'attente de décisions opérationnelles qui tardaient à être prises pour la réalisation du pont, ils finirent par ne s'occuper que des constructions nouvelles de Sansovino (*Sansoviniane Fabbriche Nuove*), le long du Grand Canal. Ils avaient trouvé une raison de poursuivre leurs activités et de se maintenir en place, au moins pour quelques années [3].

C'est précisément à la juridiction spéciale — qui n'exerce plus mais qui n'a pas été supprimée — que l'on fait de nouveau appel trente ans plus tard, au moment du vote définitif en faveur d'un nouveau pont en pierre qu'il faut construire à la place de l'ancien devenu trop vétuste [4]. On attend de cette institution, chargée une nouvelle fois de résoudre des questions de programme, d'ordonnancement des travaux, de propriétés, d'expropriations, qu'elle fasse preuve de zèle, qu'elle soit diligente et s'engage de façon permanente jusqu'à l'exécution complète de son mandat. C'est exactement ce que le Sénat avait imposé aux trois magistrats responsables de la restauration du palais ducal, après les incendies de 1574 et 1577. Les commissaires (*Provveditori sopra la Fabbrica di Palazzo*) constituèrent une référence explicite pour le Sénat quand il s'est agi du pont[5]. Le maître d'œuvre est la Seigneurie : promoteur et propriétaire du sol sur lequel sont implantés le vieil édifice en bois et les quais qui le longent, la Seigneurie est détentrice du droit de l'affecter à l'usage commercial ; elle restera par la suite propriétaire des lieux construits, et aura donc la possibilité de les vendre aux enchères et d'en tirer périodiquement le revenu des loyers.

L'expérience acquise depuis des siècles dans l'île du Rialto en matière d'espaces publics couverts et découverts réservés au contrôle, à la vente et au stockage de marchandises, a été renouvelée récemment avec la nouvelle attribution, par lots successifs, à des personnes privées, de parts de l'ensemble construit par Sansovino le long du canal[6]. Mais déjà, vingt ans plus tôt, la *Drapperia* et les *Fabbriche dello Scarpagnino* situées autour de la place San Giacomo, avaient été louées par adjudication publique au meilleur offrant ; c'est, comme cela s'est fait pendant des siècles, la possibilité d'accoster le long des quais, de construire des entrepôts sous les portiques, ou d'installer des comptoirs à la *Pescheria* ou à la *Beccheria*[7].

Au-delà de la rhétorique due peut-être à l'importance de l'ouvrage, le mode de financement n'est pas non plus nouveau. Les provéditeurs s'adressent à la magistrature ordinaire compétente en la matière, les Commissaires du Sel (*Provveditori al Sale*). A Venise, ceux-ci sont institutionnellement chargés de financer les ouvrages nouveaux et l'entretien des constructions de la République ; on leur ordonne les affectations de crédits, c'est-à-dire le retrait, parfaitement enregistré, des sommes spécifiques nécessaires, prélevées sur la « petite caisse » réservée depuis toujours à ce genre de dépense[8].

Comme cela s'était déjà produit pour des entreprises du même genre, un emprunt public contracté avec la Monnaie (la première tranche est de 25 000 ducats[9]) permet de disposer des sommes nécessaires dès le début des travaux ; le remboursement est prévu et doit se réaliser grâce à la vente aux enchères des nouvelles constructions, au fur et à mesure qu'elles seront disponibles[10]. C'est ce processus institutionnel qui permet, une fois que le Sénat a fait les choix définitifs et que la construction est commencée, de pallier la difficile quête de fonds et l'irrégularité des entrées d'argent. Dès ce moment, donc, c'est-à-dire à partir du 5 décembre 1587, les provéditeurs agissent au nom de la ville qui a passé commande ; ils sont libres de procéder aux meilleurs choix et sont tenus de les soumettre au Sénat pour une délibération définitive et deviennent ensuite responsables de l'exécution. Ils ont pour tâches de coordonner les projets et de superviser l'état d'avancement des travaux, d'en centraliser ou d'en déléguer une partie. Les ordres sont donnés avec la plus grande minutie, l'exécution elle-même ne semblant pas être contrôlée ensuite avec la même rigueur. La charge dont sont investis les provéditeurs va du rôle de dépositaires de l'autorité de l'État à celui de simples interprètes de la volonté du Sénat. Dans le premier cas, l'acte est scellé par l'authentification de leur signature, ou par la possibilité qui leur revient de « faire justice » dans les contentieux les plus divers[11] ; dans le second, par l'intervention des *Pregadi,* lorsqu'il s'agit de problèmes vraiment importants[12].

Une fonction presque notariale de garant de l'exactitude des actes revient au secrétaire du bureau, employé permanent, à qui incombe aussi la tâche d'émettre les mandats de paiement, sur la base des polices régulièrement souscrites qui lui sont présentées[13]. Pour en souligner les rapports institutionnels et les fonctions, celui qui assume cette charge est aussi (ou a été) secrétaire du Sénat. Mais la complexité de ses responsabilités est telle qu'il peut employer, à l'occasion, deux adjoints rétribués. La procédure est compliquée, le dispositif bureaucratique n'est ni rapide, ni autonome, ni spécialisé. Le conducteur de travaux (*proto della fabbrica*) est l'interlocuteur et l'intermédiaire des provéditeurs : de par sa formation et ses compétences, il ne joue pas seulement le rôle d'architecte chargé du projet, ni de maître maçon qui se contente de superviser la construction, d'organiser simplement les commandes et de distribuer l'argent ; car ce n'est pas la présence des secrétaires qui empêche les techniciens d'assumer aussi des fonctions bureaucratiques. Le conducteur de travaux prépare à l'avance les mesures, les épures, les plans et rédige les termes des contrats, les devis et les polices de paiement ; il fixe les conditions de réutilisation des matériaux de démolition : en somme il dresse le plan concret et exécutoire des travaux. Ce qui peut parfois (soit du fait de la pesanteur de l'organisation et du contrôle croisé des juridictions, ou du caractère particulier du pont) nécessiter également un contrôle des *Savi alle Acque* ou des *Provveditori di Comun* [14].

Construction de grandes dimensions, le pont représente un investissement absolument extraordinaire, si on le compare à d'autres transformations urbaines de la même époque : au premier bilan, en 1593, 245 537 ducats ont été dépensés pour l'achat, le transport et la manutention des matériaux nécessaires, pour la rétribution des ouvriers qui creusent, déblaient boue et détritus, pour la mise en œuvre de la grandiose structure nouvelle et de ses annexes, les boutiques, y compris les 130 000 ducats dépensés pour les expropriations (surtout celles qui ont précédé la construction des immeubles près de San Marco) [15]. La dépense globale annuelle pour l'entretien des lieux de passage public urbains avoisine 6 % de ce chiffre : en 1582, la somme dépensée par les *Provveditori di Comun* pour remettre en état (*aconciar*) ponts, portes et voies publiques avait été de 16 750 ducats. Vu le désordre des sources, il est difficile de comparer ce bilan avec celui d'autres chantiers publics à Venise : nous savons cependant que la restauration du Palais Ducal, entre les deux incendies de 1574 et 1577, a coûté au total, au dépositaire de l'*Ufficio al Sale*, 20 660 ducats en trois ans. Après cette date, c'est-à-dire après la seconde destruction par le feu, les devis d'Andrea Palladio et de Paolo del Ponte et Andrea da Valle pour la réfection des deux façades du Palais, tous deux refusés, s'élevaient respectivement à 40 000 et 50 000 ducats ; de fait, les sommes affectées à la suite de plusieurs délibérations entre 1577 et 1582 atteignent 88 585 ducats. Enfin, dans les mêmes années qui virent la réalisation du pont, le devis établi pour la construction des nouvelles prisons, près du bassin de San Marco, au-delà du rio di Palazzo, s'élevait à 80 000 ducats [16].

Il n'existe pas, dans les actes, de délibération explicite nommant Antonio da Ponte *proto della fabbrica*. Malgré la demande insistante de l'un des provéditeurs, Marcantonio Barbaro, d'élire quelqu'un qui soit à même de diriger les travaux en fonction d'un résultat déterminé à atteindre, le choix du Sénat est pragmatique, plutôt qu'en faveur d'un projet spécifique et de celui qui l'a rédigé [17]. Cependant, beaucoup de responsabilités reviennent pendant cette période à Antonio qui avait parmi d'autres présenté des projets et mené à bien des réalisations ; il avait en particulier été chargé d'« assécher les deux côtés » du canal par des palplanches (*serrar per parti*) pour jeter les fondements des culées (il s'était proposé d'entreprendre l'opération à ses frais, sans remboursement en cas d'échec)[18]. Il avait acquis une expérience considérable en matière de construction publique : ex-fonctionnaire de l'*Ufficio al Sale,* dans les premiers documents relatifs au pont, il est appelé *proto di Palazzo*[19].

FIG. 2. — Antonio da Ponte, projet d'un arc unique pour le pont du Rialto, et rangée de maisons et de boutiques, première hypothèse d'un raccordement avec le Campo di San Bartolomeo, 1588, projet redessiné aujourd'hui (dessin Paolo Rosso)

Une fois votées la modification d'alignement du nouveau pont par rapport au vieil édifice en bois et la disposition des maisons et des boutiques[20], une fois sa proposition adoptée, les compétences qui lui sont demandées sont plus larges et plus diversifiées que celle de simple maître d'œuvre : compétences techniques pour l'organisation du travail et pour mettre fin aux controverses sur l'emploi de la main-d'œuvre ou en référer aux provéditeurs ; compétences administratives, inhérentes à une comptabilité spécifique, difficile et complexe ; il doit faire appel à des collaborateurs pour partager ses tâches : d'autres responsables chargés de fonctions diverses sont désignés, les *sotto-proti.* L'un s'occupe d'épures, de mesures, de la qualité des matériaux ; l'autre est chargé de stipuler les clauses des contrats fixant les quantités et les délais ; le troisième doit choisir le meilleur offrant. Un autre responsable tient à jour les cahiers de comptes, tel autre prépare ou vérifie les polices de paiement, tel encore inspecte les locations[21].

Les procédures, qui ne sont pas toutes prévues au début, sont élaborées au fur et à mesure de l'exécution des travaux ; personne n'est capable de maîtriser le « chantier » dans son ensemble, c'est-à-dire le lieu même de la construction (qui exige une « garde » nocturne adéquate[22]) et l'organisation qu'il nécessite. Car le chantier réclame le concours d'une immense main-d'œuvre, d'hommes qui travaillent à l'intérieur de l'excavation ou sur les échafaudages, parfois même la nuit, à la lueur des chandelles, ou qui posent les pierres vives de l'arche et les briques de chant. Le chantier s'étend au « ponton » flottant sur le canal, fourni par l'Arsenal en 1589[23], servant de point d'appui aux travaux de cintrage et destiné à « faciliter le travail » (*accomodar*) des ouvriers ; il s'élargit parfois à quelque vieil immeuble du voisinage qui, endommagé par les travaux, nécessite une restauration[24].

Les maîtres d'œuvre, les structures, les dispositifs qui définissent l'ensemble architectural du pont du Rialto sont donc nombreux et complexes. Ils se combinent en une forme institutionnelle, presque une personne juridique, qui existe avant que le « chantier » ne soit ouvert, avant que le pont ne se matérialise, et dont l'activité ne s'arrête pas à la pose de la dernière colonnette de la balustrade ou aux chiffres romains gravés sur les impostes ; elle connaît aussi les démolitions, les procédures d'expropriation, de vente des boutiques et maisons ; elle se préoccupe du remboursement des dettes contractées avec la Monnaie, dresse les rapports avec les particuliers et avec les fournisseurs, et prend en charge les travaux d'entretien... Pour cette entreprise publique extraordinaire, le Sénat vénitien a ouvert temporairement un chapitre spécial de son activité gouvernementale.

II. — Le contentieux

Les rapports qui s'établissent entre les différents protagonistes de la construction du pont institutionnellement concernés, ou impliqués dans leurs intérêts individuels, ne sont pas simples ni toujours faciles.

« Redresser » l'axe du pont qui « actuellement est gauchi » est un choix difficile et coûteux pour la Seigneurie, et le simple fait de « faire taire » les propriétaires des boutiques et des entrepôts devant être démolis aboutit à des contestations sans fin, et les accords sont difficiles. Sur l'ancienne construction précaire mais rentable, les « propriétaires » avaient acquis leur droit à plus d'un titre : achat, droit de préemption *jure sanguinis,* acte de partage, sentence compromissoire ; depuis des années, chaque marchand exerçait quotidiennement son activité, ayant obtenu l'usage du lieu, ou d'une partie de ce lieu, contre paiement d'un loyer, ou par bail emphytéotique, ou après avoir été investi *sive proprio,* et en tirait bénéfice. C'est pourquoi, pour les premières expropriations, on a énormément de peine à se repérer dans le labyrinthe des parcelles, des *commissarie,* des héritages, et à fixer les modalités d'indemnisation (directement, par sentence, par procuration, contre délivrance d'une quittance, avec des bons du Monte Nuovissimo et du Monte di Sussidio, par un acte bancaire *per contenta*)[25].

D'autre part, le fait de procéder à l'acquisition de terrains, à l'expropriation d'immeubles, à la rectification des limites ou des bornes, est une opération préliminaire nécessaire à la réalisation d'un projet d'intervention sur la ville et de réévaluation foncière qui, tout en restant prudent quant aux modes d'exécution, s'impose à Venise par ses dimensions et par sa localisation. Il modifie inévitablement les rapports personnels et les usages de l'espace urbain. Pour cette raison, il engendre parfois des controverses à propos du respect des règles juridiques et financières, et le recours à des intermédiaires entre les maîtres d'œuvre délégués et les particuliers titulaires d'un droit acquis.

Certains choix de stratégie urbaine — comme celui de modifier l'alignement du nouveau pont par rapport à l'ancien pont de bois, et d'en élargir l'accès du côté de San Marco en respectant la perspective avec les campaniles de San Giovanni Elemosinario et de San Bartolomeo — peuvent en effet se heurter à de multiples problèmes tels qu'une boutique acquise, ou la difficulté de fixer le tarif exact d'indemnisation de quelques mètres carrés de terrain[26]. Des décisions d'investissement immobilier, comme celle de rénover les bâtiments délabrés sur la rive opposée à l'île du marché, en y construisant des « immeubles » destinés à la spéculation, risquent de s'enliser devant un mur mitoyen entre deux propriétés, d'être bloquées par la répartition des parts de propriété et des dépenses inévitables qui en découlent.

Dans tous les cas, c'est aux provéditeurs que l'on fera appel pour une médiation, obligés d'entamer « maintes discussions » avec chaque propriétaire (les religieuses de l'Arsenal, les Malipiero, les Lippomano, les Vendramin, les Corner), avec les locataires (le libraire de la *Salamandra,* le verrier Girardi, le parfumeur du *Cavalletto d'oro*), ou avec les « interprètes » du patriarche et du chapitre. Ils les laisseront ensuite, du moins en ont-ils l'intention, « s'entendre » avec le *proto della fabbrica* pour les questions de détail[27].

Mais le processus conçu pour respecter les procédures, ne pas léser les droits acquis ni les promesses légitimes, se brise devant de nouvelles éventualités. Des changements survenus « en dépit » de ce qui a été établi — l'appropriation indue d'un lieu collectif par exemple — pourront obliger les provéditeurs à régler un contentieux long et difficile. Acceptée dès le début, cette éventualité fait partie de leur manière de procéder. L'interruption des travaux, la suspension ou l'annulation d'ordres donnés pour des inspections et pour des expertises, un accord anormal de répartition de propriétés, la présentation à la Seigneurie d'une nouvelle maquette ou d'un plan correctif, des remboursements non prévus au budget, des amendes et des obligations de restauration, deviennent alors les étapes nécessaires de l'évolution quotidienne de la construction[28].

Les difficultés continuent une fois la construction terminée : pour trouver rapidement de l'argent, le Sénat avait autorisé les provéditeurs à procéder aussitôt à la vente aux enchères des boutiques sur le pont, ainsi que des boutiques et des maisons situées sur la *ruga.* Mais deux lots successifs ne parvinrent pas à atteindre les sommes espérées et leur vente fut annulée dans la crainte que l'opération ne réussisse pas à s'autofinancer et que les nouvelles constructions ne suscitent pas l'intérêt escompté, ou pire qu'elles ne réaniment pas le marché[29]. On se résigna à étaler le recouvrement de la dette avec la Monnaie : les maisons, et surtout les boutiques furent louées, et leur vente ne reprit que sept ans plus tard ; en 1600, un tiers de la somme dépensée avait été récupéré, et tout n'avait pas été vendu[30].

Il y eut aussi de mauvaises exécutions. Là encore, les provéditeurs durent trancher et réunir des témoignages et des avis nécessaires. En 1591, des erreurs commises par le tailleur de pierre ou des imprécisions dans les épures fournies par les conducteurs de travaux obligent à débloquer de nouveaux fonds destinés au nivellement d'une imposte qui n'avait pas été construite parfaitement plane. Une enquête est ouverte et les procès durent de 1593 à 1596, avec l'intervention de l'*Avogaria di Comun*. Un forgeron est accusé de vol et de fraude à l'encontre de son associé, titulaire du contrat. Sur le contremaître Banelli, dont la carrière rapide suscite peut-être des jalousies — passant de simple tailleur de pierre à une charge *de facto* dans la construction du pont — pèse le soupçon d'une certaine complicité[31]. Une supplique, une dénonciation ou un quelconque ragot peuvent donner lieu à une proposition d'accord à l'amiable ou à une intimation de paiement.

Les conflits existaient, ils duraient longtemps, mais ils devaient pouvoir se résoudre sans trop mettre en jeu l'honnêteté des corporations ou d'un magistrat public.

III. — Géographie des approvisionnements

L'examen des sites d'approvisionnement des matériaux destinés aux travaux préliminaires et à la construction du pont n'apporte pas de grandes surprises : l'implantation géographique des fournisseurs, et le tracé des liaisons entre le lieu d'origine des matériaux et le chantier ne sont pas modifiés. Points d'extraction habituels pour les constructions de la cité lagunaire, ils sont tous compris à l'intérieur de son *dominium* : la pierre vive vient d'Istrie (en particulier de la carrière d'Orsera et d'autres carrières situées aux environs de Rovigo) et — seulement lorsqu'il s'agit de petites quantités de marbre *bronzino* pour les bas-reliefs sur les façades du pont — de la Predara, près de Vérone ; les pierres cuites proviennent des régions de Padoue et de Trévise ; la chaux *negra* de Padoue, la chaux blanche de la région de Trévise ; les sables de Treporti et Lio Mazor (la langue extrême de terre ferme, qui clôt la lagune nord) ; le bois — pieux d'orme et de mélèze — vient du bassin de la Brenta et des montagnes au-dessus de Bassano, les pointes de pin et les planches de sapin viennent aussi des bois de Montello ou de ceux du Frioul[32].

Le principal lieu de déchargement et, parfois, de premier entrepôt des pierres est San Pietro in Volta, où il est nécessaire de mettre en place un nouveau débarcadère à l'entrée du port ; parfois, au contraire, on rassemble les pierres à Malamocco, aux environs du deuxième accès portuaire à la lagune[33]. Provenant par mer des carrières situées de l'autre côté de l'Adriatique, la pierre vive y est parfois dégrossie, puis travaillée à Venise dans les deux ou trois ateliers de tailleurs de pierre qui ont signé des contrats pour le pont. Transportée en ville par les fournisseurs qui l'ont acquise « pour leur compte »[34], elle s'entasse provisoirement sur les *campi* de Venise et, seulement plus tard et en quantités limitées, « selon les besoins », elle est livrée, en *peata* et à force de bras, sur la Riva del Vino ou directement à la base des piles : « l'enclos » construit au Rialto pour y entreposer les matériaux[35] est de petite dimension, enserré entre les quais et les constructions qui lui font face. Les entrepôts des marchands de bois sont plus nombreux en ville et dispersés dans les *sestieri* situés au voisinage d'un quai pouvant être atteint avec des moyens de transport adéquats : à San Giovanni et Paolo on en compte au moins deux, aux Zattere trois différents ; mais il en existe aussi à San Vidal, à San Francesco, à Barbaria, à San Basilio et à la Madonna del'Orto. Les provéditeurs achètent des troncs de mélèze, des pointes d'orme, des planches de sapin en fractionnant les commandes entre les étrangers susceptibles d'importer personnellement les matériaux qui leur ont été commandés, et les autochtones, peut-être à la tête

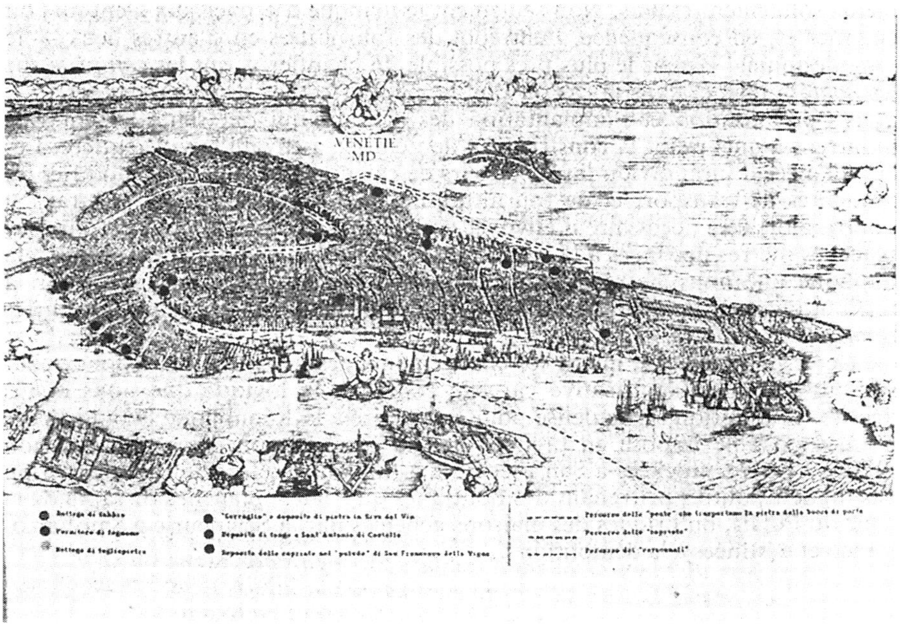

FIG. 3. — Boutiques des fournisseurs et parcours pour l'approvisionnement des matériaux utilisés sur le chantier du Rialto ; reconstruction par D. Calabi d'après la vue de De Barbari.

d'une plus grande entreprise[36]. Il en va différemment pour la chaux et les briques qui sont fournies presque exclusivement par le noble vénitien Filippo Vendramin qui les extrait sur le lieu d'origine et se charge de leur transport[37]. Quant aux ferrures, aux clous vendus au poids, aux crocs pour les batteurs de pieux, le conducteur de travaux et ses collaborateurs les achètent sans commande préalable, selon les besoins, chez deux ou trois forgerons qui en disposent ou sont en mesure de les fournir rapidement, principalement à San Samuele[38]. Pour le plomb, le contrat prévoit une fourniture *ad hoc,* en permettant même le prélèvement du matériau déjà utilisé, et une remise en œuvre spécifique[39].

Ainsi, malgré les limitations imposées par les travaux au trafic sur le Canal, malgré les préoccupations diverses lorsqu'on décide comment faire les fondations dans l'eau, le va-et-vient des embarcations autour du pont, suscité par le chantier lui-même, semble plus intense que jamais. *Burchi* et *peate* effectuent les livraisons par lots successifs, échelonnées dans le temps. Les charpentes doivent être transportées du « marais » du couvent de San Francesco della Vigna, où elles sont construites[40], pendant que des barques transportent en sens inverse, à Sant'Antonio di Castello, la boue qui a été retirée du fond du Canal, sous le contrôle des *Soprastanti alle acque* et du contremaître Antonio da Ponte[41]. Le pont du Rialto demande donc, au XVIe siècle, un déploiement extraordinaire d'activité, mais il ne bouleverse ni l'ampleur ni la répartition traditionnelles des entreprises et des ressources, ni les habitudes et les comportements solidement établis. Non seulement le manque d'espace aux alentours du chantier et, en conséquence, l'entrepôt des fournitures en d'autres lieux — le long des quais, « mais le plus près possible du chantier », sur les *campi*, « sur les *peate* » même, dans le cas où l'on ne peut effectuer le déchargement [42] —, mais l'organisation et l'implantation des artisans qui ont signé les contrats montrent à quel point la construction de l'édifice a envahi la ville entière. Les défaillances d'un individu ou les erreurs de calcul peuvent parfois modifier les modalités de transport et de fourniture des matériaux fixées par contrat : il pourra alors être nécessaire d'envoyer en Istrie un « expert » pour extraire sur place les pierres destinées à la voûte [43] et en contrôler le chargement. Mais cela implique une nouvelle répartition des dépenses : un poste différent pour la rétribution occasionnelle de porteurs et de portefaix, une organisation différente des transports.

De ce point de vue, même les changements des lieux d'approvisionnement sont une donnée significative car elle confirme la logique que nous avons décrite. L'attention scrupuleuse pour faire faire à la République le plus d'économies possible impose en premier lieu la récupération des matériaux non utilisés : pierres stockées à San Pietro ou à Malamocco lors de la construction des digues, poutres provenant des boutiques du Rialto, planches de sapin de la cour du Palais, ou briques des maisons achetées par la Seigneurie à San Bartolomeo et destinées à la démolition [44].

IV. — Le travail

Les procédures auxquelles les responsables de la construction recourent pour fournir sur place les matériaux finis, semi-ouvrés ou devant encore être travaillé, sont la commande groupée (pour des quantités importantes) et l'achat au détail (fournitures en quantités limitées, dont les besoins sont urgents ou occasionnels). Les responsables lancent des appels d'offres lorsqu'il s'agit de travail partiel sur des périodes définies, et procèdent au recrutement direct pour une prestation sur un temps court et qui se renouvelle de semaine en semaine.

Les deux « creuseurs de canaux » (*cavacanali*) qui remportent l'adjudication d'ouverture de chantier appartiennent à une association de professionnels. Ils parcourent la ville en offrant leurs services ; il ne s'agit donc pas vraiment d'une corporation [45]. On leur demande de construire les ensembles de palplanches — un par rive — nécessaires pour délimiter et assécher la zone des fondations du pont. En vingt jours, ils doivent assécher les deux batardeaux et les dégager de la boue provenant des couches superficielles ; pour cela, ils doivent savoir établir un programme, mener leur entreprise et diriger les compagnons recrutés. Cependant, peu entreprenants, ils ne savent pas adapter leur travail aux difficultés propres à chaque cas particulier, à la largeur et à la profondeur inhabituelles du canal, et à la nécessité évidente d'obtenir une étanchéité presque parfaite. Les contractants ne respectent pas les délais, travaillent sur une rive à la fois, et dès les premiers essais d'étanchéité, l'eau réussit à l'infiltrer.

Antonio da Ponte intervient pour sauver ce qui a été fait. Il offre de construire une troisième rangée de digues, et pour cela il utilise le procédé le plus simple et le moins onéreux. Il ne détruit pas, ne révolutionne pas, mais il réutilise en améliorant[46]. Il réussit. « Charpentier » (*marangon*) de formation, da Ponte avait deviné et senti, plus qu'il ne savait pas science, comment réagissaient les bois réunis pour former la structure, lorsqu'ils subissent les effets de la pression de l'eau et des courants.

Une fois écartés ceux qui s'étaient révélés incapables de remplir le contrat, de nouveaux terrassiers travaillent sous la dépendance directe de da Ponte, mais en tant que simples exécutants. L'ancienne association des « creuseurs de canaux » n'a pas formé d'individus capables de jouer pleinement leur rôle dans le chantier, exceptionnel, du pont. Se contentant d'une pratique élémentaire — enfoncer et creuser —, l'entreprise reprend les habitudes ancestrales, sans les perfectionner ni les adapter : sans imagination. Pour élever le second ensemble de palplanches sur la rive opposée, la corporation des *marangoni*, ayant tiré profit des enseignements apportés par la première construction, inspire davantage confiance. Les compagnons qui avaient démoli le vieux pont remportent l'appel d'offres lancé par les provéditeurs ; ils s'engagent à fournir un travail achevé pour une somme convenue d'environ 600 ducats[47].

D'autres charpentiers préparent les cintres du pont. Des dizaines de pieux, débarqués et déposés loin du chantier, sont rabotés, sciés et équarris, montés en charpentes, assemblés et cloués selon les mesures et les indications fournies par le conducteur de travaux et établies dans les clauses du contrat[48]. Ils arrivent, au moment voulu, par les canaux de la ville, sont déposés près des piles du pont, et mis en œuvre par les charpentiers eux-mêmes. Chaque charpente coûte 38 ducats.

La fourniture des pierres pour les culées, l'imposte et la voûte du pont fut confiée à deux contractants, une compagnie de trois associés (unis peut-être pour l'occasion), et un individu, tous inscrits à la corporation des tailleurs de pierre, un *arte di consumo*[49]. Les provéditeurs ont préféré répartir les charges : peut-être pour réduire, de cette manière, à leur propre avantage, les risques de non-respect du contrat ou les complications ; peut-être parce que les responsabilités qui incombaient à chaque signataire étaient trop lourdes. Chaque mois, ceux qui avaient remporté l'appel d'offres devaient consigner, en conformité avec les modèles (forme et mesure) qui leur avaient été fournis, respectivement 5 000 et 2 000 pièces[50]. Les contractants travaillent eux-mêmes, mais ils font surtout travailler : ils prennent en charge et paient l'extraction de la pierre des carrières, son transport à Venise, son stockage et l'équarrissage en ville. Ils administrent, pour eux-mêmes et pour les compagnons de corporation qu'ils ont choisis et qu'ils paient, les arrhes et les acomptes remis périodiquement par les provéditeurs ; ils avancent, inévitablement, quelques sommes et doivent pour cela être reconnus solvables. Ils gagnent, certes : dans les dix premiers mois, les mandats émis à leur profit sont respectivement de plus de 3 500 et de 1 000 ducats, ce qui représente beaucoup d'argent en circulation. La corporation des tailleurs de pierre compte dans ses rangs des individus capables de gagner autant par leur travail que par leur capacité d'initiative ; ils sont prêts aussi à prendre des risques.

Un autre corps de métier, les porteurs (*bastasi*), devait se charger, avec un contrat précis, de l'enlèvement des blocs taillés du lieu où ils étaient travaillés et de leur décharge sur les quais, à proximité du chantier, tandis qu'un *peater* devait s'occuper de rassembler une équipe chargée de les transporter [51]. Comme cela se faisait régulièrement pour tout édifice à Venise à partir du XVe siècle, de larges *battuti* — véritables forêts de pieux enfoncés dans le silt argileux du fond lagunaire — devaient tasser et rendre homogène le plan de fondation : par milliers, troncs d'aulne, de sapin, de mélèze et de pin, de longueur et de diamètre variables. Ils sont surmontés d'une plate-forme de plusieurs épaisseurs sur laquelle on fonde les piles et les culées de pierre du pont. On ne peut prévoir le détail, comme dans le cas des blocs de pierre, ni les interstices se trouvant entre les pieux, ni la quantité des pièces de colmatage et de bandage nécessaire, à trouver quand il le faudra, sur place. Un seul fournisseur ne suffit pas. Près de trente « marchands de bois » sont concernés par les besoins du chantier [52]. A trois ou quatre exceptions près, le fournisseur des matériaux destinés au *battuto* ne les vend pas pour la fabrication des caissons étanches, les batardeaux, ni pour les boutiques provisoires, construites pour dédommager les personnes expropriées, ni pour les baraquements qui abritent les hommes et les outils. Les fournitures oscillent entre un maximum de 3 000 pilots et un minimum de 16 *ponti* ; les sommes versées par les provéditeurs vont de 1 200 ducats et plus à 2 ducats ! Calculer une moyenne serait arbitraire [53].

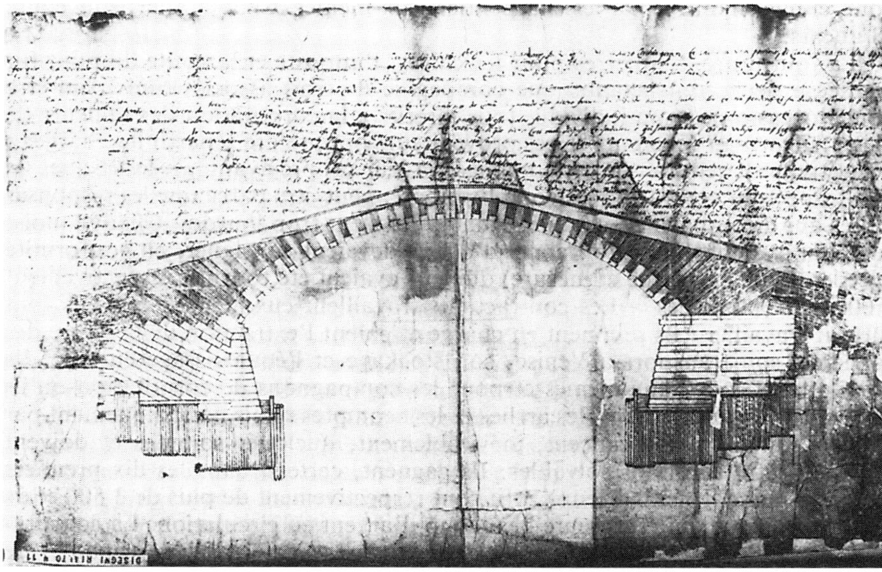

FIG. 4. — Francesco Zamberlan (?), les piles du pont du Rialto : à gauche la pile construite par Antonio da Ponte, à droite la manière dont elle devait être construite en 1588 (ASV, *Provveditori sopra la fabbrica del Ponte di Rialto*, dessin n° 11).

C'est la structure même du groupe qui détermine le comportement des provéditeurs et la multiplicité des commandes. Le fournisseur de grosses pièces peut négliger les petites ou, inversement, celui qui négocie le petit matériel n'est pas en mesure d'assurer la livraison des équipements lourds. Les marchands de bois ne constituent pas une corporation, mais une association commerciale traitant avec un revendeur ou un commerçant et, s'il le veut, ce commerce peut ne durer que le temps de la négociation. Association occasionnelle, fragmentée et discontinue, elle autorisait une activité libre permettant de répondre à n'importe quelle demande du bâtiment civil de la ville. Les bois possédés en propre par la Seigneurie furent le plus souvent destinés à l'Arsenal.

La corporation des maçons (*mureri*) garantit les moyens et la préparation nécessaires à la réalisation des *battuti* selon les indications du conducteur de travaux[54]. Les contractants doivent tailler les pieux et les enfoncer, poser horizontalement les plates-formes, en utilisant leurs propres outils ; en outre, obligés de recruter d'autres compagnons, ils actionnent les machines mises à leur disposition par les provéditeurs pour puiser l'eau des infiltrations ; ils doivent dégager la boue à l'aide de douze pelles fournies par un marchand de fer payé sur les deniers du chantier. Ils doivent remettre un travail fini ; dans leur cas aussi, il s'agit d'un contrat chiffré à environ 500 ducats.

Si le contrat pour les fondations comprend le creusement en profondeur, ce ne sont pas les mêmes hommes, ni la même corporation, qui transportent et déchargent la boue loin du chantier. Ce travail est confié aux mariniers (*burchieri*), groupés en association qui, comme celle des « creuseurs de canaux », n'est pas classée au nombre des corporations. Quatre de ses membres contractent l'engagement[55] de fournir les barques (*burchielle*) en plus de la manutention qu'ils garantissent, et ils s'occupent de recruter et de rétribuer la main-d'œuvre qui leur sera nécessaire. L'association n'a pas de compétence particulière, mais elle reproduit les gestes accomplis de tout temps avec un zèle infatigable. Dans les deux premiers mois, plus de 200 barques chargées de boue quittent les quais, soit en moyenne trois ou quatre par jour, pour le prix de deux ducats par voyage. Un gain très modeste, en rapport avec la simplicité du travail. C'est avec le lent va-et-vient d'embarcations semblables que de nombreuses parties de la ville ont été assainies et transformées. Les piles sont dressées une à une dans la cavité à l'intérieur des batardeaux, au-dessus des fondations. La mise en place de la première pile, vers Rialto, est confiée au premier groupe de tailleurs de pierre, responsable de la fourniture des pierres ; ils doivent se tenir, avec les ouvriers qu'ils ont embauchés, à la disposition du contremaître. La pile de Rialto sert de modèle à celle qui lui fera pendant sur la rive opposée, vers San Bartolomeo, dont l'exécution a été adjugée à un autre groupe capable, lui aussi, de diriger, de recruter, de prévoir[56]. C'est une autre équipe encore qui répond à l'appel d'offres pour exécuter l'imposte et l'archivolte à trois bandes de la grande arche du pont, avec les matériaux préparés par les fournisseurs agréés et, pour l'occasion, par un nouveau tailleur de pierre[57]. C'est un travail bien défini, qui demande de la précision, car bande et archivolte sont les premières parties de l'édifice à émerger de la boue et de l'eau, à s'offrir aux regards, les premières à illustrer les *maestà*, ou façades, du pont. D'autres encore — bien que les noms ou les surnoms attestent des liens de parenté avec les précédents — construisent l'arche entière, en pierres et en briques[58]. Ce travail de précision est lent : poser de chant les pierres, mesurer, niveler, cimenter. Il est confié à des maîtres maçons, rétribués au mètre carré, calculé sur la base de l'intrados. Ils remettent l'arche terminée, avec corniche et revêtement, les bas-reliefs exécutés par les sculpteurs (le groupe de l'Annonciation, les saints Todaro et Marco, la colombe et le mascaron en clef de voûte).

Il est de nouveau fait appel aux tailleurs de pierre de l'imposte et de l'archivolte : ils signent le contrat pour l'édification des façades des deux rangées de boutiques qui doivent être construites sur le pont [59], en ordre rustique, avec des balèvres ; il faut faire preuve de toute l'habileté que la tradition de la corporation autorise et garantit.

Maçons, charpentiers, sculpteurs, tailleurs de pierre rivalisent ensuite pour obtenir une série de petites adjudications pour les parties en pierre et en briques, les couvertures en bois, les comptoirs, les corniches, les consoles, les gouttières, les numéros des boutiques ; pour la frise, la corniche, l'architrave des grandes « voûtes » ou « entrées » au milieu du pont [60].

Les contractants et les adjudicataires ne sont pas les seuls à animer le chantier. Bon nombre d'ouvriers travaillent sous la dépendance directe du conducteur de travaux et de ses assistants. Aides, manœuvres, portefaix, charpentiers participent au travail des « maîtres » indépendants. Le chef de chantier se charge lui-même de payer ces journées de travail, témoins du labeur et de la fatigue de ces hommes. C'est à lui que sont adressées les nombreux mandats de paiement souscrits par les provéditeurs : sommes minimes et considérables à la fois : 1, 20, 50, 100, 200 ducats. C'est lui qui les répartit individuellement ou qui les encaisse à titre de remboursement de l'argent qu'il avait avancé de sa poche [61].

Sur le chantier, hommes et traditions sont à l'opposé d'un encadrement strict et rigide. Chaque corporation cherche à atteindre son but, et elle agit de façon autonome, utilisant des outils traditionnels et familiers. La phase ultime — le pont édifié — fut toujours étrangère aux réalisations individuelles, spécifiques et partielles : elle les domina. Les chefs de chantier agirent sans provoquer de gêne puisqu'ils ne possédaient pas l'habileté manuelle des maîtres, et parce que l'indépendance des corporations ne l'aurait pas permis.

V. — Les machines

Les ensembles de pieux furent construits le plus souvent à la force des bras : celui de Rialto à l'aide des béliers qu'on soulevait et qu'on laissait retomber à la main, celui de San Bartolomeo au moyen de cables et de poulies ; maîtres, aides et tâcherons enfoncent des centaines de pieux, de planches de mélèze et de sapin pour former trois rangées de palplanches l'une dans l'autre, encastrées, clouées, serrées par des renforts d'angle et séparées par des espacements remplis de boue [62]. Instruments simples et énergie humaine : on ne dépensa pas beaucoup d'argent pour les construire et pour se procurer la main-d'œuvre.

Même pour l'opération plus complexe du pompage de 1 500 à 2 000 mètres cubes d'eau, le chef de chantier pouvait compter sur le rendement de machines connues depuis des siècles : les pompes hydrauliques — ou pompes « à eau », ou pompes « à sécher » comme on les désigne sur les mandats de paiement [63].

Une fois la bouche inférieure munie de clapets immergée, directement dans l'eau ou dans un caisson de canalisation, une fois actionnée la tige solidaire de l'extrémité — à la main ou au moyen d'un mécanisme capable de la soulever et de l'abaisser alternativement, soit par une bielle-manivelle, soit à l'aide de vis-treuil — la pompe fonctionne. Dans sa course ascendante le piston aspire et sous l'effet de la pression atmosphérique, une fois le clapet soulevé, l'eau entre et monte grâce à la dépression. Arrivé en fin de course, le piston est repoussé dans la position de départ ; sous l'effet de la compression, les clapets de sa base supérieure s'ouvrent, l'eau les traverse et passe dans la partie supérieure de la pompe. Avec une série de mouvements alternés, elle gagne le conduit d'évacuation. Instrument familier à l'équipage des bateaux ou des galères pour vider les sentines, utilisé par les Égyptiens, les Grecs et les Romains, plus simple que la machine « de Ctesibius » décrite par Vitruve (aspirant et refoulant en même temps)[64], la pompe hydraulique, ou *antlia*, ou siphon, figurait déjà dans les traités. Giorgio Agricola l'avait depuis longtemps décrite et dessinée pour évacuer l'eau des mines[65]. Agostino Ramelli, à la même époque, en récapitulait plus d'une cinquantaine, de mécanismes variés, actionnées par l'homme ou par le courant d'un fleuve : pompes doubles, en série, en chaîne, accouplées à des mécanismes aussi ingénieux que compliqués[66]. Basée, comme l'avait expliqué Giuseppe Ceredi, sur la « seconde raison naturelle de Pline »[67] — « il n'est pas d'espace qu'aucun corps n'occupe » — la pompe était certainement le noyau même des innombrables maquettes qui remplissent les « chambres secrètes » des *Provveditori di Comun*. Maquettes de machines, construites pour faire étalage d'habileté et acquérir des brevets, des privilèges, ou des emplois, mais pas plus efficaces que leur ancêtre plus élémentaire, et donc destinées à rester là, toujours inutiles et cachées[68]. Les simples pompes employées à Rialto, construites et réparées sur place par les maîtres et par les ouvriers, pouvaient élever l'eau à une hauteur utile de presque sept mètres sous le seul effet de la pression atmosphérique, puisqu'on est au niveau de la mer, assez donc pour les besoins de la construction.

Quand les machines travaillent, porteurs et creuseurs de canaux aident à l'assèchement au moyen de seaux, passés de main en main, du fond jusqu'au sommet de la palissade[69]. Une combinaison élémentaire d'efforts considérée par Agostino Ramelli comme une machine, dans la mesure où elle fournit un travail avec facilité et rapidité, sans devoir être construite ni réparée[70].

Une fois le fond vaseux atteint, les ouvriers l'attaquent avec pioches et bêches, travaillant jour et nuit. Ils jettent les boues en remontant d'échafaudage en échafaudage — les mêmes qui avaient été utilisés pour le pompage[71] — boues que les embarcations vont décharger au loin. Ce système est aussi ancien que la ville surgie sur les lagunes et les *velme*, et indissolublement lié à sa croissance et à sa construction. Le capitaine et ingénieur Girolamo Maggi, qui conseillait de l'adopter pour « vider » rapidement les canaux de navigation[72], le considérait comme essentiel et efficace.

Une fois atteint le niveau auquel le conducteur de travaux juge qu'on peut bâtir les fondations (qui, à Rialto, se situe à cinq mètres au-dessous du niveau moyen des marées), les « maîtres » commencent à mettre en place le *battuto*. Les pilots s'enfoncent avec peine, sous les coups de maillet : « On les voyait frapper sur les pieux avec acharnement, et souvent ils frappaient vingt-cinq fois dessus » témoignent des passants attirés sur le chantier par la fosse profonde, par le trafic et par les bruits [73].

Un va-et-vient de brouettes poussées par les hommes déversent les pierres et les briques sur le lit de fondation [74]. Leur propre poids aide à la descente des rampes jetées entre les échafaudages contre les parois intérieures des palissades. Même les cintres, d'une certaine manière, fournissent un travail : pendant près d'un an, en effet, ils soutinrent le poids des pierres, des briques et des mortiers de l'arche, au fur et à mesure que celle-ci s'élève et qu'elle s'étend des impostes à la clef de voûte. Cette arche suit la courbe du « rembourrage », c'est-à-dire du « sol », échafaudage de mélèze soutenu à son tour par des systèmes de chaînes et de charpentes du même bois, mortaisés dans le tablier du pont, et disposés en séries parallèles sur sa largeur. Il n'y eut pas de pompage ultérieur, et les échafaudages ne furent pas étayés par des pilots enfoncés dans le lit du Grand Canal ; en revanche on apporta de l'Arsenal, dans l'espace compris entre les deux piles, un grand ponton soutenant un large *pagiol* ou plateau servant de point d'appui et de service flottant [75].

Les premières pièces de charpentes reposent donc directement sur les saillies construites sur les impostes, s'appuyant les unes aux autres au moyen d'étriers et de ligatures de fer. De cette manière, le poids, uniformément réparti sur les chevrons, passe dans les anneaux dans lesquels, avec les poinçons et les tirants, ils convergent, subissant de façon équilibrée soit des résultantes égales et contraires, soit la réaction des impostes. Une structure assez semblable avait été conçue par l'expert Felice Brunello pour le projet du pont de bois qu'il proposait de construire [76], ou par Vincenzo Scamozzi pour « l'armature » de l'arche du pont du Rialto : une structure où les tirants d'une première série de trois faîtes sert de chevrons à deux supports situés au-dessous [77]. Les deux systèmes, tout comme celui adopté par Antonio da Ponte, étaient des perfectionnements ou des adaptations, pour ce cas particulier, de procédés déjà en usage, par exemple pour les arches du *Ponte Sisto* romain, construites plus d'un siècle auparavant.

Le conducteur de travaux veille à la réalisation scrupuleuse, et suit de près le travail des menuisiers chargés de scier, raboter, modeler et d'assembler les pièces des faîtes. La connaissance des réactions du bois, des pierres et de l'eau leur permet, cas par cas, de suppléer au manque d'un savoir spécifique pour la construction des ponts. C'est de ce manque de savoir que se plaignait, encore en 1716, Gautier, alors qu'il tentait le premier de conférer aux notions pratiques éparses la cohérence d'un système de connaissances [78].

Une fois les cintres placés, on utilise les « treuils » (*argani*) et les « moufles » (*taglie*) [79]. « Monter les pierres vives de la voûte », font état les mandats de paiement : mais comme les pierres, d'après le contrat, doivent être déchargées sur les quais et qu'il n'est pas fait mention de déplacements ultérieurs de barques, des porteurs les transportent à force de bras sur le grand ponton, d'où elles sont élevées par les machines. Plutôt que d'en constituer des variantes élaborées, les treuils et les moufles de Rialto fonctionnent de la même manière que les *ergata* et les poulies des machines « agissant par traction » décrites par Vitruve [80]. Au Moyen Age, elles furent le symbole d'un travail « en chantier », comme le montre la multitude des Tours de Babel et des Temples de Salomon dans les miniatures des bibles et des bréviaires.

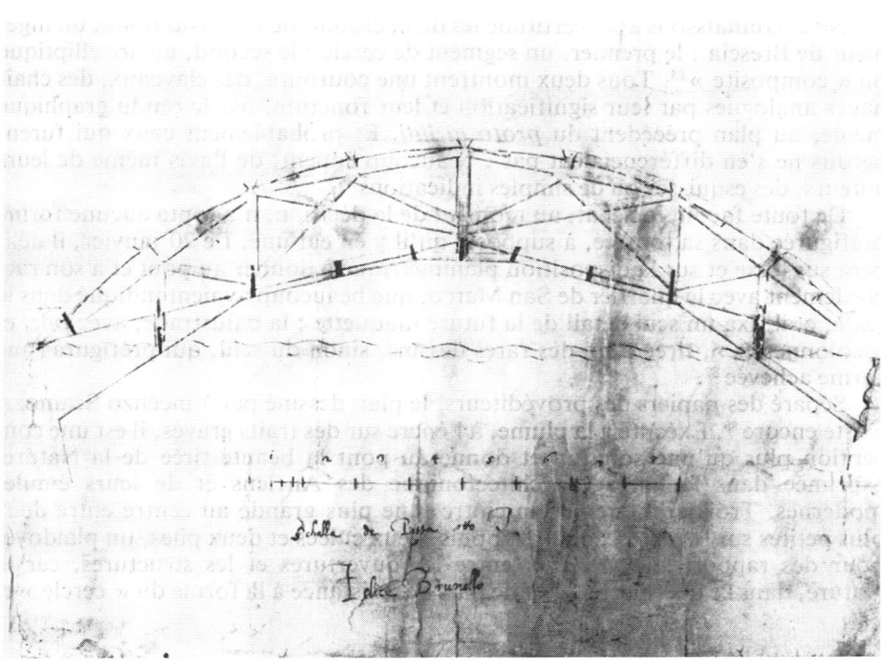

FIG. 5. — Felice Brunello, projet en bois pour le pont du Rialto (ASV, *Provveditori sopra la fabbrica del Ponte di Rialto*, dessin n° 9).

VI. — Les dessins

Parmi les papiers des *Provveditori sopra la fabbrica del Ponte di Rialto,* figurent 25 dessins ; 12 concernent la construction du pont [81]. Il y a peu d'autographes, beaucoup sont anonymes et aucun ne préfigure une forme achevée.

Les premiers avis parvenus aux provéditeurs sur le choix à faire de trois arches ou d'une seule arche sont peu nombreux : cinq en tout, et trois seulement des chefs de chantier et experts qui intervinrent en accompagnant leurs commentaires de croquis [82]. Il ne reste peut-être aujourd'hui qu'un seul de ces dessins, celui de Giacomo di Guberni, *proto ai lidi*. Il servait à montrer la corde, la flèche, la courbure de l'arc, la position des claveaux, les systèmes de fondation [83]. Il était tracé pour rassurer les esprits quant à la stabilité de l'édifice et constituait une illustration complète, suivie des nombreuses raisons données dans le texte en faveur du parti choisi.

Et il reste encore très peu de dessins joints aux nombreux avis envoyés dans une seconde phase aux provéditeurs qui avaient lancé un nouvel appel d'offres et fait des demandes pour mieux évaluer les qualités et les défauts de l'une et de l'autre solution [84].

Nous connaissons avec certitude les deux croquis de Dionisio Boldi, un ingénieur de Brescia : le premier, un segment de cercle ; le second, un arc elliptique ou « composite »[85]. Tous deux montrent une courbure, des claveaux, des chaînages analogues par leur signification et leur fonction, par le rendu graphique même, au plan précédent du *proto ai lidi*. Et probablement ceux qui furent perdus ne s'en différenciaient pas : beaucoup étaient, de l'avis même de leurs auteurs, des esquisses ou de simples indications[86].

De toute façon, le Sénat, au moment de la décision, n'adopta aucune forme préfigurée dans sa totalité, à supposer qu'il y en eut une. Le 20 janvier, il délibéra sur l'axe et sur la disposition planimétrique à donner au pont et à son raccordement avec le quartier de San Marco, que beaucoup avaient indiqué dans le texte, et il fixa un seul détail de la future maquette : la balustrade, avec *pozi* et « colonnettes », tirée d'un des rares dessins, sinon du seul, qui préfigurât une forme achevée[87].

Séparé des papiers des provéditeurs, le plan dessiné par Vincenzo Scamozzi existe encore[88]. Exécuté à la plume, à l'encre sur des traits gravés, il est une conception plus qu'une solution et donne au pont la beauté tirée de la Nature, exprimée dans le langage architectonique des Anciens et de leurs émules modernes. Trois arches en plein cintre, une plus grande au centre entre deux plus petites sur les côtés ; quatre appuis, deux culées et deux piles, un plaidoyer pour des rapports harmonieux entre les ouvertures et les structures, car la Nature, dans ce qu'elle crée, donne plus de résistance à la forme du « cercle » et

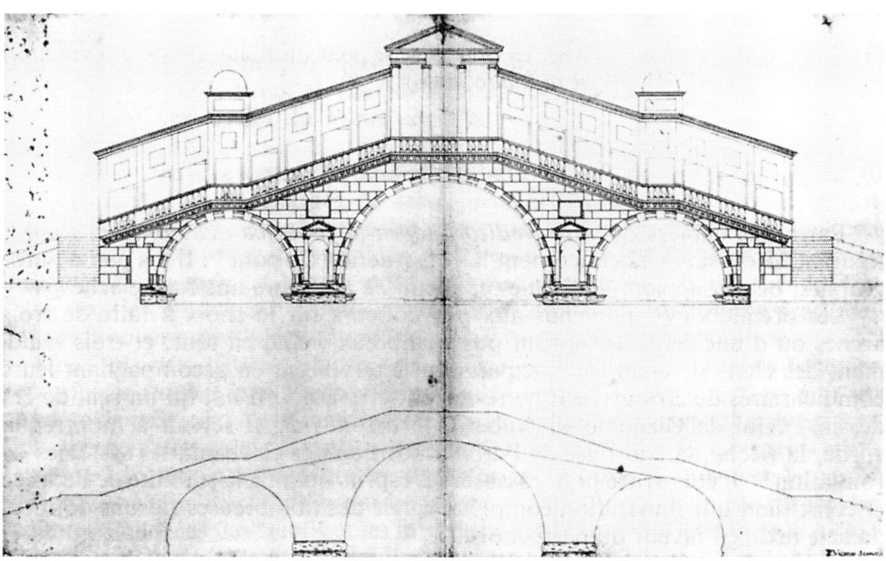

FIG. 6. — Vincenzo Scamozzi, projet à trois arches pour le pont du Rialto, élévation, 1587 (Londres, RIBA, Burlington Devonshire VIII-10).

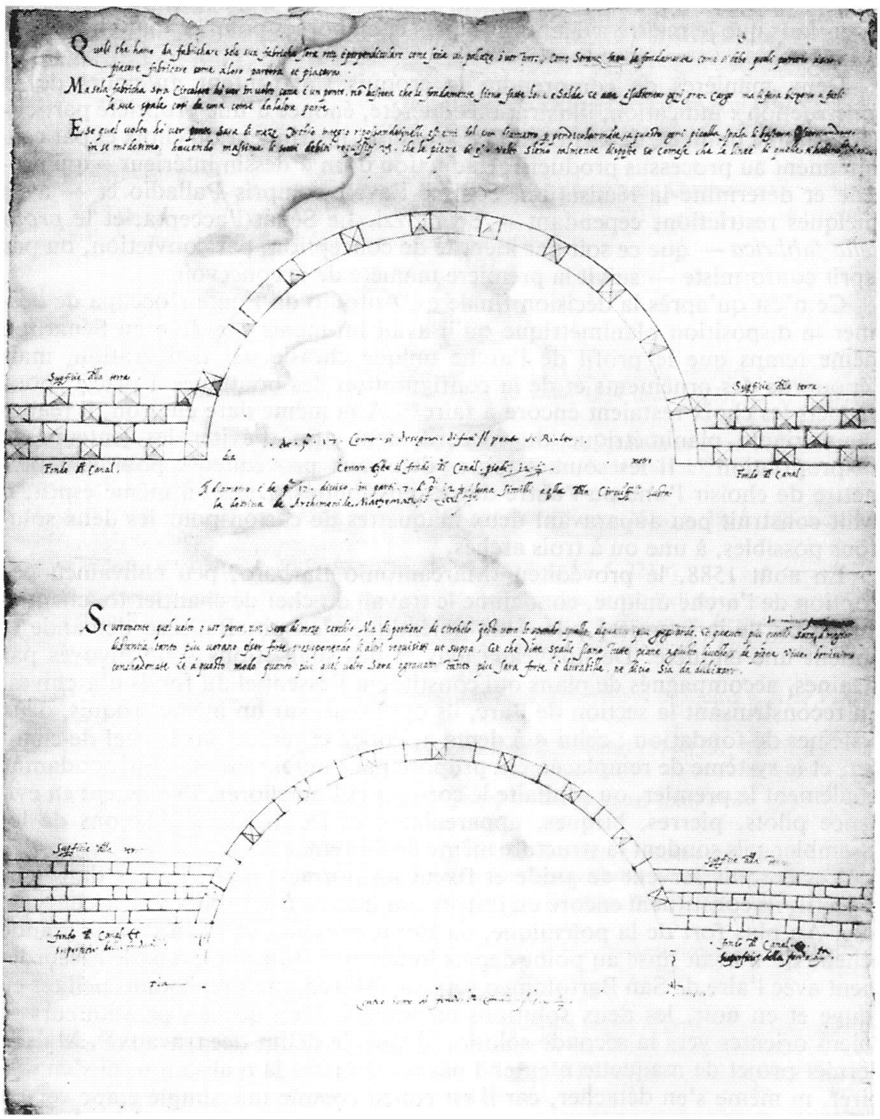

FIG. 7. — Marc'Antonio Barbaro (?), dessin où l'on voit comment devait être construit le pont du Rialto, en plein cintre, entre des culées de pierre avec des appareillages plans (ASV, *Provveditori sopra la fabbrica del Ponte di Rialto*, b. 4).

à tous les animaux un nombre pair de points de soutien, car elle crée l'unité en liant de façon équilibrée des parties singulières et distinctes. Le projet de Vincenzo Scamozzi est un morceau « in memoriam » d'Andrea Palladio, des inventions que le maître avaient produites et élaborées pour le pont entre 1551 et 1570[89].

Deux manières de comprendre le croquis s'affrontent au cours de la construction : indication, illustration concrète, énoncé d'une propriété particulière, simple point de départ pour la discussion, instrument du choix et par cela immanent au processus productif, traduction d'un « dessin intérieur » qui précède et détermine la réalisation, comme l'avait compris Palladio et — avec quelques restrictions cependant — Scamozzi. Le Sénat l'accepta, et le *proto della fabbrica* — que ce soit par identité de conception, par conviction, ou par esprit conformiste — suivit la première manière de le concevoir.

Ce n'est qu'après la décision finale qu'Antonio da Ponte s'occupa de dessiner la disposition planimétrique qu'il avait lui-même suggérée au Sénat, en même temps que le profil de l'arche unique choisie par délibération, mais dépourvue des ornements et de la configuration des boutiques à placer, pour lesquels les choix restaient encore à faire[90]. A la même date environ, il réalise une variante planimétrique du raccordement afin d'éviter les lenteurs de l'expropriation[91]. Il les soumet toutes deux aux provéditeurs pour leur permettre de choisir l'une ou l'autre de ces possibilités. Dans un même esprit, il avait construit peu auparavant deux maquettes de carton pour les deux solutions possibles, à une ou à trois arches.

En août 1588, le provéditeur Marcantonio Barbaro, peu convaincu par l'option de l'arche unique, condamne le travail du chef de chantier touchant la fondation de la première pile. Une polémique s'ensuit. Barbaro demande et obtient une enquête. De nouveaux avis et des suggestions sont envoyés par dizaines, accompagnés de plans qui constituent l'essentiel du fonds d'archives. En reconstruisant la section de l'arc, ils opposent, sur un même croquis, deux systèmes de fondation : celui « à dents », conçu et réalisé par le chef de chantier, et le système de remplacement proposé par l'auteur du plan qui condamne totalement le premier, ou souhaite le corriger et l'améliorer. Ils mettent en évidence pilots, pierres, briques, appareillages et les différentes façons de les assembler ; ils sondent la structure même de l'ouvrage[92].

Les croquis servent de guide et fixent les normes, mais, limités dans leur objectif, ils constituent encore un instrument interne à la production architecturale. Au plus fort de la polémique, da Ponte présente une maquette à grande échelle qu'il avait mise au point depuis longtemps. Elle comprend le raccordement avec l'aire de San Bartolomeo vers San Marco, mais reproduit, peintes en rouge et en noir, les deux solutions envisagées, bien que les provéditeurs se soient orientés vers la seconde solution depuis le début des travaux[93]. Mais le dernier projet de maquette n'entend pas transcender la réalisation, ni s'en inspirer, ni même s'en détacher, car il est conçu comme une simple étape vers la décision adoptée.

Avec le refus systématique d'un plan complet initial, le pont s'éleva par tronçons : épures, expositions, retailles, maquettes partielles (rarement en bois, plus souvent en carton) des cintres, des boutiques, des maisons sur le raccordement, des moulures furent fournies aux maîtres d'œuvre pour l'usage et l'utilisation des matériaux, au fur et à mesure de l'avancement du chantier.De fait, jamais une forme ne fut choisie a priori ; on préféra élaborer progressivement la construction. La perte partielle ou la dispersion de l'ensemble des croquis ne doit pas étonner : ils avaient peu de valeur en eux-mêmes et furent dépassés par la réalisation du pont.

Traduit par Annie Duchesne

NOTES

1. ASV, *Senato Terra* (plus loin, *Sen. Ter.*), 17 janv. 1551 (a.s. 1550), registre 37, fol. 88 r. ; pour la reconstruction du marché : ASV, *Sen. Ter.*, 13 janv. 1514 (a.s. 1513), r. 18, f. 100 v.-101 r.

2. Le vote est l'occasion de reprendre le programme d'aménagement et de restauration de la ville déjà confié à deux *Savi* nommés en 1535, quand Andrea Gritti était doge (ASV, *Sen. Ter.*, 2 sept. 1535, r. 28, f. 66 r.).

3. ASV, *Sen. Ter.*, 27 sept. 1554, r. 39, f. 165 r. ; *ibid.*, 3 nov. 1554, r. 39, f. 188 v. . *ibid.*, 14 nov. 1554, f. 189 r.

4. « Ridotto a termine ». ASV, *Sen. Ter.*, 5 déc. 1587, r. 57, f. 223 r.

5. S. Sinding-Larsen, « Christ in the Council Hall. Studies in the Religious Iconography of the Venetian Republic », dans *Acta ad Archaeologiam et Artium Historiam Pertinentia*, Rome-Spolète, 1974, V, 4-5 ; G. Lorenzi, *Monumenti per servire alla storia del Palazzo Ducale di Venezia*, Venise, 1868, doc. 786.

6. D. Calabi, P. Morachiello, « Sansovino e le Fabbriche Nuove di Rialto », *Arte Veneta*, XXXVIII (1984).

7. D. Calabi, P. Morachiello, *Rialto : le fabbriche e il ponte, 1514-1591*, Turin, Einaudi, 1987.

8. ASV, *Provveditori sopra la fabbrica del ponte di Rialto* (plus loin, *Prov. fab. Rialto*), 1er févr. 1588 (a.s. 1587), *busta* 5.

9. ASV, *Sen. Ter.*, 7 janv. 1588 (a.s. 1587), r. 57, f. 246 r. ; *Mensa patriarcale*, b. 10, f. 111 r. ; BNMV, *Documenti e scritture originali ed in copia relative alla fabbrica del ponte di Rialto in Venezia e a quella del ponte di San Michele in Vicenza*, ms. It. VII, 2207 (= 9549), f. 6 r. ; BMCV, *Curiosità*, s. coll. f. 1 r.

10. ASV, *Sen. Ter.*, 2, 11 mai 1590, r. 60. L'enchère n'atteint pas la somme espérée. Une nouvelle mise aux enchères est effectuée en mars 1591 : ASV, *Sen. Ter.*, 16 mars 1591, r. 61, f. 26 v. ; *ibid.*, 28 mars 1591, f. 13 r. ; *Collegio*, 28 mars 1591, r. 53, f. 24 r.

11. Voir, à titre d'exemple, l'examen de contrôle auquel est requis le provéditeur Jacopo Foscarini pour mettre fin à une controverse entre le conducteur de travaux et un maître tailleur. ASV, *Prov. fab. Rialto*, 26 mai 1590, b. 4 ; *ibid.*, *Processi*, 7, 14, 16 juil. 1591 ; 11, 13 août 1591 ; 1er, 18 sept. 1591 ; 18 nov. 1591 ; 16 janv. 1592 (a.s. 1591).

12. Par exemple quand il s'agit de prendre une décision sur la façon de construire les batardeaux, pour assécher les zones du canal à l'intérieur desquelles fonder les culées : ASV, *Sen. Ter.*, 12 mars 1588, r. 58, f. 7 v.

13. Antonio da Ponte rappelle, par exemple, avoir consigné deux polices qu'il avait rédigées à l'attention de Camillo Ziliol, secrétaire des provéditeurs : ASV, *Prov. fab. Rialto*, 15 janv. 1588 (a.s. 1587), b. 3.

LES CHANTIERS URBAINS

14. ASV, *Sen. Ter.,* 10 sept. 1588, r. 58. Cf. en outre, ASV, *Savi ed Esecutori alle Acque,* b. 117, ff. n.n. ; 29 mars 1594, ff. 23 r. et v. ; *Provveditori di Comun,* 22 mars 1593, b. 17, Actes, Livre n° 33, f. 109 r.

15. ASV, *Prov. fab. Rialto,* 16 janv. 1593 (a.s. 1592), b. 3 ; cf. aussi : D. Calabi, « La Direzione del nuovo ponte di Rialto e il negotio degli stabili di San Bartolomeo », *Bollettino dei Musei Civici Veneziani,* XXVII, 1982, 65.

16. Pour les dépenses des *Provveditori di Comun* : BMCC, Cod. Cicogna, n° 2509, MCCCCXCII ; cf. aussi Besta, p. 335 ; pour la restauration du palais ducal, un total de 109 245 ducats en neuf ans a été voté par le *Consiglio dei X* ; cf. G. Lorenzi, *Monumenti per servire alla storia del Palazzo Ducale di Venezia,* Venise, 1868 ; pour les prisons : BNMV, ms. It. VII, 295 (= 10047), *Scrittura inviata ai Provveditori sopra la restauration del Palazzo,* le 6 févr. 1581 (a.s. 1580), f. 11 v.

17. ASV, *Prov. fab. Rialto, Pareri,* 28 janv. 1588 (a.s. 1587), b. 3, acte de Marcantonio Barbaro ; R. Cessi, A. Alberti, *Rialto : l'isola, il ponte, il mercato,* Bologne, 1934, pp. 205-206. Dans les premiers actes des provéditeurs, da Ponte est appelé « *proto del palazzo* » : ASV, *Prov. fab. Rialto,* 13 févr. 1588 (a.s. 1587), 21 avr. 1588, b. 5.

18. ASV, *Prov. fab. Rialto, Pareri,* s.d., b. 3.

19. Les délibérations de l'ASV, *Provveditori al Sale,* b. 10, et du *consiglio dei X, Misti,* 1581, 1587-1588, renseignent sur son activité très variée ; sur Antonio da Ponte en général, cf. T. Temanza, *Vite dei più celebri architetti e scultori veneziani che fiorirono nel secolo Decimosesto,* Venise, 1778, pp. 499-518 ; G.B. Cecchini, « Della vita e delle lodi di Antonio da Ponte Architetto Veneziano », *Atti della R. Accademia di Belle Arti di Venezia,* 1859-1860, pp. 16-112 ; B. Baldoni, P. Martinelli, *Antonio da Ponte, Proto al Sale : « l'acconciar » e le nuove « fabbriche ». Ponte di Rialto e Prigioni,* thèse de doctorat soutenue auprès de la Section d'Histoire de l'Architecture de l'IUAV, 1982-1983.

20. ASV, *Sen. Ter.,* 7 janv. 1588 (a.s. 1587), *op. cit.,* ; cf. aussi D. Calabi, P. Morachiello, *Rialto..., op. cit.,* ch. xvii.

21. En avril 1588, Benedetto Banelli, appartenant à la corporation des tailleurs de pierre, commence à remplacer da Ponte dans la rédaction des polices : ASV, *Prov. fab. Rialto,* 30 avr. 1588, b. 5 ; cf. en outre : *Avogaria di Comun. Processo a Bortolo Luran,* 13 oct. 1594, interrogatoire de Beneto Banelli, ff. 28 r.-31 r. A l'automne 1588 Antonio Contin, lui aussi de la corporation des tailleurs de pierre, neveu du *proto,* avait également fait son entrée sur le chantier, comme second aide : *Prov. fab. Rialto, Mandati* b. 5 : 24 sept. 1588 ; 3, 8, 22, 29 oct. ; 5 nov. 1588 ; 17 déc. 1588 ; sur Antonio Contin et sa famille, L. Tessarin, *I Contin, una dinastia di proti nel '500 veneziano,* thèse de doctorat soutenue auprès de la Section d'Histoire de l'Architecture de l'IUAV.

22. ASV, *ibid.,* févr. 1588 (a.s. 1587), b. 6.

23. ASV, *Prov. fab. Rialto,* 8 avr. 1589, b. 4 ; *ibid.,* 14 mai 1589, b. 4 ; 8 juil. 1589, b. 5 ; 30 mars 1590, b. 5 ; en outre : *ibid., Fedi di spesa e pagamenti,* 30 mars 1590, b. 8.

24. ASV, *Prov. fab. Rialto, Mandati,* 22 avr. 1589, b.5 ; *ibid., Fedi di spesa e pagamenti,* 3 juin 1589, b.6.

25. ASV, *ibid., Mercati,* 1er févr. 1588 (a.s. 1587), b. 4 ; *Mandati,* 8, 10, 12, 13, 21, 24, 26 févr. 1588 (a.s. 1587), b. 5 ; 5, 11, 15, 27 mars 1588, 21, 24 mai 1588, b. 5.

26. ASV, *Sen. Ter.,* 7 janv. 1588, *op. cit.* ; 20 févr. 1588 (a.s. 1587), r. 57, f. 275 r. ; *Collegio,* 20 févr. 1588 (a.s. 1587), Notat. r. 49, f. 184 v. ; BMCC, Cartella, « Curiosità », f. 7 ; cf. aussi : D. Calabi, *La direzione..., op. cit.*

27. ASV, *Prov. fab. Rialto,* 5 juil. 1588, b. 4.

28. ASV, *ibid.,* 19 déc. 1588, 14 mai 1589, 8 oct. 1589, 27 avr. 1591, b. 4.

29. ASV, *Sen. Ter.,* pour les ventes aux enchères : 16 mars 1591, r. 61, f. 26 v. ; 28 mars 1591, f. 13 r. ; *Collegio,* 28 mars 1591, r. 53, f. 24 r. ; pour les résultats et la révocation : 29 juin 1590, r. 60, 20 oct. 1594, r. 64.

30. Pour les nouvelles mises aux enchères : *Collegio,* 14 févr. 1599 (a.s. 1598), Not. R. 60, ff. 34 v., 60 v., 86 r., 167 v. ; 21 févr. 1600 (a.s. 1599), Not., r. 61, ff. 24 v., 33 r., 44 v., 95 v., 15 r., 97 r., 98 r.

31. ASV, *ibid., Processi,* 7, 14, 16 juil. 1591 ; 11, 13 août 1591 ; 1er, 18 sept. 1591 ; 18 nov. 1591 ; 16 janv. 1592 (a.s. 1591), b. 4 ; *Avogaria di Comun. Processo a Bortolo Laran,* 7 août 1593-6 févr. 1596 (a.s. 1595), ff. 1 r.-69 r.

32. Par exemple, pour l'achat et le travail de la « pierre vive », voir ASV, *ibid.,* b. 5. Pour les pierres *bronzine,* voir les écritures des Recteurs de Vérone : ASV, *ibid.,* 29 mars et 2 avr. 1590, b. 4. Pour le bois et les ferrures, *ibid.,* 13 févr. 1588 (a.s. 1587)-août 1588, b. 5. Pour la chaux, les pierres cuites et le sable, *ibid.,* 20 et 22 févr. 1591 (a.s. 1590), b. 8 ; 1er mars 1591, b. 9.

33. ASV, *Prov. fab. Rialto,* 12 mars 1588, b. 4 ; *ibid.,* 24 mai 1588, b. 5.

34. Les tailleurs de pierre sont payés systématiquement (aux mêmes dates) avec deux mandats distincts, pour l'achat de la pierre « en leur nom » qui vient de l'Istrie, et pour le travail de cette pierre « dans cette ville ».

35. ASV, *ibid.,* 13 févr. 1588 (a.s. 1587), b. 5.

36. Nous sommes amenés à ces considérations par les mandats de paiement cités : Trivellin da Bassan, ou Frattolin Furlan témoignent, par leur nom, de leur lieu d'origine, lequel est aussi un lieu d'importation du bois bien connu ; quant à Iseppo Porto, fournisseur aussi de Palladio, nous savons qu'il vient de Vicence.

37. ASV, *Prov. fab. Rialto,* 21, 29 avr., 11 juin, 5 juil., 12 août 1588, b. 5.

38. AVS, *ibid.*

39. ASV, *ibid.,* 13 oct. 1590, b. 4 ; *ibid., Fedi di spese e pagamenti,* 1er févr. 1588 (a.s. 1587), b. 6.

40. ASV, *ibid.,* 9 mai 1589, 22 juin 1589, b. 5

41. ASV, *ibid.,* b. 5 : par exemple le 23, le 26 et le 30 avr. 1588.

42. ASV, *ibid.,* 29 déc. 1588, b. 4.

43. ASV, *ibid.,* 19 juin 1589, b. 4.

44. ASV, *ibid.,* 12 mars 1588, b. 4 ; 13 févr. 1588 (a.s. 1587), 5 mars 1588, 31 avr. 1588, b. 5 ; 12 oct. 1590, b. 4.

45. ASV, *Prov. fab. Rialto,* 28 janv. 1587, a.s., c.à.d. 1588 ; b. 4. ASV, *Prov. fab. Rialto,* 6, 13, 24 févr. 1587, a.s. (1588) ; 26 mars 1588 ; 16 avr. 1588 ; b. 5.

46. ASV, *Sen. Ter.,* 12 mars 1588, r. 58, f. 7 v. ASV, *Prov. fab. Rialto,* 21, 26 mars 1588 ; 2, 9 avr. 1588 ; b. 5.

47. ASV, *Prov. fab. Rialto,* 6 juil. 1588 ; b. 4. ASV, *Prov. fab. Rialto,* 16, 30 juil. 1588 ; 20, 27 août 1588 ; 6, 12, 19, 24 sept. 1588 ; 5, 8, 15, 22 oct. 1588 ; b. 5.

48. ASV, *Prov. fab. Rialto,* 8 avr. 1588 ; b. 4.

49. ASV, *Prov. fab. Rialto,* 1er, 26 févr. 1587, a.s. (1588), b.4. ASV, *Prov. fab. Rialto,* 1er, 15, 26 févr. 1585, a.s. (1588) ; 1er, 16, 22 mars 1588 ; 2, 8, 21, 28, 29 avr. 1588 ; 14, 24 mai 1588 ; 6, 11, 18, 25 juin 1588 ; 12, 14, 27, 30 juil. 1588 ; 3, 9, 18, 19, 22, 27 août 1588 ; 2, 10, 12, 28 sept. 1588 ; 3, 5, 7, 8, 24, 26 oct. 1588 ; b. 5.

50. Le contrat dit textuellement 5 000 et 2 000 « *miara* », c'est-à-dire milliers ; mais nous croyons qu'il s'agit de la répétition, due au copiste, d'un même nombre en chiffres et en lettres : sinon on obtiendrait une quantité de pierres exagérée.

51. ASV, *Prov. fab. Rialto,* 10 sept. 1588 ; b. 4.

52. ASV, *Prov. fab. Rialto,* 9, 31 avr. 1588 ; 14, 16, 21, 24 mai 1588 ; 11, 18, 25 juin 1588 ; 1er, 2, 7, 9, 16, 28 juil. 1588 ; 5, 13, 18, 19, 25 août 1588 ; 10, 24, 26, 29 sept. 1588 ; 8, 15 oct. 1588 ; b. 5.

53. Trois commandes atteignent 1 000 ducats, une est supérieure à 500, six tournent autour de 100, sept sont inférieures à 25 ducats.

54. ASV, *Prov. fab. Rialto,* 16 févr. 1587, a.s. (1588) ; b. 4. ASV, *Prov. fab. Rialto,* 28 mai 1588 ; 4, 11, 18 juin 1588 ; 1er, 5, 12 juil. 1588 ; 8 oct. 1588 ; b. 5.

55. ASV, *Prov. fab. Rialto,* 18 févr. 1587, a.s. (1588) ; b. 4. ASV, *Prov. fab. Rialto,* 23, 26, 30 avr. 1588 ; 24 mai 1588 ; 1er, 4, 11, 18 juin 1588 ; 1er, 6 juil. 1588 ; 10, 17 sept. ; 3, 8, 15 oct. 1588 ; b. 5.

LES CHANTIERS URBAINS

56. ASV, *Prov. fab. Rialto,* 29 déc. 1588 ; b. 4.

57. ASV, *Prov. fab. Rialto,* 17 sept. 1588 ; 20 déc. 1588 ; 4 mars 1589 ; b. 4.

58. ASV, *Prov. fab. Rialto,* 1er sept. 1589, b. 4.

59. ASV, *Prov. fab. Rialto,* 30 juil. 1590 ; b. 4.

60. ASV, *Prov. fab. Rialto,* 21 août 1590 ; 1er nov. 1590 ; 5 déc. 1590 ; 2, 20 mars 1591 ; 10 mai 1591 ; 2 juil. 1591 ; b. 4.

61. ASV, *Prov. fab. Rialto,* 16, 22, 31 avr. 1588 ; 14, 21 mai 1588 ; 4, 11, 18, 25, 28 juin 1588 ; 1er, 9, 16, 23, 30 juil. 1588 ; 5, 13, 20, 27 août 1588 ; 3, 10, 17, 24, 26 sept. 1588 ; 3, 5, 8, 15, 22 oct. 1588.

62. ASV, *Prov. fab. Rialto,* 2, 31 avr. 1588 ; 23 juil. 1588 ; b. 5.

63. ASV, *Prov. fab. Rialto,* 21 mai 1588 ; 4 juin 1588 ; 4, 13 août 1588 ; 15, 29 oct. 1588 ; 14, 19 nov. 1588 ; 17 févr. 1588, a.s., c.à.d. 1589 ; b. 5.

64. *De architectura,* L.X, chap. VII.

65. A. RAMELLI, *Le diverse et artificiose machine,* Paris, 1588, fig. I-III, VI-XLII, XLV-XCV.

67. G. CEREDI, *Tre discorsi sopra il modo d'alzar acque da' luoghi bassi. Per adacquar terreni. Per levar sorgenti ; e piovute dalle campagne, che non possono naturalmente dare loro il decorso. Per mandare l'acqua da bere alle Città, che n'hanno bisogno, e per altri simili usi,* Parme, 1567, pp. 14-15.

68. *Ibid.,* p. 19.

69. ASV, *Prov. fab. Rialto,* 19 sept. 1588 ; b. 5.

70. A. RAMELLI, *op. cit.,* fig. CXI.

71. ASV, *Prov. fab. Rialto,* 3 oct. 1588 ; 5 nov. 1588 ; b. 5.

72. *Vocabolario degli Accademici della Crusca,* Florence, 1863, I, pp. 133-134.

73. ASV, *Prov. fab. Rialto,* 12 août 1588, b. 3.

74. ASV, *Prov. fab. Rialto,* 3 oct. 1588, b. 5.

75. ASV, *Prov. fab. Rialto,* 8 juil. 1589 ; 30 mars 1590 ; b. 5.

76. ASV, *Prov. fab. Rialto,* 15 janv. 1587, a.s., c.à.d. 1588.

77. V. SCAMOZZI, *L'Idea dell'architettura universale,* Venise, 1615 (Bologne, 1982), VIII, pp. 345-346 ; A. RONDELET, *Essai historique sur le Pont de Rialto,* Paris, 1836, pp. 22-23.

78. GAUTIER, *Traité des Ponts,* Paris, 1714.

79. ASV, *Prov. fab. Rialto,* 21 août 1590, b. 4. ASV, *Prov. fab. Rialto,* 16 mars 1590 ; 13, 19 avr. 1590 ; 4, 18 mai 1590 ; b. 5.

80. *De architectura,* L.X, chap. III-V.

81. ASV, *Prov. fab. Rialto,* Dessins.

82. ASV, *Prov. fab. Rialto,* 20, 23 déc. 1587 ; b. 3.

83. ASV, *Prov. fab. Rialto,* Dessin n° 1.

84. ASV, *Prov. fab. Rialto,* 28, 29, 30 déc. 1587 ; 1er, 2, 5, 6, 7, 12, 15, 16, 17, 21, 22, 27 janv. 1587, a.s., c.à.d. 1588 ; b. 3.

85. ASV, *Prov. fab. Rialto,* Dessins n° 2 et 3.

86. ASV, *Prov. fab. Rialto,* 11, 12 août 1588 ; b. 3.

87. ASV, *Sen. Ter.,* 7 janv. 1587, a.s., c.à.d. 1588 ; r. 57, f. 246 r.

88. Londres, R.I.B.A., Burlington-Devonshire VIII - 10.

89. Cf. D. CALABI, P. MORACHIELLO, « Rialto : " sacrario " da conservare o da rinnovare », dans L. PUPPI, G. D. ROMANELLI (sous la dir. de), *Le Venezie possibili,* Catalogue de l'exposition, Venise, 1985, pp. 55-70, et aux indications bibliographiques qui s'y trouvent.

90. ASV, *Prov. fab. Rialto,* Dessin n° 10.

91. ASV, *Prov. fab. Rialto,* 26 août 1588 ; b. 3.

92. ASV, *Prov. fab. Rialto,* Dessins n° 6, 7, 11, 12.

93. ASV, *Prov. fab. Rialto,* 13 mars 1589 ; 20, 27 août 1589 ; b. 5. ASV, *Prov. fab. Rialto,* 13 mars 1589 ; b. 7.

8
The Pul-i Khwājū in Isfahan: a combination of bridge, dam and water art

H. Luschey

INTRODUCTION

Beginning with the seventeenth-century travellers such as Chardin, Tavernier and E. Kaempfer, a vast amount of literature and eulogy has been written concerning the bridges of Isfahan, especially the Pul-i Khwājū.[1] But *one* element, the fact that this bridge, with its dam and water regulation, represents the phenomenon of a "water-art", has been largely overlooked. It was hinted at by Chardin, and then clearly understood by E. Kaempfer in his interpretation and illustration of the bridge (Pl. I). Later, Le Bruyn and Curzon[2] interpreted the form and function of the bridge, and then, especially, A. U. Pope:

This bridge, far from being a mere transit utility, is indeed a place for tarrying. It has become an end in itself, and with imagination and sound architectural sense exploits the site to the utmost. Permitting and safeguarding traffic, it also opens up for use and enjoyment all the advantages of an admirable location[3] (cf. Pl. IIIa–b).

I have looked through many books on Iran and on Isfahan, expecting that someone would have seen the outstanding value of this bridge as "water-art" and as a *Gesamtkunstwerk*; but the texts are either conventional or else enthusiastic over the architectural form, and the pictures give inadequate attention to the role of the water in combination with the actual bridge. E. Höltzer, a gifted German engineer working for the British Telegraph Union in the eighteen-seventies and eighties, did photograph the phenomenon of the storm flood running through the Khwājū bridge,[4] but the harmonic form of the "water art" can only be found in Kaempfer. Thus the publication of my photographs may be useful for a better understanding of the Pul-i Khwājū (Pls. Ia, II, IV, Va, VIa, b).

DESCRIPTION

Before going into details, however, we should review the general appearance of the bridge (Figs. 1 and 2). The main roadway is flanked on both sides by walls and arcades for the pedestrians and onlookers. The lower story of the bridge, at the level of the heavy piers, offers another opportunity for people to pass over the stream or else to sit on the stairs before the piers or under the shady arches of the bridge. This is best illustrated in a photograph by Pope (Pl. IIIa), showing a good number of people on the platforms, steps and in the arcades, and also shown in a drawing given by Jane Dieulafoy (cf. n. 3 above) (Pl. IIIb).

My own photographs, taken on March 30th 1963, give a different impression (Pls. I–VI). The water is softly flowing over the central platforms and steps which extend across the entire width of the stream, transforming the whole expanse into a great cascade. The steps are visible under the flow of water, and the play of water over the softly-carved planes of the middle platforms creates waterfalls of great harmony.

The bridge and its dam, which form an artifical lake in front of the pavilion of Ā'īne-Khāne and the palace of Haft-Dast, were planned by a unknown architect for the pleasure of Shāh 'Abbās II and his court in 1650,[5] and the conception of "water art" was for his pleasure and that of the people.

HISTORICAL DOCUMENTS

We possess historical documents for the understanding of the technique and meaning of the architecture of the Pul-i Khwājū by such authors as Chardin, Kaempfer, Le Bruyn and Curzon. Chardin, with whom we often visit in imagination Isfahan through his precise descriptions of various buildings, offers us on this occasion only a general text, and inadequate etchings for the understanding of the bridge.[6] Kaempfer, however, offers us detailed text and pictures (Pl. Ib and VII), which we will now review.[7] With some variations, the following translation is based on the edition of W. Hinz.

I feel that the structure is more superb compared with other bridges, especially when we compare the crowning forms—the arcades on both sides, the variations of the doors and vaults, and the elegance of the pavilions in the middle and at the ends. The bridge has twenty-four arches built of carefully-hewn, square stones. In the middle of each arch are wooden boards by which the river may be dammed, thus bringing it to the height of a lake before the nearby royal garden of Sa'ādatābād, "the Garden of Happiness". This provides the pleasure for the people looking on—the water flowing through the arches and running over the stone steps, these artificial impediments, then falling down to the depths with a heavy murmur—as you can see in the etching presented here.

His Latin text offers the following:

Arcus habet viginti quatuor, ex lapide quadrato politoque factos, in medio valvis ligneis interceptos, quibus sisti flumen potest, ut in horto adjacente Regio, bāgi Sehadèt abaad, i.e., horto beatitudinis, in altitudinem lacus extumescat. Non minorem proflues unda aspicientibus voluptatem parit, dum per arcus elapsa, mox, per saxea graduum superbissimorum impedimenta, in profundum praeceps cum gravi murmure devolvitur, velut ex annexa icone and ovulum patescit.[8]

On May 1st 1704, Corneille Le Bruyn, after giving a description and the name "Pont d'Hassan-Abad" to the bridge and the situation of the Talār (A'īne-Khāne) and the Serail or Harem (Haft-Dast), noted:

Lorsque le Roy s'y trouve, il fait arrêter le cours de la Riviere, par des Digues de bois, dans les Cannaux du Pont d'Hassan-Abaet, pour faire venir l'eau contre le talael, proche duquel il a deux ou trois méchantes Barques, dans lesquelles il va se divertir, à la rame, avec ses Concubines.[9]

Curzon, apparently aware of the text of Le Bruyn, wrote in 1892:

The bank of the river from the Pol-i-Khadju upwards was lined with gardens, and by means of sluice-gates at the lower bridge, the king was in the habit of damming the river, till it formed a great lake before the talar, known as the Aineh Khanhe, upon which he disported himself in boats with his ladies, and which at night was made the scene of some illuminations.[10]

Finally, Galdieri, in his study concerning the role of water in Isfahan, his "Isfahan quasi lagunare", helps us with a painting from the middle of the nineteenth century showing the Ā'īne Khāne and Haft Dast, military tents before these, and boats on the artificial lake.[11]

These "water pleasures" have forerunners in the "water plays" at the Allāh-Verdi-Khān bridge of the time of Shāh 'Abbās I, held in June or July and described by Pietro Della Valle.[12] Apparently, for the purpose of "water plays" the damming of the river was done by means of wooden boards, which on each occasion were placed between small pillars under the arcades beneath the bridge.

TECHNICAL ASPECTS

It is quite astonishing, considering all the studies on the Pul-i Khwājū, that the special technique of guiding the water in order to create an art form has not been sufficiently discussed. The plan and dimensions of the Pul-i Khwājū by P. Coste did not adequately represent the problem of the dam and the regulation of the water.[13] Pope wrote an encomium of the bridge[14] and published a picture of people sitting on its steps (Pl. IIIa); yet in regard to the technical aspects, he simply reproduced the drawing of Coste, which we again present in Fig. 1. M. Ferrante, in his brief article published in 1974, had a good understanding of the artistic components and the technical finenesses of the bridge and the corresponding balance between the bridge and Ā'īne-Khāne.[15] In 1974, the German architect K. Würfel again studied the bridge and published new, carefully drawn plans and descriptions.[16] Finally, in the *Catalogue* of the London Exhibition of 1976, *Isfahan, city of light*, a new section and plan of the bridge were

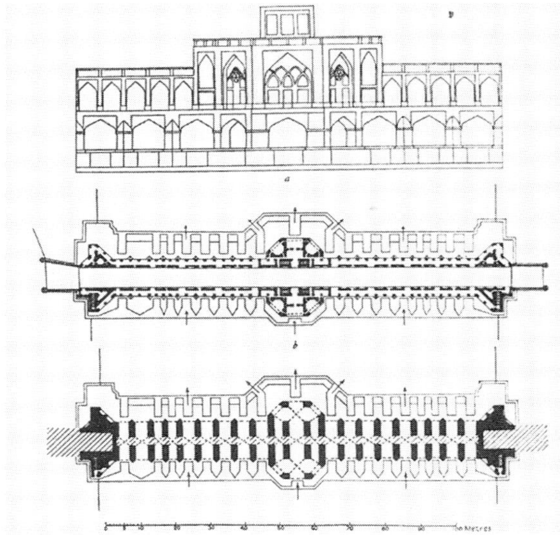

Fig. 1. Plan and elevation of the Pul-i Khwājū by Coste. After Pope, Survey.

Fig. 2. Cross-section of the Pul-i Khwājū. Author's sketch with use of a drawing in Isfahan, city of light.

presented by B. Shirazi. Nonetheless, the decisive factor, viz. the exact position of the wooden barrage boards, remains unclear. Thus I feel it is necessary to explain this technical point in a drawing (Fig. 2 and Pl. VIa), since it is important to note that the wooden barrage boards are not placed in a vertical position, as shown in previous drawings, but at an angle. It is this angular placement which is essential for the understanding of the "water art" involved. To quote from my own unpublished *Erkundungsbericht,* XIV, 1963:

. . . the closing of the water tunnels through slanting set wooden boards brought the level of the water approximately two metres higher and guided the mass of water not through the tunnels, but in bright, foaming cascades over the steps on the eastern side of the bridge, hence symmetrically over the whole width of the river—a splendid spectacle, quite baroque in style . . .

Furthermore, other fine points can be observed. The platforms, over which the waters glide before the middle pavilion to the steps, have a slightly convex surface which softly divide the waters into fan-shaped forms, and then the steps themselves, not exactly flat, but also slightly convex, giving the flooding a finer fall.

Even now, the people sit on chairs at the edge of the floods, delighted at the refreshing elements and at the art of the ruler who 300 years ago gave this building to the Isfahanis.

Slightly later, in 1965–8, Ferrante gives the following description:

En correspondance avec les arches il y a, entre les soubassements, des rainures inclinées pour y introduire des planches en bois dans le but d'utiliser le pont comme un barrage capable de rehausser d'environ deux mètres le niveau de l'eau en amont.

En aval, dans les périodes de crue, l'eau descend sur les gradins formant autant de petites cascades; le pont et les innombrables cascades produisent l'effect scénographique d'une sorte d'immense fontaine.[17]

Thus the two of us have made nearly the same observation; however, I had the opportunity also to take photographs illustrating the "immense fontaine". If we look at Ferrante's instructive drawings of the centre of the bridge with the water-channels in the basement and the pavilion on the higher storey (Figs. 3 and 4), here the fan-like platforms are clearly visible; these, having no technical function, represent only an artistic accent. For the water art begins precisely at the level of these platforms only when the flow of water is barred.

There is one further technical point. Visible in my photographs (Pl. VIb), looking through the length of the bridge to the north, are slits in the walls on the pedestrian level and in all places, two rectangular holes in the interior of the eastern pillars (Fig. 2 and Pl. IIIb). In a discussion with K. Würfel, I have learned that here at a still higher level were heavy horizontal beams supporting wooden boards for a second damming of the river. This technique, one should note, is similar to the setting of mediaeval town or castle doors, and may explain certain details visible in Kaempfer's drawing (Pl. Ib). Some of the arcades of the bridge appear to be closed by walls, leaving the steps before them dry. At Pul-i Khwājū, there are walls, but these are wooden walls composed of several boards with horizontal cross pieces.

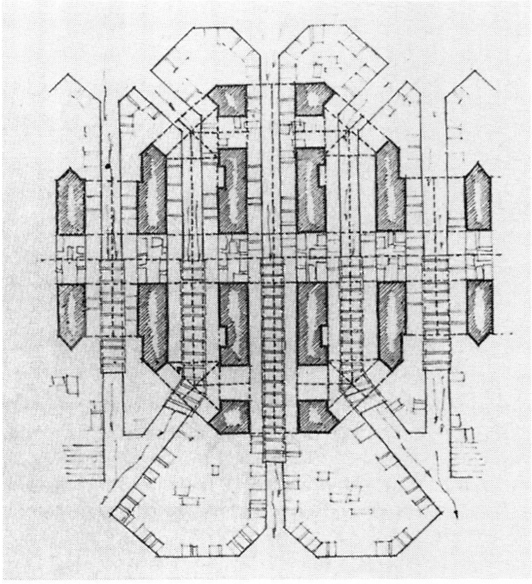

Fig. 3. Structures in the basement under the middle pavilion. Drawing by M. Ferrante.

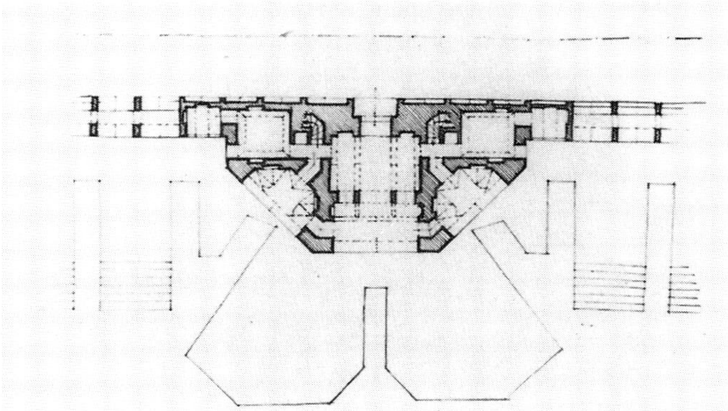

Fig. 4. Plan of the middle pavilion, level of the street. Drawing by M. Ferrante.

There remains a structural problem: did the central platforms of the bridge originally have steps like the other postaments? The etching of Kaempfer (Pl. Ib), with the water flow in the centre, does not enable us to decide this. But Chardin and Le Bruyn indicate steps, the first with some misunderstanding; in the end, Coste[18] is the most accurate. On the other hand, Höltzer in *ca.* 1880, in the first photographs of the bridge, shows the central parts such as we know them today, with a vertical drop. It is possible that the restoration dated 1290/1873–4,[19] which replaced most of the Safavid tiles, especially in the lower level, removed not only the balustrades of the upper level (reconstructed after Kaempfer in Fig. 2 and Pl. II), but the steps in the centre too. The other steps are connected with the platforms in interlocking joints, which cannot be removed without destroying the whole structure. Moreover, it is improbable that the steps were only abutments on to these central platforms in such a way that they could easily be removed. The structure of these platforms shows the original surface and construction technique as known from the pillars and walls of the bridge itself (Fig. 2).

At present, the problem is not soluble, and nobody seems even to have observed and discussed it. Pope's drawing after Coste's plan (Fig. 1) is, one must say, very sketchy and rather avoids the problem, which may perhaps have been felt. Certainly, the function of the central platforms, with or without steps, remains the same, viz. the fan-like distribution of the waters. Yet the variation of the whole composition of the bridge is richer and more baroque, if the central fan like platforms have vertical contours and no steps, as have the other parts.

After observing the Pul-i Khwājū in descriptive terms, it should now be examined from the historical and aesthetic point of view.

PROBLEMS AND TRADITIONS

In order to understand the Pul-i Khwājū in the light of architectural and aesthetic tradition, we must look first at the history of bridges in combination with dams in Iran, and second, at the tradition of water displays in the garden art of Iran. Concerning the first problem, it has long been noted that there is a clear tradition stemming from the Sasanian dam-bridges, as at Shushtar (Pl. Vb)[20] and Dizful, running through the mediaeval Islamic period, as exemplified in the Band-i Amīr,[21] to the bridges of Isfahan. The Sasanian bridge of Shushtar, moreover, also contains elements of stairs, although these appear to play a more technical role, giving the dam more substance and vigour, one, which may again be seen in the Band-i Amīr in the tenth century. This development has already been discussed by M. Dieulafoy, Pope and Gh. Kouros; the latter has also published descriptions of a large number of mediaeval bridges.[22] The beginning of this tradition appears nevertheless to lie in the Achaemenid period, from which dam building has been reported.[23]

The nature of the country requires such help in order to store water for the irrigation of fields and gardens, and to bring it to cities and palaces. This technical aspect, the age-old problem of how to use the river for the irrigation of the surrounding countryside, was clearly seen by Kaempfer, who observed that miles before the city of Isfahan, the river was dammed and the water guided into channels:

The river is dammed up long before the city and divides in several channels, partly for the irrigation of the fields before the city and partly for the different wishes of the city—filling the pools in the palaces and houses. There it may be raised as a fountain for pleasure, but mostly it remains in the cisterns, needed for religious ablutions.

In his Latin text:

Cataractis enim longe ante urbem intercepte, varieque divisa, tum per campos subarbanos irriganti causa circumducitur, tum in ipsam urbem multis distinctionibus diffluens, aedificiorum impluvia atque intima civium gynaecae penetrat, in aliis ad voluptatem tubis hydraulicis saliens, in plurissimis ad necessitaem stagnans, cisternis excepta, quia his pro eluenda corporis illuvie in Legis observatiame opus est quotidie.[24]

Tavernier gives us a more precise account and measurements:

A trois ou quatre lieues au-dessus d'Ispahan on l'a (rivière Senderu) détourné par plusieurs canaux pour arroser les terres et les jardins, qui sans cela ne produisaient rien.[25]

P. Coste, in his famous plan of 1840, presented a drawing of the "canaux d'irrigation" running east-west through Isfahan parallel to the river, along the southern bank through Julfa, and those near the Hazār-Jarīb; but the place where these channels branched off from the river was outside the scope of the plan.[26] The contributions of A. K. S. Lambton in her *EI²* article "Iṣfahān" and in her article "The regulation of the waters of the Zayande Rud"[27] primarily concern the old legal texts. Thus Kaempfer, whom we have often quoted, remains our primary source, where he states that the channels for Isfahan and the south bank begin *longe ante urbem*. Adam Olearius, who stayed in Isfahan in 1637, wrote about the use of water in the residences and, in particular, in the royal garden which he referred to as the "Tzarbagh".[28] As one can easily see from his plan,[29] the "Tzarbagh" is, in fact, the Hazār-Jarīb: "The garden is square . . . and the water comes from the river Senderuth through subterranean channels" and runs through groves and fountains", . . . "and in the middle [of the garden], there is a lake with a fountain 12 ells (*sc.* 7 m) in height".[30] Olearius's mis-identification was tacitly corrected by Galdieri, who emphasized that the Hazār-Jarīb, with its terraces, cascades and fountains, continued to the slopes of the Kūh-i Ṣūfī in the south.[31]

Here Le Bruyn observed "une chute d'eau, qui tombe des Montagnes".[32] Thus the mountain springs also provided water for the garden.

We have discussed these problems of irrigation in order to achieve a better understanding of the function of the Pul-i Allāh-Verdi-Khān and the Pul-i Khwājū. As far as I can now see, the primary function of the Pul-i Khwājū did not lie in the irrigation of gardens or, if this was a function, it was only such at special times. A contrary opinion has been presented by Pope, Boulanger and Würfel, who state that the bridge served too for "watering many square miles of richest meadow-land . . . on both sides".[33] But that the purpose was primarily artistic is further supported by the Pul-i Jūbī, the aqueduct-bridge, situated between the Pul-i Allāh-Verdi-Khān and the Pul-i Khwājū. The Pul-i Jūbī (dated 1654–5) guided the water several metres above the river level and brought it to the Haft-Dast and Ā'īne-Khāne, the now-vanished palaces on the south bank (Pls. VII–VIII).[34] Thus the Pul-i Khwājū was simply not needed for irrigation, serving instead for carrying traffic across the river, for taming the high water, for people to tarry on it and for the formation of a lake and a waterfall.

We must now review our second concern—the role of water in the garden art of Iran. In a country such as Iran, it would be unjust to view water only as a luxury. Because of the harsh climate, the people have had to develop artifical oases, such as gardens, beside the natural oases, in order to survive. The oldest regular water channels with water pools are found in Pasargadae in the park in which the Palaces of Cyrus were built. 'Alī Sāmī was the first to observe these channels,[35] and more thorough studies were conducted by D. Stronach and a reconstruction drawn by F. Krefter.[36] Here we see the beginning of the Persian gardens with their precise planning and water channels in a rectangular pattern. The garden of

Semiramis may represent an even earlier example. These gardens, near the mount of Bīsūtūn, with a spring in the centre, are documented in Diodorus (II, 13) and may have survived to the time of Alexander the Great (Diodorus, XVII, 110).[37] The garden carpets and the miniatures also illustrate the role of water in garden art, and may best be seen in D. Wilber's *Persian gardens and garden pavilions*.[38]

The artifical waterfall, an element of the seventeenth-century garden, is documented in Isfahan in the shape of the Chahār Bāgh[39] and remains an important element in the eighteenth- and nineteenth-century Iranian gardens. Perhaps earlier examples, again in the Iranian tradition, may be visible in the Mughal gardens, as demonstrated by S. Crowe.[40]

We have one example of a little waterfall in the garden of Ā'īne-Khāne near the Pul-i Khwājū. M. Ferrante suggested that the palace and bridge form a single artistic composition.[41] The water falls softly from the platform of the Talār, flowing to the river, which in itself may create a second waterfall, a hundred times greater, at the Pul-i Khwājū (Pl. VIIIa–b).[42]

CONCLUSION

The art of Isfahan of the seventeenth century, so modern in its time, was always combined with the knowledge of the older traditions. Tavernier, Chardin and Kaempfer saw the three bridges in Isfahan within a historical perspective (Pl. VII).[43] Kaempfer's first bridge, the Pons Maranbuun or Pul-i Marnān,[44] the westernmost, was the oldest, and resembled in the lower parts the Shahristān bridge, possibly a Sasanian foundations with mediaeval superstructures.[45] The second bridge, the Pul-i Allāh-Verdi-Khān, built in the time of Shāh 'Abbās I as an integral part of his new town planning, was a pioneer work combining the elements of bridge and dam.[46] Finally, the Pul-i Khwājū, which replaced an earlier bridge,[47] brought the older tradition of bridge building to a newer and more sophisticated level.

If we try to compare the art of the Pul-i Khwājū with European Baroque art, we find that there are no adequate comparisons. Although European art traditions from Antiquity onwards includes such elements as fountains and waterfalls, the water art of the European Renaissance and Baroque remained within the more restricted concept of garden art and fountains in cities, such as one can see in the gardens of the Villa d'Este in Tivoli and the Fontana di Trevi in Rome.[48]

But an entire river was never used as a tamed element in the hand of an architect; the Tiber would revolt against such a task, as well as the Salzach at Salzburg or the Neckar at Heidelberg! Such rivers could not be tamed. Only the Chateau at Chenonceaux in France, with its six arches built on the foundations of an older bridge in 1512, presents a similar effect; but the waters of the river Cher, an affluent of the Loire, flow quietly.[49]

The Zayandeh-rud in Isfahan, however, had a special character. In the summer it nearly vanished, only to return in the spring with high spirits and vigour. Here an ingenious Iranian architect of the Safavid period saw his chance to surprise the river god in the summer by building his bridge and dam. Then, the following year after the snows melted in the mountains, the great masses of water came and found themselves tamed and transformed into art. Even now one can easily recreate the image and sounds of music, similar in spirit to the water music, which Handel composed for London, being performed in the pavilion in the middle of the Pul-i Khwājū.

But behind the artistic form also lies a religious feeling for the element of water and for the Zayandeh-rud—the "Living River"; perhaps the old veneration of the Goddess of Life and Water, the Anahita of the Sasanian period, has found here a certain echo.

A last remark. The distribution of the waters in a diagonal flow is an interpretation of the architecture of the middle pavilion. The spiritual significance of the axis as intended by the architect becomes visible in the flowing of the waters. We can see this tendency towards a spiritual, "flowing" architecture open to the universe in the Palace of the Hasht Bihisht, dated to about 1670,[50] which is open on three sides to the garden. The same tendency can be seen in the architecture of the Madrasa-yi Mādar-i Shāh of 1708, with its new diagonal axis in the central room and its open octagonal courts in the corners of the main courtyard.[51]

The architecture of Isfahan in the seventeenth century has a baroque character, and although it is in rivalry with contemporary European art, it remains on its own ground. The Pul-i Khwājū of 1650 represents a high point in the "Iranian Baroque", creating great possibilities for the future, a future which, unfortunately, was cruelly destroyed by the Afghan invasion of 1721.

[1] J. Chardin, *Voyages en Perse* (Amsterdam, 1711), VIII, pp. 220ff. with etchings; J. B. Tavernier, *Les six voyages de J. B. Tavernier en Perse et aux Indes* (Paris, 1679), pp. 455ff.; Engelbert Kaempfer, *Amoenitates exoticae* (Lemgo, 1712), pp. 166ff. with etchings (Pol-i Hassanabad is another name for the Pul-i Khwājū).

[2] C. Le Bruyn, *Voyages par la Moscovie, en Perse et aux Indes Orientales* (Paris, 1725), IV, pp. 130ff. with etchings; G. N. Curzon, *Persia and the Persian question* (London, 1892, repr. 1966), II, pp. 44ff.

[3] *A survey of Persian art* (New York, 1938–9, 2nd ed. 1967), III, pp. 1237ff., Fig. 439f., pl. 495; idem, *Persian architecture* (New York, 1965), p. 237, fig. 324; Jane Dieulafoy, *La Perse* (Paris, 1887), fig. p. 327.

[4] *Persien vor 113 Jahren. I Esfahan,* ed. M. Assemi (Tehran, 1976), pp. 98f., pls. 120–5; on E. Höltzer, see J. Scarce, "Isfahan in camera", *AARP* (London, April 1976).

[5] L. Honarfar, *Gan̄gīne-ye āsār-e tārīḵī-ye Esfahān (Historical Monuments of Isfahan)* (Tehran, 1344/1966), pp. 582ff; A. Godard, in *Aṯhār-é Irān,* II (1937), pp. 142ff.

[6] Chardin, VIII, pp. 220ff.

[7] Kaempfer, p. 167.

[8] *Idem, Am Hofe des Persischen Grosskönigs,* ed. W. Hinz (Leipzig, 1940, 2nd edn. Tübingen-Basel, 1977).

[9] Le Bruyn, IV, p. 134.

[10] Curzon, II, pp. 47ff.

[11] E. Galdieri, "L'acqua nell'antico aspetto di Isfahan attraverso le pitture parietali degli ultimi due secoli", in *Gururā-jamañjarikā, Studi in onore di Giuseppe Tucci* (Rome, 1974), pp. 343ff. An offprint of this important study I owe to the author.

[12] *Reise-Beschreibung* (Geneva, 1674), Part II, p. 19 (letter of March 17th 1617); Ingeborg Luschey-Schmeisser, *The pictorial tile cycle of Hašt Behešt in Iṣfahān and its iconographic tradition,* IsMEO, Reports and Memoirs XIV (Rome, 1978), p. 185. For Pul Allāh-Verdi-Khān, see Ferrante, in *Travaux de restauration de monuments historiques en Iran,* ed. by G. Zander, IsMEO, Reports and Memoirs VI (Rome, 1968), pp. 451ff.; K. Würfel, *Isfahan* (Küssnacht-Zürich 1974), pp. 175ff.; *Isfahan, City of Light,* Exhibition Catalogue, London, 1976, p. 79.

[13] *Monuments modernes de la Perse* (Paris, 1867), pl. 3.

[14] *Survey,* III, p. 1237, figs., 439–40, pl. 495.

[15] *Op. cit.,* pp. 451ff.

[16] *Isfahan,* pp. 175ff.

[17] *Op. cit.,* p. 456.

[18] Chardin, VIII, *ad* p. 220; Le Bruyn, p. 132, fig. 104; Coste, *Monuments modernes,* Pls. 50, 51; Flandin et Coste, *Voyages en Perse. VI. Perse moderne* (Paris, 1851), pl. 47.

[19] Höltzer, *op. cit.,* pls. 120–5, especially pl. 120; Godard, in *AI,* II (1937), p. 144; *Isfahan, city of light,* plan p. 76.

[20] M. Dieulafoy, *L'art antique de la Perse* (Paris, 1884), V, pp. 105ff., pl. 12; Bridge at Dezful: pls. 10, 11; C. Reuther, in *Survey,* II, pp. 570ff., pl. 261A.

[21] Pope, Survey III, p. 1228, pl. 261B (tenth century); E. F. Schmidt, *Flights over ancient cities of Iran* (Chicago, 1940), pp. 22f., pl. 24.

[22] Dieulafoy, "Notice sur la construction des ponts en Perse", in *Annales des ponts et chausses,* LIII (1883), pl. 21, figs. 1–5 (not available the author); Pope, *Survey,* III, pp. 1226ff.; Gh. Kuros, *Āb au fann ābyārī dar Īrān bāstān* ("Water and the art of irrigation in ancient Iran") (Tehran, 1971).

[23] K. Bergner, "Bericht über unbekannte achaemenidische Ruinen in der Ebene von Persepolis", *AMI,* VIII (1936), pp. 1ff., pls. 1–8; Schmidt, *Persepolis I* (Chicago, 1953), p. 57.

[24] Kaempfer, p. 165; Hinz, p. 194.

[25] Tavernier, p. 551.

[26] Coste, *Monuments modernes,* pl. 3. His plan is used by N. Ardalan and L. Bakhtiar, *The sense of unity* (Chicago-London, 1973), fig. 123, under the legend "Irrigation", the southern part is excluded. Compare also Höltzer, *op. cit.,* pp. 12f. on the quantity of watering-channels coming from the Zayande-rud above Isfahan.

[27] *BSOS,* IX (1937–9), pp. 663ff.

[28] *Vermehrte neue Beschreibung der Muscowitischen und Persischen Reyse* 1656, repr. Tübingen 1971), p. 561.

[29] *Op. cit., ad* p. 552.

[30] *Op. cit.,* p. 561.

[31] "L'acqua . . .", p. 353.

[32] Le Bruyn, IV, p. 107.

[33] Pope, *Survey,* III, p. 1237; Boulanger, *Guide bleu, Moyen-Orient* (Paris, 1965), p. 812; Würfel, *Isfahan,* p. 175.

[34] Le Bruyn, IV, p. 134, figs. 105, 106; Flandin and Coste, *op. cit.,* pl. 62A, B. (Dessin de Flandin); Flandin, *Relation de voyage* (Paris, 1851), I, pp. 356f. ("plutôt un aqueduc qu'un pont, il se trouve en face du palais de Hapht-Dest, et des cannaux, ménagés dans sa maçonnerie, conduisent des eaux d'Ispahan dans cette habitation"); Godard, *op. cit.,* p. 144; Honarfar, *op. cit.,* p. 114 (date 1654–5); Ferrante, *Travaux de restauration,* 463ff., figs. 35–9; *Galdieri.* "L'acqua . . .", pp. 346, 350, figs. 7, 45, 46.

[35] *Pasargadae* (Shiraz, 1956), p. 75 with plan.

[36] D. Stronach, *Pasargadae* (Oxford, 1978), pp. 107ff., figs. 48, 49. A reconstruction of the whole garden with all the palaces in air-view, executed by F. Krefter, is published by Ingeborg Luschey—Schmeisser in "Kunst, Kultur und Geschichte der Achämenidenzeit und ihr Fortleben", *Akten des Symposion zum 75. Geburtstag von Walther Hinz, Nov. 1981* (= 10. Ergänzungsband von AMI, Berlin, 1983), p. 281, fig. 9.

[37] Paradeisos of Semiramis: see H. Luschey, "Bisutun, Geschichte und Forschungsgeschichte", *Arch Anz* (1974), 114ff. with literature; cf. also W. Eilers, *Semiramis* (Vienna, 1971).

[38] Tokyo, 1962.

[39] Best described and illustrated by Chardin, VIII, 169ff., and best interpreted by Wilbert, pp. 102ff., with his famous and often reused re-reconstruction fig. 30, which is still problematic, cf. Luschey, "Der königliche Marstall in Iṣfahān und Engelbert Kaempfers Planographia des Palastbezirks", *Iran,* XVII (1979), pp. 71–9.

[40] Sylvia Crowe and Sheila Haywood, *The gardens of Mughal India* (London, 1972).

[41] *Travaux de restauration,* p. 453.

[42] After Flandin and Coste, *op. cit.,* Pls. 44 and 43 (Dessin de Flandin).

[43] Tavernier, pp. 455ff.; Chardin, VIII, 220, with etchings; Kaempfer, pp. 165ff., with etchings (Hinz, pp. 195ff.); Le Bruyn, p. 134, fig. 107, seems to be the first person to illustrate the Shahristān bridge.

[44] Kaempfer, p. 166 (Hinz, p. 195), "Regis Thamasi conditus"; Curzon, II, pp. 44f., "Shah Tahmasp"; Godard, *op. cit.,* p. 144, no date; Honarfar, p. 837, with figure, no date.

[45] Godard, *op. cit.,* fig. 52; Pope, *Survey,* III, pp. 1230ff., fig. 437; Honarfar, p. 8, with figure; Würfel, *Isfahan,* pp. 175ff., with figure.

[46] Kaempfer, p. 166 (Hinz, p. 196); Godard, in *AI,* II, p.95; Ferrante, *op. cit.,* pp. 443ff., fig. 1–12; Würfel, *Isfahan,* p. 180.

[47] Ferrante, *op. cit.,* p. 451.

[48] D. R. Coffin, *The Villa d'Este at Tivoli* (Princeton, 1960); C. Lamb,

Die Villa d'Este in Tivoli, ein Beitrag zur Geschichte der Gartenkunst (Munich, 1966), giving the tradition of water art from Roman Antiquity to Moorish Spain. On the Fontana di Trevi, see L. Bruhns, *Die Kunst der Stadt Rom* (Vienna, 1951), pp. 493ff., fig. 440; G. B. Colonna, *Römische Brunnen* (Berlin, 1941).

[49] G. Kaufmann, *Die Kunst des 16. Jahrhunderts*, Propylaeen-Kunstgeschichte Bd. 8 (Berlin, 1976), pp. 366f., Abb. 367, pp. with literature; G. Boekhoff, *Paläste, Schlösser, Residenzen* (Brunswick, 1971), pp. 142f.

[50] Pope, *Survey*, III, pp. 1196ff., fig. 421; Ferrante, *op cit.*, pp. 399ff., with 30 figures; Wilber, *Gardens*, p. 111, figs. 45, 46; Luschey-Schmeisser, *The pictorial tile cycle of Hašt Behešt*.

[51] Pope, *Survey*, III, p. 1213; Godard, *L'Art de l'Iran* (Paris, 1962), p. 361, fig. 226; Würfel, *Isfahan*, pp. 166ff, with several figures.

Pl. Ia. The Pul-i Khwājū in spring 1963. Author's photograph.

Pl. Ib. The Pul-i Khwājū 1684/85. Etching in E. Kaempfer, Amoenitates exoticae, *1712.*

Pl. II. The Pul-i Khwājū 1963. Author's photograph. The missing balustrade in photo-montage.

Pl. IIIa. The Pul-i Khwājū before 1938. Photograph of A. U. Pope.

Pl. IIIb. Vaults of the Pul-i Khwājū. Drawing in Jane Dieulafoy, La Perse, *1887.*

Pl. IVa. The water art at the middle pavilion, view to the south. Author's photograph.

Pl. IVb. The water art at the middle pavilion, view to the north. Author's photograph.

Pl. Va. The Pul-i Khwājū. Author's photograph.

P. Vb. The Sasanian bridge and dam of Shushtar. After M. Dieulafoy.

Pl. VIb. Longitudinal view under the bridge from south to north.

Pl. VIa. Water channel under the bridge with the sluice-gate.

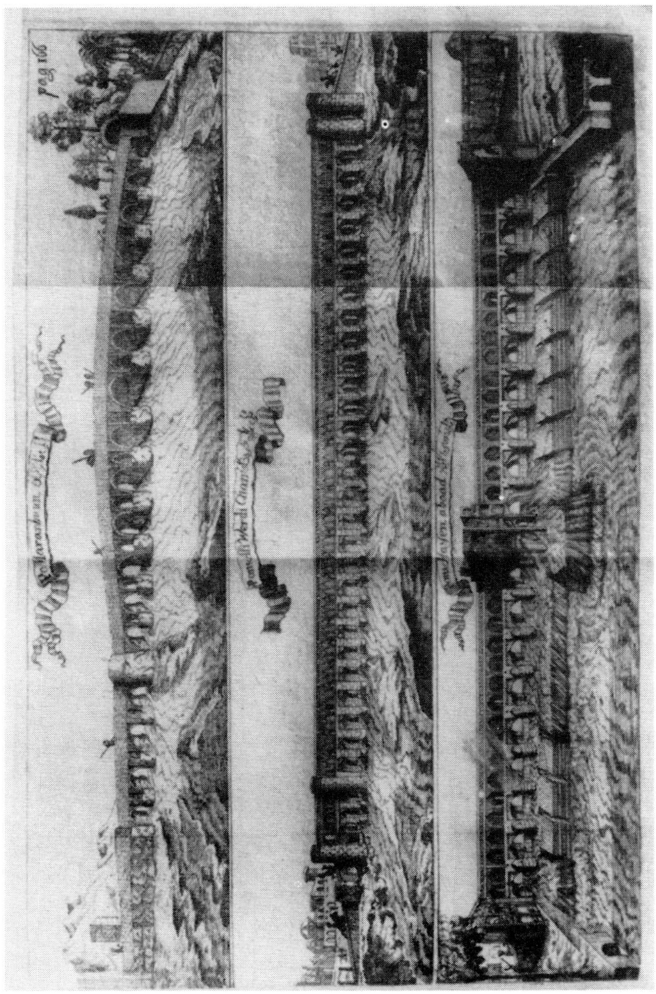

Pl. VII. The bridges of Isfahan 1684–5. Etching in Kaempfer, Amoenitates exoticae.

Pl. VIIIa. Ā'īne Khāne and Pul-i Khwājū. Drawing of Flandin.

Pl. VIIIb. Ā'īne-Khāne. Drawing of E. Flandin, 1840.

9
Berwick-on-Tweed Bridge

J. Summerson

On the accession of James VI of Scotland to the English throne, the garrison and the fortifications of Berwick became redundant and although at least one voice was raised in a plea for Berwick's continued maintenance as a nursery of valour and an internationally renowned model of discipline, common sense prevailed.[1] King James ordered the garrison to be kept on full pay only till Christmas 1603. After that it would be reduced to a company of a hundred foot under a captain.[2] The governorship of Berwick was already in abeyance; Lord Willoughby had died in 1601 and had not been replaced. Nor, apart from the treasurership, were the lesser administrative offices continued. Of the rank and file, some of the veteran soldiers, gunners and artificers remained in Berwick to be pensioned or put on half pay. In the king's first orders for reducing the establishment the office of Surveyor of Works was, with many others, to be extinguished.[3] In fact, however, it survived, nominally in the person of the aged, and probably absent, William Spicer but actually in that of his deputy, James Burrell. Spicer dying, it seems, early in 1604, Burrell succeeded him.[4] The officers of master-mason and master-carpenter did not survive. The former had been held by Spicer, conjointly with the surveyorship, and now lapsed; the latter was held by Leonard Fairley who was paid up to Christmas 1603 but not thereafter.[5]

In the spring of 1604, Berwick received a new charter from the king who supplemented it with an order to the Captain of Berwick making it clear that the mayor, bailiffs and burgesses were to take possession of the royal buildings in the town.[6] These included the palace and its appurtenances, the governor's house, Burrell's tower, the Maison Dieu and various other buildings—everything, it appears, except the actual fortifications. The responsibilities of a crown surveyor of works were thus reduced almost to nothing. From 1604 only one building project occupied the Berwick scene; but it was a majestic one—the building of a stone bridge across the Tweed.

A stone bridge to supersede the old and insecure timber structure had long been in the minds of those concerned with Berwick. It was mentioned in the course of a conference on the fortifications held as early as 1564.[7] The resurgence of the idea at

[1] SP 14/5, no. 4. [2] SP 14/5, nos. 5, 11 and 19. [3] SP 14/5, no. 5.
[4] See above, p. 626. [5] See vol. iii, p. 417, of this work.
[6] J. Scott, *Berwick-upon-Tweed; the History of the Town and Guild* (1888), p. 329.
[7] See above, p. 657.

a moment when the military importance of Berwick had evaporated but when the union of England and Scotland under one Crown rendered a new bridge symbolically and practically valuable was largely due to one man—George Home, Earl of Dunbar. An able and often ruthless administrator, Home came immediately into prominence in the new reign as the king's chief adviser on Scottish affairs. He had been High Treasurer of Scotland since 1601 and in 1603 became Chancellor of James I's Exchequer. In the following year he was created Baron Home of Berwick in the English peerage and in 1605 Earl of Dunbar in the Scottish peerage.[1] At Berwick, James granted him in absolute right all the lands which did not belong to the freemen, and in 1607 resurrected the office of Governor in his favour. He was always at Berwick when he was not at Whitehall.[2] He was instrumental in obtaining the speedy grant of the charter of 1604 and was the chief agent in promoting the building of the stone bridge.

It was probably on a hint from Dunbar that on 21 March 1605, the Lord Treasurer directed the Treasurer of Berwick to pay to James Burrell 16d. a day for life as surveyor of Berwick bridge.[3] This was in addition to the small fee of 12d. a day he was already receiving as Surveyor within the reduced establishment.[4] Further, on 30 March he received the grant of a pension of £40 a year.[5] His prime responsibility was for the maintenance of the old bridge whose survival was a necessity until a new stone bridge should be built. On 15 February 1607/8 the old bridge met with disaster. Ice coming down the river on a strong current carried away ten of the timber piers. The bridge was under repair at the time with three hundred volunteers, beside hired men, working on it. None was hurt. Burrell put a lighter to work on a ferry service till the bridge could be mended. Recounting all this in a letter to the Earl of Salisbury he stated his view that the only real solution to the bridge problem was the erection of a new bridge in stone.[6] To support this view he enclosed a statement of the expenditure on the bridge in the preceeding thirty-seven years, amounting to £5372 15s. 2½d.

A few days after Burrell's communication to Lord Salisbury, the mayor of Berwick, Robert Jackson, wrote to the Earl of Dunbar, then at Court, asking for his help. Dunbar replied promptly that he would petition the Privy Council for money for a new stone bridge. He coupled the proposal with that for a church which had been lacking in Berwick since the old church had been removed for the fortifications in Henry VIII's time. He himself had already collected £1000 for this purpose. The result of Dunbar's petition was that the king granted, by indenture dated 23 May 1608, to James Baylie, a burgess of Berwick, as accountant, £10,000 in debts and arrears of rents between 1485 and 1600, together with a moiety of £18,000 of debts remaining due to the Crown in any one year between 1485 and 1585.[7]

The money coming in under this arrangement was administered by Dunbar himself in whom the capacity of Paymaster was joined with that of Governor. Up to 1611, £3347 4s. 2½d. had been collected, of which sum £894 had been spent, with

[1] J. Scott, *op. cit.*, pp. 189–90; article on G. Home, Earl of Dunbar, in *D.N.B.*
[2] E 351/3487. [3] SP 38/8 (21 March 1604/5). [4] E 351/3483.
[5] SP 38/8 (30 March 1605). [6] B.M., Lansd. MS. 166/86.
[7] SP 14/27, no. 22; SP 14/32, no. 52; E 351/3585.

£774 11s. 9d. owing to workmen.[1] However, on 29 January 1611, the Earl of Dunbar died suddenly at Whitehall. This was a great shock to Berwick. He had dominated the place and given it hope for the future after the loss of its old affluent status as a garrison town. A great house he had been building for himself in the town was still unfinished.[2] Above all, he had been the champion of the new stone bridge. On 6 February, Sir William Bowyer, the Captain of Berwick, wrote anxiously to Salisbury stressing the late governor's belief in the national as well as local importance of the undertaking and the extent of his preparations.[3] He wrote again (4 March) expressing the view that the bridge was so important to the town that the townsmen if they could, 'wold rather build it them selves than want it'. He hoped that in the building of it there would be none of the abuses which prevailed at Berwick in Queen Elizabeth's time, when money was squandered on needless officers and clerks and idle overseers. The town itself, he thought, would furnish all the administration necessary with only one 'special clarke' in charge. As for James Burrell, who in this year was Mayor of Berwick for the second time, Bowyer was not entirely confident of his capacity to rule so exceptional an undertaking;

> ffor Mr Borrell trewlly my Lord I know him to be a very honnest man and yet I thinke in all his tyme he never had hand in Suche a worke I meane of Suche diffecultye or Import by Reasone of the nerenes of the Sea and the quicke Retorne of the watter. It will require the Best experimented and Ingenious freshe sperits to work without lose as not to do and undoe. Your L will fynd his Sufficiencye by confference who I think can do better than he can speake and yet in this perchance promis more then he shalbe able to perfforme being a worke of Suche hassard.[4]

Bowyer added that the Lord Chancellor and Lord Advocate of Scotland would soon be on their way to Court and would have in their party a certain James Murray who was 'the surveyor and builder' of the great house which the Earl of Dunbar had been building in Berwick. He was honest and 'of a notable Ingenious speritt' and might be worth Salisbury's consideration.[5] Murray was in fact the Master of the King's Works in Scotland, an office he had held since 1607.[6]

Enclosed with this letter was an estimate signed by James Burrell, for a masonry bridge of six pillars, seven arches and two landstalls taking the bridge only over the deepest part of the river, the remainder to be built of timber; the figure was £5440 10s. 4d.[7] A few days later the recorder and an alderman, with Burrell in their company, left for London to attend the Privy Council.[8] On the evening of their departure the old bridge again collapsed, losing two of its timber bays and prompting further letters from Bowyer and the mayor, begging Salisbury to consider the urgency of the new project.[9] A further estimate was submitted for a stone bridge of thirteen arches, 'as appeareth by the modell', to cost £5000.[10] But the final estimate, sent to London by the hands of the recorder and an alderman in

[1] E 351/3375. [2] SP 14/61, no. 27 [3] *Ibid.*
[4] SP 14/62, no. 5. [5] *Ibid.*
[6] H. Colvin, *Biographical Dictionary of British Architects 1600–1840* (1978), p. 567.
[7] SP 14/62, no. 5. [8] SP 14/62, no. 15.
[9] SP 14/62, nos. 17, 18 and 19. [10] SP 14/62, no. 20.

April, was for £8462 8s. 4d., over and above the money already provided. This estimate was signed by Burrell as 'Bridge Maister'.[1]

It would be clear to Salisbury that if the bridge was going to be built at all it would need more ready money than could be got from the collection of old debts. Accordingly, he conveyed to James Baylie, the accountant for the collecting process, the king's order to surrender the indenture of 1608 for cancellation. The king was 'unwilling that so important a work should rely on old debts slowly recoverable'. While he regarded the figure in the estimate as 'overmuch' he nevertheless directed the Lord Treasurer, by warrant dated 16 May 1611, to pay to the mayor, bailiffs and burgesses of Berwick, £8000 in instalments. First, £2000 for the work in hand, including the £774 odd already owing; second, £1000 a year, or more at the discretion of the Lord Treasurer and Chancellor of the Exchequer. Timber would be available from the royal forest of Chopwell, near Gateshead. The cash total of £8000 was under no circumstances to be exceeded.[2]

No time was wasted. Only eleven days after the date of the warrant Lord Salisbury sent to the mayor, bailiffs and burgesses of Berwick a list of orders for the conduct of the works. These, in brief, were as follows. No more money was to be expected. Workmen and materials were to be obtained at the best rates and work undertaken at the best times of year, with materials punctually assembled. Two selected burgesses were to have the daily oversight of materials; and the mayor and eight selected aldermen were to see the workmen paid weekly and to sign the books. George Nicholson, paymaster of Berwick, was to be present at the pays, and to sign the books; also to be made acquainted with all proceedings for the king's satisfaction. The Earl of Salisbury was to be regularly informed of the state of the works so as to acquaint the king therewith and to receive an annual report. Finally, James Burrell, described here as master mason, was to have 2s. 6d. a day for overseeing the works, 'as he received in the time of the Earl of Dunbar', with 1s. 4d. a day for a clerk to keep the books.[3]

George Nicholson's position needs explaining. The Earl of Dunbar had, as we have seen, combined the office of governor with that of paymaster and in 1609 delegated the paymastership to Nicholson who, after Dunbar's death, continued to hold it in his own right.[4] In 1611 he had taken upon himself to advise the king that no successor to Dunbar would be needed as the garrison was, in the course of nature, dying off.[5] The king took the same view but effected a further economy by merging the paymastership and surveyorship. Nicholson executed both offices till his death in or around 1631.[6] James Burrell was thus eliminated from the garrison establishment and salaried only in respect of the bridge and out of the money paid by the Crown to the Corporation. The pay proposed for him by Salisbury was the same as what Spicer had had for exercising the joint offices of master mason and surveyor.

Work on the bridge started on 19 June 1611, and a detailed account of the expenditure was duly forwarded by the mayor and corporation to Lord Salisbury at

[1] SP 14/63, no. 42. [2] E 351/3585. [3] *Ibid.*
[4] E 351/3490, 3491 [5] J. Scott, *op. cit.*, p. 191. [6] *Ibid., loc. cit.*

the end of September.[1] The document gives a complete list of the labour force, of which the following is a numerical summary:

Masons	31	Sawyers	2
Slaters	2	Smiths	9
Carpenters	12	Quarrymen and wallers	22
Ships carpenters (mending the lighters)	6	Labourers	76
		Boys	9

In the larger classes the wages are carefully graded. The most costly workers were the leading mason, Lancelot Bramston or Branxton and the leading carpenter, Roger Richardson, both at 1s. 8d. per day. Labourers were paid 6d. and 5d. Burrell had his 2s. 6d. a day as prescribed in the warrant, with a clerk, John Shell, at 13d—3d less than the prescribed wage. Further expenses included carriage of stone by water from the quarry at Tweedmouth, cooperage, cobblers for mending bellows, press and conduct money, iron, steel, and rope. The total spent in the period of the account (14 weeks) was £317 5s. 0½d. Attached to the account is a neat drawing, perhaps by Shell, the clerk, of the masonry in hand at the first pier, signed by Burrell; with a postscript stating that he has devised a new type of 'ginne' (or engine; obviously in this case a pile-driver) hoisting a ram of nine or ten cwt.. and operated by five men, instead of the usual twenty.[2]

It appears that the administration of the work required more clerical help than what was allowed for in Salisbury's directions, for the Berwick corportion raised a levy on its members, differentiated according to status, to pay for extra functionaries.[3]

The mayor and corporation forwarded a second account to Lord Salisbury on 21 January 1612.[4] By this time £562 5s. 3d. had been spent and it was realized that the balance of less than £700 would not be sufficient to see the bridge though the spring and to lay in materials for the summer. Salisbury was, accordingly, petitioned to advance a further £1500 at once, together with 100 tons of timber.[5]

Actually, a further £600 was paid in the Michaelmas term, 1611, and sums of £400 and £500 in the following Easter and Michaelmas terms. After that, there were five regular payments of £500 and, in 1616, two of £1000.[6]

Work on the bridge proceeded without incident, the corporation regularly sending one of their number to the Court in London to report progress and carry back the cash. By May 1617 the whole of the sum of £8000 had been spent. Salisbury, now dead, had ruled this to be the outside limit; but the bridge was by no means finished. Salisbury's successor could not extend the aid without royal authority and the corporation therefore petitioned the king. With some pride, they stated that 'not one penny of what came to yor peticoners hands hath bene any way mispent' and they enlarged on the greatness of what they had in hand—'a great worke, yea such a one as beinge finished, they do not thinke that any Kinge in the worlde ever did a more worthie and memorable worke, then this wilbe to yor Matie'

1 SP 14/66, no. 55. 2 SP 14/66, no. 55 (i). 3 J. Scott. *op. cit.*, pp. 413–14.
4 SP 14/68, nos. 21 and 21A. 5 SP 14/68, no. 22.
6 351/3585, B.M., Lansd. MS. 166/84 is a summary of the account for 1613.

Their humble suit was for further annual grants up to £4000 or £5000.[1] The former sum was almost immediately sanctioned, the first grant of £1000 being paid in the Easter term, 1618, the second in the Michaelmas term and two more in the following years.[2]

By the beginning of 1620, when £3000 had been paid out, the Council began to feel some anxiety about the rate of progress. A letter was dispatched (7 February 1620) to local justices who were commissioners of the Border, instructing them investigate the expenditure of money and look also into the spoils of timber from Chopwell Wood.[3] The justices duly reported in a letter of 5 March.[4] Their opinion, reached with the help of an independent surveyor whom they took with them to Berwick, was generally favourable. They commented doubtfully on the large sums spent on journeys to London to obtain the grants; but the construction work, they said, was 'workmanlike and substanciallie well wrought' and some recent flood damage to the brandreths of three piers could be repaired for £20 or thereabouts.[5]

Berwick bridge came again before the Council on 5 April.[6] They had now before them an abstract of the corporation's accounts up to 31 December 1619, prepared by the auditors of the Exchequer.[7] In this document the auditors assumed that it might be desired to take the bridge works out of the hands of the corporation. If that were done, they would endeavour to make up the account in detail; otherwise they recommended that the account should run until completion of the bridge.[8]

The Council was concerned at the expensive running of the bridge works but agreed that it must be finished. The question was how soon and for how much money. Letters were again sent to the local justices asking them this time for an estimate of the cost of completion. At the same meeting of the Council, it was ordered that the commissioners of the Treasury direct the Bishop of Durham to strike tallies for £1000 to be employed on the bridge in the current year. This was the final instalment of the £4000 promised.[9]

The Bishop of Durham becomes, at this stage, the main agent in the business. Richard Neile was a man of fifty-eight who had occupied the see for some three years, having previously held those of Lichfield and Lincoln. He was an unscholarly prelate but of great business capacity, close to the king and with considerable experience of building.[10] He knew the situation at Berwick very well, having been there earlier in the year and, in fact, having already communicated his own views and rough estimate to the Council through one of the clerks.[11] On 17 July 1620, the Council wrote asking him, as a 'person of speciall quallity and place' in the North, to take the finishing of the bridge under his special care so that neither money nor materials would be misapplied.[12] Six days later they wrote again asking him to make it his business to deal with a case of assault on the mayor and sabotage of the bridge which had been reported from Berwick.[13]

[1] SP 14/109, no. 70. [2] E 351/3585. [3] SP 14/113, no. 6. [4] *Ibid.*

[5] *Ibid.* 'Brandreths' were the horizontal timber grids laid over the piles and carrying the masonry, but here the term may include the masonry starlings.

[6] *A.P.C., 1619–21*, p. 170. [7] SP 14/113, no. 6 (ii–iv). [8] SP 14/113, no. 45.

[9] *A.P.C., 1619–21, loc. cit.* [10] Article on Richard Neile, *D.N.B.*

[11] SP 14/112, no. 57. [12] *A.P.C., 1619–21*, p. 254. [13] *Ibid.*, p. 262.

The bishop undertook the Council's commission to set the affairs of Berwick bridge in order with characteristic vigour. After some conference in London with the Chancellor of the Exchequer and also, apparently, with Inigo Jones, he went up to Berwick in August. At this date, seven of the fourteen arches, now proposed, starting from the north, had been turned; the eighth was in preparation. The 'landstall' on the south side was also in progress, with the first arch and pier adjoining. There were still five more arches to build. The bishop, in a long and detailed letter to Greville (22 August), expressed his disapointment.[1] He found, after a careful survey, that while the mayor and his brethren had been 'carefull and faithfull' in their proceedings, there were various factors operating against early completion. Immediate factors were the slow delivery of timber from Chopwell (where, we know from another source, there was some maladministration by the royal bailiff)[2] and the effects of exceptional floods in the previous year. More serious, in the bishop's view, was the way in which the work was conducted. 'I do utterly dislike the spinninge of it out either by dayes worke or parcellinge it to taske for soe the overseers of the worke and the workmen will delay and gaine time of us doe we what we can'. The bishop's remedy for these defects in procedure was to complete the bridge by contract in gross:

> I fell into a project of compoundinge for the whole worke to bee well and substantially performed by a sett time for a certaine some of money and soe much more Tymber as necessarily must be hadd for the perfectinge and finishinge the whole worke and if such a course may stand with his Maj. and your Hons good likinge I will proceed in it.

The contract proposed by the bishop was to be based on priced quantities. He specifies some of the prices in his letter, observing that 'Mr. Surveyor of his Maj^{s} Workes' had previously seen them and considered them more reasonable than he could have expected. He estimated that the bridge might be completed for a further £1200 and 200 tons of timber. It would be passable by vehicles by midsummer 1621 and complete with its paving and parapet walls one year later.[3]

Greville did not send a reply till 12 September and had probably been taking technical advice on the feasibility of completion by contract. He agreed in principle but warned against work being skimped and stressed the need for a capable overseer. The contract should contain safeguards particularly as regards setting of piles, making foundations and the type and sizes of stones.[4]

Meanwhile, the bishop had followed up his earlier letter with a detailed estimate on which a contract could be based[5]. This amounted to £2073 19s. 5½d. with 250 tons of timber—substantially more than the first estimate. But the figure had been reached with the advice of the other commissioners and after consultation with workmen. He pressed Greville for a prompt indication of the official view, so that he could conclude agreements before leaving Berwick on 9 October. The communication crossed with Greville's letter indicating approval in principle. It brought back

[1] The bishop's letter is quoted *in extenso* in E 351/3585 and in Lincoln Record Office MS. M.M., vi/10/1/3.
[2] SP 14/115, no. 113. [3] E 351/3585. [4] *Ibid.*
[5] A copy of part of this estimate is in Durham University Library (uncatalogued Add. MS.). The whole text is incorporated in E 351/3585.

a prompt reply. The king had been consulted and was agreeable to the the extra cost in money and timber. His Majesty expected that with the bishop's care the work would be finished 'in all pointes fitt for soe Royall a Monument of blessed union betweene the two Kingdomes.' The letter concluded with the rather less than princely hope that strict bargaining 'in that cheap country' might have the effect of pricing the work down to within the bishop's first estimate.[1]

The way was now clear for drawing up a contract. The bishop must have delayed his departure for a few days, for the document was executed on 12 October. It is in the form of an agreement between the Bishop of Durham and five other members of the bridge commission of the one part and, of the other, James Burrell, alderman and surveyor, the designer of the bridge, and Lancelot Branxton, 'mason and burgess', the chief mason. Burrell and Branxton were to erect the five missing arches and six piers towards the south side and to make the bridge fit for traffic by midsummer 1621; and to complete the parapet walls and cobble paving in the following year. For this they were to receive £1750, to be paid in instalments as the work advanced. Burrell and Branxton were to enter into a bond of £2000 with the king to keep the terms of the contract 'without fraud or delay'.[2]

To ensure the technical feasibility of the contract, the bishop had with him at Berwick the Newcastle bridge master, John Johnson. He then installed Johnson at Berwick, paying him 2s. a day out of his own purse, to supervise the work. With the sharp incentives contained in the contract and the presence of Johnson's critical eye, the bridge began to move rather faster. It is not clear whether completion of the arches was achieved by midsummer but it certainly was by Michaelmas. Then came disaster. In October 1621 the Tweed was flooded as never within living memory. Floating hay-stacks and loose timber bore down the old wooden bridge, part of which broke away and with the accumulated debris launched itself against the stone bridge. The new masonry being still green and the centering of some of the arches still in place the assault was irresistible and the whole of the year's work was undone in a few hours.[3]

Neither the domestic state papers nor the acts of the Privy Council tell us what action was taken at this unhappy juncture. There was no general provision against storm and flood in the contract, though as Burrell was held responsible for making good any damage which might occur in the winter of 1620–21 it may be assumed that no liability lay against him in respect of subsequent mishaps. All we know is that on 25 June 1622 a warrant issued to pay to the mayor, bailiffs and burgesses of Berwick £3000 for finishing the bridge, any surplus after completion to be set towards the building of a church.[4] The final account for the bridge shows that the works were re-started on 8 April 1622, £873 13s. 6½d. being spent between that date and 21 March following. Thereafter, expenditure steadily declines. By 1625–6 it is down to £85 13s. 8½d. and we may assume that the bridge was then virtually complete. The account was not closed, however, till 24 October 1634.[5]

The total cost of the bridge in materials, freight and wages between 1611 and 1634 is given in the account as £12,522 1s. 9d. In addition, however, there was a

[1] E 351/3585. [2] *Ibid.* [3] *Ibid.* [4] *Ibid.* [5] *Ibid.*

sum of £1000 paid by the Bishop of Durham to Burrell and Branxton, which we must suppose related to the work performed by them under contract before the disaster of October 1621. Further, there was an award of £20 a year for five years to Sir William Bowyer who had been Mayor of Berwick continuously from 1620 to 1623 and again in 1625. This recognition of his special care of the bridge was evidently obtained for him by the bishop, whose directions as to the award were dated 14 July 1627.[1] Bowyer did not live long to enjoy it; he was buried at Berwick on 12 November following.[2]

There was an allowance to the corporation of nearly £500 for twenty-three years expenses. This, with some costs of repairs thrown in at the last moment, brought the final account to £14,960 1s. 6d. The total of exchequer grants was £15,000. There was thus only £39 18s. 6d. left for the church. The account was declared by Sir Robert Jackson, acting for the corporation, on 20 July 1637.[3]

Of the structural history of the bridge something may be learned from the 'particular payments' which survive, with gaps, through the period 1611–20 and fragmentarily for later years.[4] Details of piling for the 'pillers' (piers) and 'brandreths' are supplied in an undated document signed by James Burrell and Lancelot Branxton the master-mason. The piles were of oak, varying in length from 6 feet to 23 feet and 12 inches to 18 inches square. They were fitted with wrought iron shoes. Eight hundred and seventy-three oak trees were used for the purpose, nearly all being felled in Chopwell forest and shipped to Berwick from Newcastle.[5]

The stone seems all to have been obtained locally. A quarry at Tweedmouth was rented for £6 13s. 4d. a year and this was the main source, though stone was sometimes fetched from 'Ethermouth'. A recurrent entry is 'wyning, loading and unloading of Burrs,' which presumably means roughly dressed stones. Other masonry items include 'scew table', 'smouth penns', 'rough penns', 'smouth coyne heads', 'cantes' and 'ashler'.[6] An uncommon masonry procedure is suggested (August 1620) by the following: 'Ashler with tenner and mortis', 'springers with tenner and mortis' and 'making 8 mortis Joyntes'.[7] In the account for 1–14 August 1618 we have a reference to architectural ornament:

> To Willm Crowe and his ptnrs for. . . the length of the 2 Dorica pillers 14 foote at x^{d} the foote xjs. viii d. the 2 Capitalles with there ordinances about them 13 foote at x^{d} the foote being wrought from the 20 July till the first August x^{s} x^{d}.[8]

A further reference to a 'Dorica piller' occurs in the account for 19 June to 20 July 1619.[9] If this Doric order was associated with one or more of the piers it no doubt

[1] *Ibid.* [2] Berwick-on-Tweed Burial Register. [3] J. Scott, *op, cit.*, p. 415.

[4] The muniments of the Corporation of Berwick contain four books of accounts for the Bridge, Br A/1 consists of weekly pay-books from 29 Sep. 1616 to 29 Sep. 1617. Br A/2 contains similar pay-books from 27 Jan to 24 Dec. 1612. Br A/3 contains paybooks from Jan. 1615 to Dec. 1620, with some accounts of 1667–9. The fourth book, which has no identification other than its title, 'Bridge Account Book,' contains pay-books from 8 Aug. 1622 to 26 June 1635. In the muniments of the Massingberd-Mundy family in the Lincoln Record Office are pay-books for 1618–19 (M.M. vi/10/2) and for 1619–20 (M.M. vi/10/3) as well as other documents relating to the Bridge (M.M. vi/10/1 and 4).

[5] Lincoln Record Office, MS. M.M. vi/10/1/2. [6] *Ibid., passim.*

[7] Lincoln Record Office: MS. M.M., vi/10/1/3.

[8] *Ibid.* M.M. vi/10/1/2. [9] Ibid. M.M. vi/10/1/3.

perished in 1621, for no recognizably Doric ornaments appear on the bridge as finally completed.

In later accounts are references indicating the conclusion of the enterprise. In December 1626 and again in July 1627 are entries for paving the bridge and finishing the 'vamor' (parapet). In September and October 1626 are various references to 'the intended tower. . . to be built on or about the middle of the bridge' and in the pay-book for 28 October to 3 November is a payment of £6 to William Crowe for building the west side of the intended tower. Where the tower was to be situated and whether it was completed is not known.[1]

In the pay-book for 14 July to 20 July 1627, the signature of James Burrell appears for the last time. It is very shaky. He was to survive, however, for a few more years. He had been admitted a free burgess of the Corporation in 1604. Besides designing the bridge and directing its construction from first to last he had served as an alderman and filled the office of mayor of Berwick in 1609 and 1611.[2] He was buried at Berwick on 20 January 1631.[3]

The bridge having been completed, its maintenance became a matter of concern. In 1633, the Corporation petitioned Charles I to allow them an annual sum of £50 for this purpose out of the money supplied for the pensions of the old garrison, now a fast diminishing body. The petition was granted. In 1645–6 the bridge suffered a breach and a petition to the House of Commons resulted in a special grant of £500. By this time the annual grant of £50 instituted by Charles I had ceased payment and the Corporation in 1653 petitioned Parliament for the sequestered estate of Lord Mordington. This they obtained, enabling them once again to put the fabric in good order, while at the same time Parliament approved an annual grant of £120. This was paid regularly until the Restoration.[4]

[1] Berwick Corporation Archives, 'Bridge Account Book'.
[2] Berwick Corporation Archives, MS. GM 7(9), f. 36; J. Scott, *op. cit.*, p. 479.
[3] Berwick-on-Tweed Burial Register. [4] J. Scott, *op. cit.*, pp. 415–16.

10
Hollow spandrels in arch bridges: a historical study

E.C. Ruddock

Synopsis
In the history of masonry bridges pride of place is usually given to the arches, with much discussion of their structural behaviour, but in fact the designers' greatest problems were always in the foundations and they also gave serious attention to the spandrels and the road. This paper is devoted to two methods of hollowing the spandrels which were introduced into British bridge building during the eighteenth century, both of which solved some problems of foundation and arch design as well as improving the spandrels.

Gautier's *Traite des Ponts* **1714**

Gautier's book, published in Paris in 1714, was the first comprehensive treatise on bridge-building in any language and was quoted as the standard authority by British engineers and architects for at least 50 years. As well as a history of bridges and practical guidance about all aspects of construction, Gautier gave empirical design rules for the proportions of arches, piers, etc. In particular, he advised that the ratio of span to thickness of an arch should not be greater than fifteen. He also included a short passage on the 'eyes of bridges', that is, voids or tunnels through the spandrels of multispan bridges, and plates of several Roman bridges, the medieval Pont St. Esprit and the Renaissance Pont Neuf at Toulouse, all with an 'eye', usually cylindrical, through each spandrel over the middle of the pier. The advantages of eyes are stated: they 'relieve the structure of much of its weight, save masonry, and make passages for floodwater'.[1]

Westminster Bridge, 1736-49

The first public proposal to make cylindrical voids in the spandrels of a bridge in Britain may have been stimulated in part by Gautier's book, but it contains a notable advance on his ideas. The design was for Westminster Bridge (Fig 1); it was made by Batty Langley and explained in two pamphlets published in 1736[2] and 1737[3]. It showed a large cylindrical

2 HOLLOW SPANDRELS IN ARCH BRIDGES

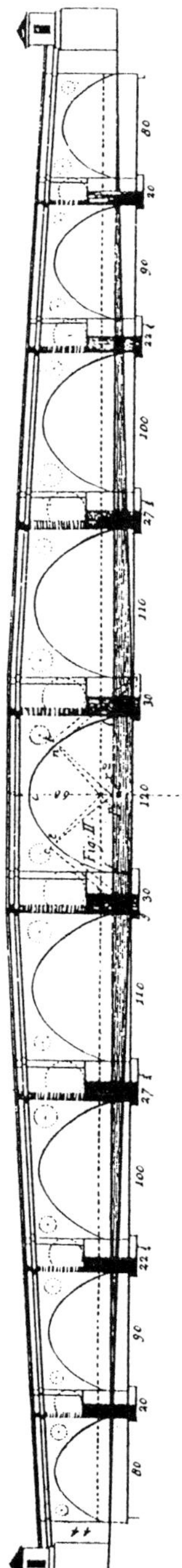

Fig 1. Batty Langley's design for Westminster Bridge[2].
(By courtesy of the British Museum)

void over the centre-line of each pier, low down against the haunches of the arches, and two further voids higher up, one over the extrados of each adjacent arch. The spandrel walls were to continue over the ends of all the voids, which shows that the motive of providing extra passages for floodwater was absent. Langley gave two reasons for the voids. One was that 'the weight of each spandrel will . . . be greatly abated', which is equivalent to Gautier's first reason for eyes; but the second reason, which appears to be the chief one, is more sophisticated and quite new:

> 'As the spandrels over the side arches [i.e. all the arches except the middle one] . . . have an unequal pressure on them, the highest side of each being most heavy, their opposite parts are unequally pressed; and if the inward arches next the shores first decay, all the others unto the middle arch will be in danger of falling down . . by making the cylinders in the upper spandrels, proportionably larger than the cylinders in the lower spandrels, their pressures will be equal, and therefore free from the danger of unequal pressure'.[2]

In a review of the published designs for Westminster, John James[4] argued, in effect, that the stability of an arch was not so sensitive to unequal loading as Langley inferred, and this was certainly true both of Langley's design and all the other published designs. But while the current theory of 'equilibrium' of an arch was much too abstruse to be understood or used by common bridge designers, Langley's idea of equalizing the loading on an arch and/or the thrusts on a pier (he seems to have considered both and perhaps confused them) by forming unequal voids in the spandrels was clear, practical and undoubtedly valid. In a design which was approaching the limits of stability it could be used to good effect.

Westminster Bridge was built to the design and under the direction of Charles Labelye. The arches were semicircular in shape, 1·07 m (3 ft 6 in) thick, and backed by great curved wedges of masonry increasing in thickness from about 1·07 m (3 ft 6 in) at the crown to almost 3 m (10 ft) at the haunches[5] (Fig 2). Labelye wrote later[6] that 'these arches taken together, do form a kind of arch, which can be demonstrated to be *in equilibrio*, in all its parts'. He believed that 'the lighter an arch is, in proportion to its piers, or . . . the heavier the piers, and spandrels are, in proportion to the arch, the firmer the arch will be', and he promised to publish full calculations of the 'thrust or lateral pressure of various kinds

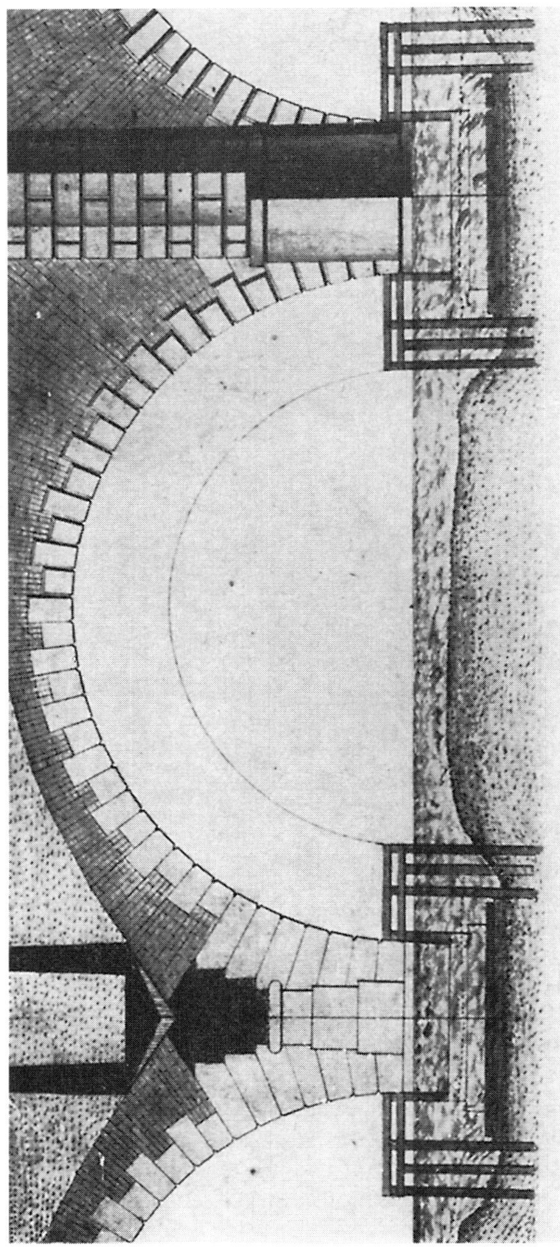

Fig 2. Drawing inscribed by Smeaton 'Copy of one of the arches of Westminster Bridge given by Mr. Labelye to Mr. Ellicott'[10].
(By courtesy of the Royal Society)

of arches', including those he made for Westminster Bridge[6]; but he never did so. Over the arches of Westminster Bridge each spandrel was divided by two walls across the bridge and two along it into nine compartments, and these were filled with gravel.[5]

When the bridge was almost finished in May 1747 a serious settlement of one of the piers began and the two adjacent arches were distorted and cracked. After supporting the arches on centres, Labelye began loading the pier with all available dense material 'as the best way to try it, and to secure it for the future', and intended to increase this loading to 2000 tonne (2000 ton). The Bridge Commissioners, however, were deluged with opinions from many other people and by the time Labelye's loading had reached 700 tonne (700 ton) they were convinced that the proper remedy was to reduce the weight over the pier, not to increase it. They therefore instructed him to remove the excess load, take down the two arches, and submit proposals for a repair by lightening the spandrel; and he did so reluctantly.[6]

Of the many schemes of repair submitted to the Commissioners, three must be mentioned. One was by Dr. William Stukeley, a noted antiquary and Fellow of the Royal Society, who afterwards claimed that his proposal was the one adopted by the Commissioners. He proposed to extend a very flat arch from crown to crown of the two damaged arches so as to form, with the halves of those arches rising from the next piers, a roughly elliptical arch spanning right over the settling pier and so relieve it of all load except the self-weight of the existing half-arches which sprang from it.[7] The space between these half-arches and the new arch would be entirely void. This scheme was actually not possible because the top of the new arch would have risen at least 1 m (3 ft) above the level of the road. The method actually used formed a void over the settled pier by a segmental arch, or 'counter arch' as Labelye called it,[6] which sprang from points on the main arches well below the crowns, thus subjecting them to very concentrated loads which they seem to have borne without any distress; the thick backings of 'secondary arches' were only partially replaced and the space between the 'counter arch' and the rebuilt main arches was left void (Fig 3).[5] Thus the total amount of load both on the main arches and on the pier and its foundation were much less than before the settlements, which accomplished the chief aim of Stukely's proposal without following it strictly.

A proposal by William Halfpenny, architect,[8] had the same

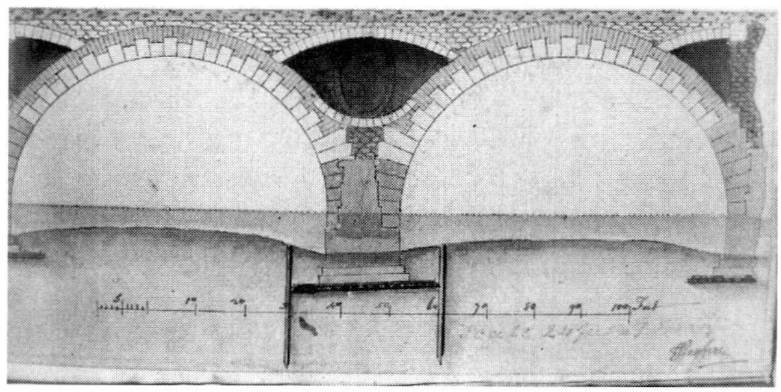

Fig 3. Longitudinal section of hollow spandrel of Westminster Bridge, drawn by Thomas Gayfere[5]. (By courtesy of the Institution of Civil Engineers)

intention, but would have achieved it by a different method. Four frames of timber would be made, extending from back to back of the rebuilt arches and tied to the masonry by iron cramps as the arches were built. Then thin brick walls would be built up on both sides of each frame spaced 100 mm (4 in) from it and the spaces within the frame and between the two walls filled with pounded charcoal, which will preserve the wood many ages'; finally the wider voids between the brick walls would be covered by small arches, that is, longitudinal vaults, to carry the pavement. The drawing of this scheme was engraved and published (Fig 4). Batty Langley knew of it[9] and John Smeaton possessed a copy.[10]

A third proposal submitted to the Commissioners was by Smeaton himself. He considered that the pier must be entirely removed and rebuilt and so confined his memorial to the problem of making a water-tight cofferdam, giving no details of the subsequent spandrel construction.[10] The substance of his proposal is therefore irrelevant to the theme of this paper, but the evidence that he had studied the failure and at least Halfpenny's scheme for repair is important.

Theory of arches, 1736–58

Men whose skill was not in bridge design but in mathematics had been writing about the form and the lateral thrust of arches since the last years of the seventeenth century. Most of the early writers were French and it was almost certainly from their works that Labelye, a French Swiss, derived the methods of calculation which he used for the arches of Westminster Bridge in the late 1730s.

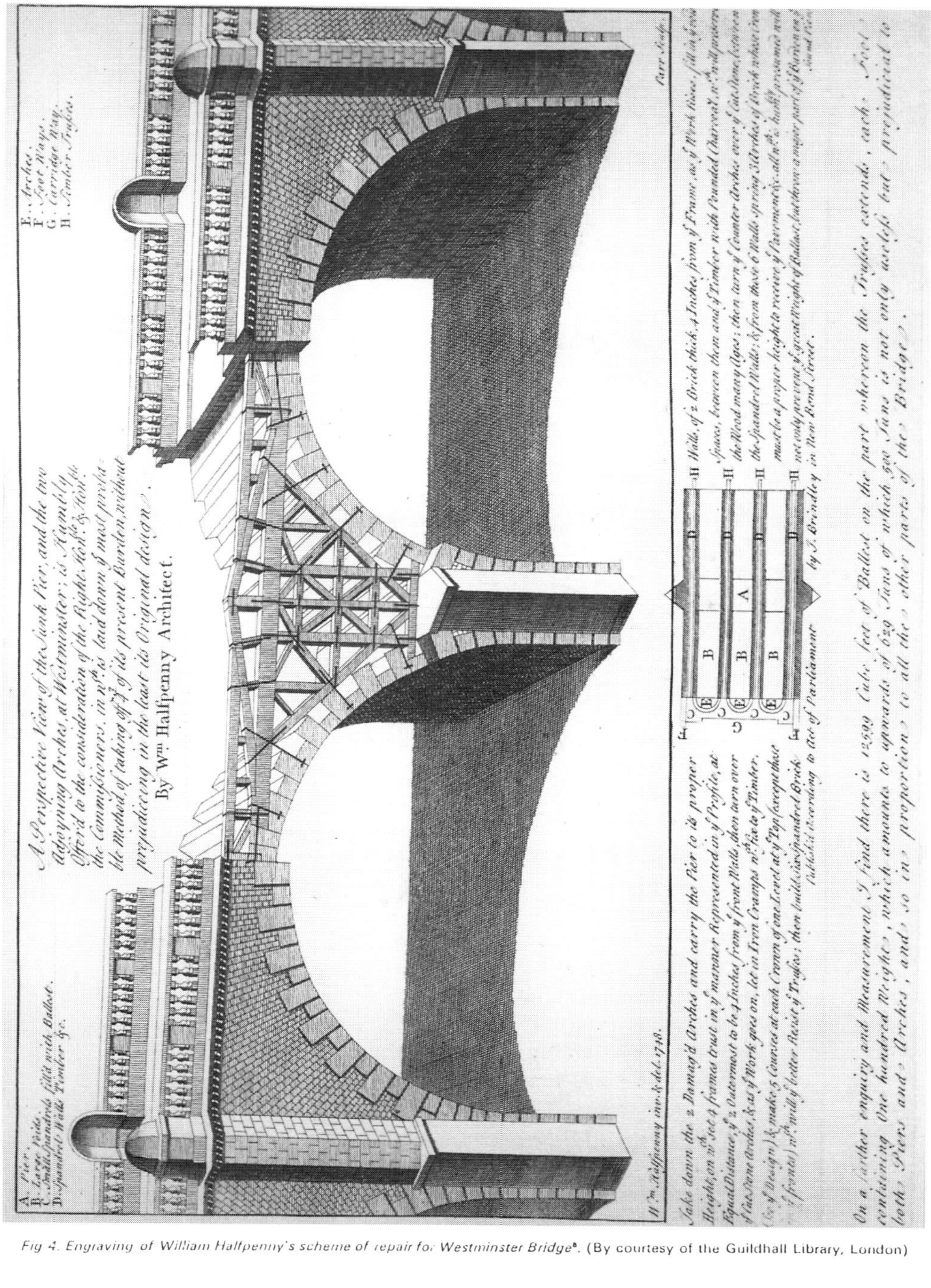

Fig 4. Engraving of William Halfpenny's scheme of repair for Westminster Bridge[8]. (By courtesy of the Guildhall Library, London)

In Britain William Emerson gave a method in his *Doctrine of Fluxions* (1743) for finding the shape of the wall which must surmount any given arch for the arch to be *in equilibrio*[11]; this is equivalent to making the line of thrust coincide with the soffit. John Muller had done the same in his *Mathematical Treatise* (1736).[12] Applications of the method to semi-circular and semi-elliptical arches and all others in which the curve is vertical at the springings showed that the top of the spandrel or the filling material over the arch should start parallel to the curve of the soffit at the crown, thus falling at first towards the haunches, but then rise steeply as the soffit approached the vertical. This was on the assumption of constant density of material in the arch and spandrel.

In later books both men tried to make this theory yield practical arch forms. In the second edition of his *Principles of Mechanics* (1758)[13]—but not in the first edition (1754)—Emerson first suggested the use of materials of varying density to make the top of the spandrel work out as something like a horizontal line, but concluded that suitable materials could not be found for the semi-circular arch. He then stated his findings that the spandrel top was approximately horizontal if the arch was a segment of 90° arc and the crown thickness 16 per cent of the radius; or if the arch was 120°, the crown thickness 7 per cent of the radius and the material about halfway down the arch on each side of half the density of that at the crown and the springings. After giving similar rules for a catenary arch, he gave a table of ordinates for a soffit curve which would ensure 'equilibrium' with a horizontal spandrel top, span of 18·3 m (60 ft) and crown thickness of 1·07 m (3 ft 6 in), and noted that this curve was particular to the given crown thickness. (There was one fairly minor error in this section which he corrected in later editions.) He added that the setting-out and cutting of stones for an irregular curve, such as that defined by the table, need not be difficult in practice, but betrayed again his essentially mathematical viewpoint in his final statement that any other curve 'must in time give way, and fall to ruin', while the curve he had derived,

> 'sustaining everywhere a quantity of pressure proportional to its strength, will never give way, so long as the piers, which are its bases, stand good; but by virtue of its figure, will stand firm and unshaken, as long as the materials the arch is made of, will last'.[13]

It is clear from the text that Emerson knew at this time that for a given soffit curve his theory would yield a range of different curves for the spandrel top, each achieving equilibrium with a different thickness at the crown of the arch, and he stated this and explained it when he published the derivation of the rules and the table in 1776.[14]

Muller, in his *Practical Part of Fortification* (1755)[15], concentrated on the problem of making the piers large enough to resist the horizontal thrust of arches of various shapes, and found the semi-elliptical curve much preferable to the semi-circle. He also derived the curve of the top of the spandrel for a semi-circular soffit from the same theory as Emerson, but because some mathematical errors made his result appear reasonably practical he found no disadvantage in arches which are vertical at the springing and had no reason to propose irregular curves as Emerson did.

Pontypridd, 1756

The story of William Edwards's four attempts to bridge the Taaf at Pontypridd in the years 1746–56 was first published in an article by his friend Thomas Morgan in *The Gentleman's Magazine* in December 1764.[16] The authorship of this article, and hence its historical authority, was established in an able survey of documents by Williams in 1945.[17]

The third bridge failed on 7 November 1754.[18] It was a single arch of 42·8 m (140 ft) span (by one account 44 m (144 ft))[18] and apparently 10·7 m (35 ft) rise, and had been completed six weeks earlier.[19] Thomas Morgan described the cause and manner of failure as follows:

> 'The quantity of matter in the *crown* of the arch was but little in proportion to that which was necessary to be laid on the *abutments* in order to make the ascent easy. The weight of this matter caused such an inequality of pressure on the arch, that in about a year's time, it crushed that stupendous pile, and it fell again to the bottom.'[16]

In 1804 B. H. Malkin, after a meeting with William Edwards's son David, described the failure thus:

> 'The arch was finished, but the parapets not yet erected, when such was the pressure of the unavoidably ponderous work over the haunches, that it sprung up in the middle, and the keystones were forced out.'[20]

An account by Thomas Telford, written in 1812, is obviously derived from Malkin's but he added a statement by William Jessop that Edwards sought advice from John Smeaton after this failure.[21] In a recent paper[22] it is shown that this statement is inaccurate but not implausible; and here it can be shown that by 1754 Smeaton had considerable knowledge relevant to the problem at Pontypridd and serious aspiration as a bridge designer.

He wrote in 1760[23] that he had watched the founding of the piers of Westminster Bridge (in the 1740s); he had also obtained drawings from the site of the internal construction of the arches and spandrels (Fig 2) and the arrangement of wedges by which the centres were struck[10]; as noted above, he had studied the problem of the settling pier and himself proposed a solution; and he is virtually certain to have read the pamphlets and designs published in 1735–7, including Batty Langley's. He had begun sketching designs for Blackfriars Bridge in 1753 and produced a full design in 1754.[10]

Of special interest is a small list of bridge dimensions and proportions[10] which he wrote out, very likely in these years of study, and clearly in an attempt to establish what were the safe proportions and forms of arches by his own observation, as an alternative to accepting the printed rules of Palladio, Gautier or anybody else.

One of the arches he listed was the middle arch of Ouse Bridge at York, described as 'two segments that meet in a very obtuse angle'. In the report which condemned this bridge to demolition some fifty years later, Thomas Harrison explained the shape as follows:

> 'The curve of this arch was . . . undoubtedly designed for a segment of a circle; but perhaps from the centers upon which it was built, giving way from the pressure of the arch, or from the piers not being sufficiently strong to resist the lateral pressure, it altered its form, and rose above the real line of the circle at the springing and sunk below it on each side towards the apex or crown of the arch.'[24]

This explanation is illogical, but the shape Harrison observed and which is shown in many eighteenth century prints was clearly the same as that seen by Smeaton, and the true explanation of the distortion must be that the high spandrels had forced the arch to deform in the same way as did the third Pontypridd (weakness of the centering might have been a contributing factor, but not the primary cause). The third Pontypridd, however, lasted only a few weeks before breaking at the crown, whereas the arch at York lasted more than two centuries.

Segmental arches of two other bridges on Smeaton's list, namely Tanfield Arch (now called Causey Arch) near Newcastle-upon-Tyne and Newton Cap Bridge at Bishop Auckland, exhibit similar distortions today (Newton Cap has shown it at least since 1825[25]) and amongst the distorted arches of medieval bridges there were probably other examples which Smeaton would have seen. It is even possible that upward breaking at the crown was a recognized mode of failure in arches. One other collapse by this mode is on record, in the bridge of Witton-le-Wear, in 1763[25].

Whether Smeaton interpreted the failure of the third bridge at Pontypridd by reference to the shape of arch he had observed at York (and probably elsewhere) is an unanswerable question. What is more or less certain is that the *cause* of the shape of the York arch was the same as that of the Pontypridd failure.

Smeaton's list when first written contained measurements of eleven bridges (two others were added later) including three with arch spans greater than 27·5 m (90 ft). One of these was the very flat cycloid arch of Santa Trinita Bridge in Florence and another (at Newton Cap) was a circular segment of 120° arc, while six of the shorter arches listed were segments of 120° or less. Therefore the large majority of the entries were relatively flat arches, and the seven 120° or less segments were arches for which Emerson's rules[13] would yield practical shapes of spandrel to keep the arch *in equilibrio*. The ratios of span to crown thickness were listed and for nine of the arches this ratio was greater than 30; but for all which had a ratio greater than 36 Smeaton also noted the existence of 'counter-arches', by which he meant second arches laid directly on top of the first to reinforce them, like Labelye's 'secondary arches' at Westminster, but of constant thickness.

The list displays a range of knowledge which Smeaton could have applied to arch design and suggests some principles of his approach. The only explicit statements of a design philosophy we have from him were written, not before 1756, but in 1760. Concerning the arches of his second design for Blackfriars Bridge, made in 1759, he claimed to have

> '. . . chose such a figure for them as having no tendency to vary their original form, either by their own gravity, or the incumbent load . . .'

and deprecated

> '. . . a great load of rubble work necessary in the haunces, in order to preserve the form of the arch when unnatural.'[23]

The arches he was defending were circular segments of 120°, a form he used in almost all his early bridge designs.[10] As mentioned above, Emerson approved this form, though only with a specific crown thickness and a varying density of material, and there is a striking similarity, both of ideas and expression, between Smeaton's statements and those quoted above from Emerson's book.

William Edwards's arch, in his fourth bridge at least, was a segment of 106°. His answer to the problem of instability shown by the third failure was to make three open cylindrical voids of 2·75 m (9 ft), 1·83 m (6 ft) and 1·22 m (4 ft) diameter, and reputedly a further hidden semi-circular void, through each spandrel, to place large timbers between the voids and the main arch, and to use charcoal for spandrel filling over and between the voids.[22] The thickness of the arch was 0·76 m (2 ft 6 in) or less,[22] giving a ratio of span to thickness of at least 56. This bridge has stood for over 200 years. It has probably never been subjected to even moderate traffic loads because the steepness of the approaches made it impractical in use; but since traffic loads were so low as to exert no influence on bridge design at the time it was built, it must be judged a successful structure because the arch has carried the weight of the spandrels and parapets without noticeable distortion.

As the mode of collapse of the third bridge bore a clear relationship to earlier known arch distortions, so the form of the fourth related clearly to earlier events, designs and theories; but as to how this came about there is no evidence except Telford's statement that Smeaton was consulted. An important and well-known forerunner to the Pontypridd voids was the large void over the settled pier of Westminster Bridge. Of the several different proposals for hollowing this spandrel one, namely Halfpenny's, suggested the use of charcoal filling,[8] though not primarily for its low density. Again, there was some similarity of purpose and a close similarity of form between the secondary cylindrical voids over the haunches of the arches in Batty Langley's design for Westminster Bridge[2] and the voids used by Edwards. Langley's motive was the equalization of lateral thrusts on each pier of his multi-arch bridge and/or the achievement of perfect 'equilibrium' (in modern terms, stability) of each arch. Edwards's motive was to obtain a distribution of loading on the crown and haunches of his single arch which would make it stable. The common concern with the stability of arches, however these two men understood the phenomenon, and their use of cylindrical voids to achieve it, establish a relationship between the two designs which it is difficult to consider fortuitous.

Thirdly, the problem of stability which Edwards solved practically was precisely the same as one of those solved theoretically by Emerson and Muller in their first books. The theoretical curve for the top of the spandrel (or road surface) for a 106° arch with the crown thickness of Pontypridd cannot be derived from the simpler rules in Emerson's *Mechanics* (1758), but it can be derived from his statement that 'the height . . . must be as the cube of the secant of the arch', a statement fully explained in the earlier *Fluxions* (1743).

The derivation requires that the crown thickness be given, and the best assumption that can be made now about the thickness of the arch, fill and road-metal at the crown of Pontypridd when first built is 1·068 m (3 ft 6 in). This is measured from Smith's survey drawing made in the 1830s,[26] and the assumption yields the curve T_1, in Fig 5. However, since spandrel height is actually used in Emerson's theory to represent the total load over any point of the soffit curve, the volume of the parapets, averaged across the width of the arch, should be added to the crown thickness. Using Smith's dimensions of the parapets, this gives a total crown thickness of 1·403 m (4 ft 6 in) and the resulting curve of spandrel top is curve T_2 in Fig 5.

The actual vertical load is also shown in Fig 5, as an equivalent spandrel height. Curve *A* is the line of the road surface given by Smith, curve *B* shows the averaged loading including the parapets and curve *C* shows the actual loading allowing for both parapets and voids. If curve *C* were smoothed it would lie very close to the theoretical curve T_2.

The closeness of Edwards's design to the requirements of Emerson's theory is remarkable. There were a number of mathematicians in Britain who could probably have made the calculations before the bridge was built if they had been asked to do so; of architects and engineers only Smeaton is at all likely to have had the skill after Labelye had left the country about 1752.[27] But the fact that neither Smeaton nor any mathematician is known to have claimed credit for such a resounding success of scientific method is strong evidence that theory was not used. A much less rigorous derivation from Langley's design and/or Labelye's repair at Westminster seems much more likely.

London Bridge improvement, 1758–9

In 1758–9 two arches of the old London Bridge were taken down and a new segmental arch of 21·4 m (70 ft) span and 7·02 m (23 ft) rise was built across the double opening, the intermediate pier being then removed to provide a wide passage through the bridge.[28] When this arch was being demolished in 1831 Edward Cooke made a sketch[29] which

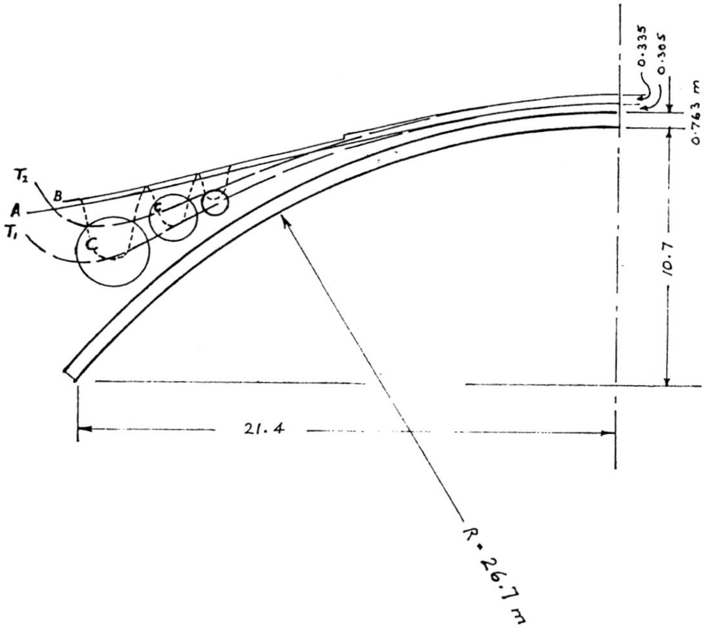

T_1 Spandrel top required by Emerson theory for crown thickness 1.068m

T_2 Spandrel top required by Emerson theory for crown thickness 1.403m

A Actual road surface (after Smith).

B Actual road surface adjusted to include weight of parapets averaged across arch width.

C Adjustments to B over voids to give actual distribution of load per unit width of arch.

Equation for theoretical lines $h = \dfrac{aR^3}{(R-z)^3}$,

where h = height from soffit to top of spandrel,
a = crown thickness,
R = radius of curvature of soffit,
z = vertical ordinate of soffit below level at crown.

Fig 5. Half elevation of Pontypridd with lines of equilibrium load and actual load

shows that there was a large hollow in each spandrel formed by two vertical walls running transverse to the bridge, one just over the springing of the arch and the other some way up the extrados, and a barrel arch from one to the other under the road surface. It cannot be inferred with certainty that these hollows were made when the arch was first built, but it seems very likely. Speculation about the precise motives of the designers, George Dance and Robert Taylor, is unjustified, but one obvious benefit was a reduction of load on the piers and foundations. The effect on the stability of the arch would be similar to that of Labelye's 'counter arch' over the settled pier at Westminster.

Other drawings made by Cooke in 1831 show some small barrel vaults within the older spandrels of London Bridge, most of them running transverse to the bridge but one being parallel to its length.[29] The drawings show no order or design in the positions of these vaults, which presumably covered void spaces. Although they were almost certainly there before 1756 it is most unlikely that any of them was made for a structural purpose. They are more likely to have been first made as storage vaults or cellars for the houses and shops on the bridge. Indeed, vaulted voids in bridge spandrels for various purposes were probably common enough. On London Bridge itself, for instance, there had been a crypt in the spandrel below St. Thomas's Chapel, probably ever since the first construction of the bridge in 1176–1209; another instance was a prison cell in one spandrel of the bridge at Inverness which was in use at least as early as 1718.[30] These voids are irrelevant to a history of the hollowing of spandrels for structural benefit, except as a practical proof to eighteenth-century designers that voids could exist, at least in some circumstances, without prejudice to the stability of a bridge.

Bristol Bridge, 1759–62

Knowledge of William Edwards's success at Pontypridd seems to have spread very slowly. The first recorded visit by an architect was that of James Bridges who went there in 1759[31] and was impressed by several things, among them the echo under the arch and the smallness and flatness of the stones. He considered that the arch-stones in a large bridge must be large and wedge-shaped.

Of four designs which he submitted the following year for the new Bristol Bridge,[31] the one he favoured was a single arch of 39·7 m (130 ft) span; but in describing it in writing he made no mention of hollow spandrels, while he proposed in an alternative design of three semi-circular arches to make 'a *concave horizontal cylinder* in each spandrel, each pier, and in each abutment', the cylinders in adjacent spandrels to be of different sizes, and he gave the reasons for these cylinders in exactly the words of Langley's pamphlet of 1736.[2] In a third design of three semi-elliptical arches he proposed cylinders in the piers and abutments, but not in the spandrels.[31]

Two years later it was John Wood the younger, with an elliptical arch of 34·8 m (114 ft) span,[31] and Ferdinando

Stratford, with a circular segment, first of 48 m (160 ft) and later of 45·8 m (150 ft),[33] who favoured single-arch designs. Stratford's design actually showed a span of no less than 61·7 m (202 ft) between the real springings of the arch on the rock deep under the river banks. Nothing is known of the spandrel construction intended by Wood, but Stratford intended the spandrels to be 'filled up with rubble stonework, laid in a particular manner to ease the pressure on the haunches of the arch', and his published drawing shows 'land-breast counter-arches' within the spandrels.[33] The form of these is not clear, but they are not at all like William Edwards's open cylinders.

Stratford sent his design to John Smeaton for his comments in November 1762 and received encouragement for the large span.[34] But the bridge eventually built was a three-arch bridge which owed nothing to Pontypridd or Batty Langley and not a great deal to any of the designs of Bridges, Wood and Stratford.

Blackfriars Bridge designs 1759–60

Concurrently with the start of the controversy in Bristol, over 50 designs were submitted late in 1759 for the new Blackfriars Bridge in London.[35] Of the 11 designs of which some details are known, only one, that of George Dance, proposed voids in the spandrels.[36] The voids were cylindrical but hidden inside continuous external spandrel walls and, according to a pamphlet by Robert Mylne,[37] were proposed only to save money. This claim to cheapness, which Langley and Bridges also made, was rightly doubted by Mylne, since the arch round a void would cost more than gravel to fill it.

Two late proposals for Blackfriars Bridge show the continuance of at least mild interest in hollow spandrels, of two very different kinds, for multi-arch bridges. One was a design by Stephen Riou in his *Short Principles of Stone Bridges* (1760)[38] with an 'eye' over each pier very similar to those in Gautier's illustrations of Roman bridges. The other was the publication in *The Gentleman's Magazine* of August 1760 of a plate by William Stukeley showing how his proposal for the Westminster Bridge repair could be extended to form a bridge of 'interlaced arches' right across the river.[7]

Developments in Scotland 1762–84

In 1762 a design for a bridge with 'eyes' came from a completely new source. The trustees for Coldstream Bridge, over the Tweed between England and Scotland, brought to Parliament in February a petition for financial aid and exhibited a design by Robert Reid, a Haddington mason.[39]

The design had five arches, and over each pier was a large masonry ring with four keystones on the vertical and horizontal diameters, and the circle within it washed black.[10] They also had an alternative design by John Smeaton, with a very different treatment of the spandrels.[10] They obtained their grant and initially adopted Reid's design, but in April or May 1763 they consulted Smeaton again and in two further designs made in July he copied Reid's ring and black circle over every pier.[10,40] Neither in Reid's nor in Smeaton's designs were these rings low enough to form useful passages for floodwater in any spate of the river which was then foreseen.

One of Smeaton's designs was adopted immediately and the spandrels were being built, under Reid's direction, by May 1765. He then asked whether the eyes were to be 'left open or filled up with rubble', and Smeaton replied that they should be filled up.[40] The question must have been with reference to the holes in the external spandrel walls (Fig 6); the spaces within were filled with material described many years later by Sir John Rennie as 'loose earth which extended down nearly to the springing of the main arches'.[41]

This is the first time Smeaton is known to have designed a bridge with rings on the spandrels. It is fairly clear that he copied them from Reid's design, presumably to please the trustees, and these eyes had nothing to do with any of the practical motives of the Romans and the French, of Batty Langley or of William Edwards; they were pure decoration.

A few weeks before Smeaton made his designs with the decorative rings, a preliminary drawing of a new bridge in Edinburgh was made by James Craig, an architect who made the plan for Edinburgh's New Town three years later. His drawing of the bridge was engraved in August 1763, and shows rings on the spandrels, with the holes closed by masonry.[42] Twenty-two designs were submitted in competition for this bridge in February 1765, and it is likely that the winning design by David Henderson, who had worked at Coldstream in 1763, had open cylinders through the spandrels.[43] The purpose must have been only to reduce weight, as the bridge was across a dry valley.

Smeaton went on to Perth immediately after the acceptance of his design for Coldstream, in August 1763. His design and estimate for the bridge at Perth were completed in June 1764,[44] and the design had rings on the spandrels just like those at Coldstream.[10] The quantities in the estimate,[34] together with later references in the Commissioners' Minutes,[44] make it clear that he intended the spandrels to be filled solid

Fig 6. Coldstream Bridge today

with rubble masonry to 1·83 m (6 ft) above the spring of the arches and with gravel over that.

The next significant events did not take place until four years later. Within that time the spandrels were built at Coldstream and filled with gravel or earth; at Bristol, Perth, Edinburgh and doubtless many other places, the building of piers and arches, and at Bristol and Edinburgh spandrels too, went on by conventional methods. At Blackfriars Bridge in London, which was well advanced, with some spandrels built, by 1764, the filling of spandrels with rubble and gravel continued.[45]

Then in May 1768 Smeaton was asked for advice on a foundation failure in the new bridge at Dumbarton. While the centres were still standing under the last two arches built, the pier between them had settled and tilted so much that both centres were crushed and both arches broken. Smeaton reported to the magistrates in a single sentence of 194 words ![34] He observed the need 'to make the superstructure as light as possible' and showed, in a sketch plan of the spandrel, external walls 0·61 m (2 ft) thick, two parallel internal walls 0·46 m (1 ft 6 in) thick, and three voids of 1·07 m (3 ft 6 in), 1·83 m (6 ft) and 1·07 m wide (Fig 7). He believed that his

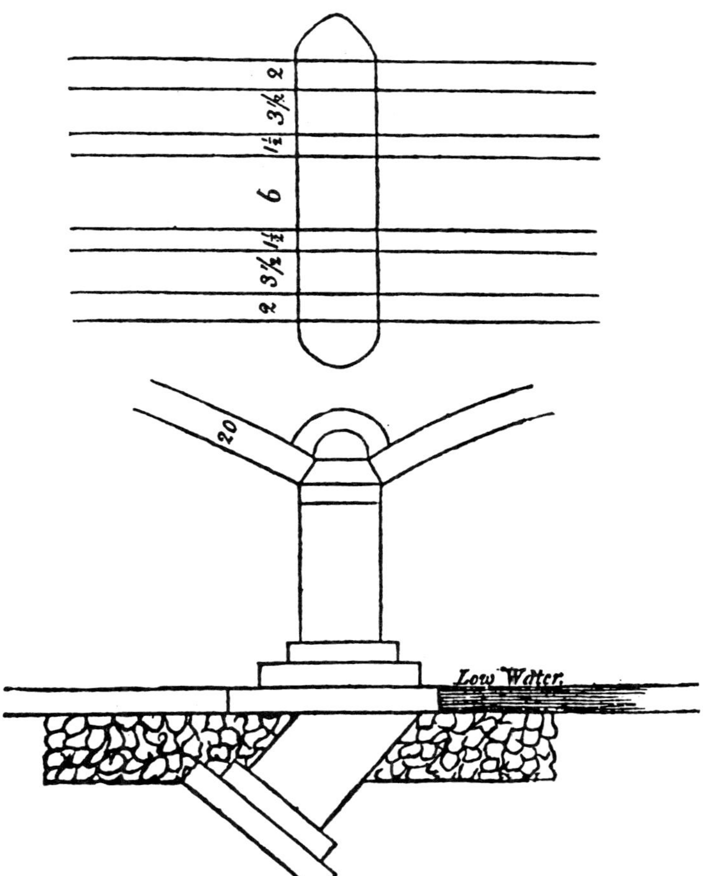

Fig 7. Sketch of damaged pier in Smeaton's report on Dumbarton Bridge[34] (1768)

advice was followed, although this is unconfirmed and there is some doubt about it. It is suggested that William Mylne, designer and builder of the Edinburgh bridge, had some part in the repair at Dumbarton,[46] and within a month or two of Smeaton's report there Mylne designed a new bridge over the Clyde at Glasgow.[47] The river bottom was very soft, under a thin crust of gravel,[34] and over each pier Mylne designed a large open cylinder which must have been placed there chiefly to reduce weight, since most of the area of the cylinders was above the level shown as 'highest water after great rains'. The bridge was built with these voids and stood until 1833.

Two things happened in the following year, but unfortunately evidence is lacking as to which of them happened first. On 3 August 1769, there was a serious collapse of part of the south abutment of the Edinburgh bridge. Smeaton, who was at Perth inspecting his unfinished bridge there, was called on for advice, which he gave later in the month and again early in 1770.[10,34,43] Also in 1769, but at an unknown date, he made a drawing showing a method for building the spandrels of Perth Bridge with longitudinal voids and parallel walls,[10] as he had recommended for Dumbarton, but with thicker walls and narrower voids. The voids, 1·32 m (4 ft 4 in) wide, were arched over with pointed vaults of good rubble masonry and the whole tied over the points of the vaults with iron 'chain-bars' (Fig 8[10]).

There is no positive evidence that Smeaton's decision to use this method was stimulated by the failure of the Edinburgh bridge, a structure groaning from the great weight of its gravel filling and bulging from the lateral pressure inside its spandrel and approach walls; but the conjunction of dates is very suggestive. The Perth spandrels were built according to his design in 1770.[44]

At Perth, therefore, as at Coldstream, the external rings which suggest Roman or French 'eyes' were purely decorative; but the internal construction of the spandrels at Perth was a solution of the structural problems which had troubled Batty Langley and William Edwards. The voids served to reduce weight on the piers and foundations (Langley) and the parallel walls formed stiffening diaphragms which would make it virtually impossible for any arch to change its shape or become unstable (Langley and Edwards). This stiffening was not in fact new. It was provided in most of the arch bridges of earlier times by the external spandrel walls which, because the bridges were so narrow, were not much farther apart than any two of the parallel walls at Perth. The

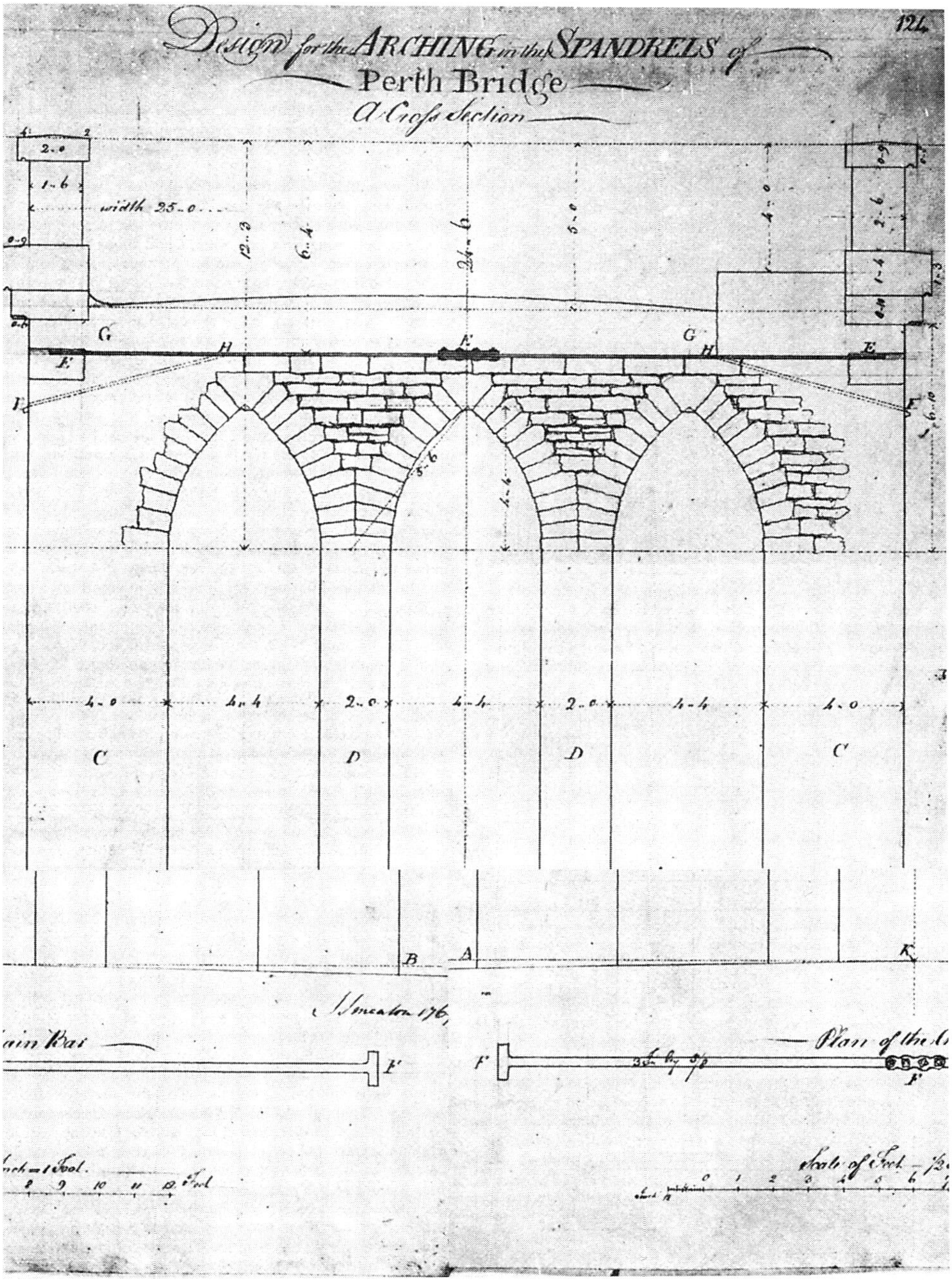

Fig 8. Design for hollow spandrels of Perth Bridge[10] (1769) (By courtesy of the Royal Society)

internal walls suggested at Dumbarton and used at Perth are the first known attempt in Britain to extend the stiffening deliberately across wider arches, and to combine it with a lightening of the spandrels. The method corresponds in general form to the proposal of William Halfpenny for hollowing the spandrel of Westminster Bridge,[8] but Halfpenny's walls, of timber and brick, would have been lighter (and more expensive) than walls of rubble masonry.

In his report with John Adam and John Baxter on the failure at Edinburgh, Smeaton recommended 'any method of arching in the spandrils between the great arches that will effectually save weight', and this in order 'as well to relieve the piers, with their foundations [one of Langley's motives], as to prevent the lateral pressure of the earth against the spandril walls'.[34] The latter is a motive not stated before, unless something of the kind is implied in the vague note on Stratford's single-arch design for Bristol quoted above.

The choice of a 'method of arching in the spandrils' at Edinburgh was presumably made by William Mylne in concert with his brother Robert who stood security for him, and it followed his design for Glasgow rather than Smeaton's method at Dumbarton and Perth. The cylindrical voids were over 6·1 m (20 ft) in diameter, perhaps the largest ever made, and ringed by barrel arches of hewn masonry. They were entirely hidden by the external spandrel walls.[43]

It was probably the accident at Edinburgh that induced the designers most closely involved to make hollow spandrels in a number of subsequent bridges although they were more expensive and at least one bridge committee raised objections.[44] John Baxter designed an open cylindrical void over each pier of Kenmore Bridge in 1772[48] and Robert Mylne did the same over the middle pier of his two-arched bridge at Inveraray in 1773.[49] In both of these there may have been some thought of providing extra flow passages for floods, but David Henderson, in his high viaduct over the Pease Burn in Berwickshire in 1783,[50] used open cylinders solely to reduce weight.

The methods spread to others who were not involved in the Edinburgh story. Alexander Stevens designed a middle arch of 30·5 m (100 ft) span for his Drygrange Bridge over the Tweed in 1773 and wisely drew 'cylindrical arches cross the great arches', but before the contract was concluded some months later these were altered to 'cylinder arches to be . . . built horizontally or longitudinally along the great arches.'[51] He also used longitudinal walls and voids in Bridge of Dun near

Montrose in 1785 and by 1792 even a layman, writing of Stevens's estimate for the Teviot Bridge near Kelso, knew the importance of hollow spandrels:

> 'Andrew Ker gave in an estimate for the same place about £20 less in appearance; but in reality much dearer. The former [Stevens's design] to have the haunches made up with small cross arches—the latter in the old abominable way—I mean with earth and rubbish subject to be affected with every change from wet to dry.'[53]

The bridge which was probably the last built in Scotland with cylindrical holes is much nearer in form to Pontypridd. This is Millhaugh Bridge over the Almond in Perthshire, a single high arch of 18·3 m (60 ft) span with its spandrels pierced by a circular void over each haunch. It was built about 1787[54] by an unknown designer, but his motive in using voids can only have been that of obtaining a stable arch—William Edwards's motive at Pontypridd. Like Pontypridd, Millhaugh Bridge was too narrow, too steep, and consequently bypassed before the end of the nineteenth century.[55]

The story of the developments in Scotland ends where it began, at Coldstream Bridge, and the last event shows clearly that hollowing of the spandrels was absolutely necessary in the larger bridges. When Sir John Rennie examined Coldstream Bridge in 1828 the bulging and cracking of the external spandrel walls was so severe that he ordered them to be taken down and rebuilt, and the internal filling to be replaced by a system of parallel walls with voids between, the voids spanned by stone slabs over which was laid 225 mm (9 in) of clay and then the gravel of the roadway.[41] Bridging the voids by slabs was by then as common as arching over them and it obviated the need for the chain-bars Smeaton had used at Perth. The parallel walls of Coldstream Bridge were photographed by the County Surveyor's staff (Fig 9) before their removal in 1960 to allow strengthening of the arches with reinforced concrete. There also exists a photograph of the voids in one of the spandrels of Perth Bridge, taken during the removal of the charges placed in that spandrel during the 1939–45 war.

Developments in Wales and western England after 1763

In south Wales and the west Midlands of England there is a sequence of bridges which follows the form of Pontypridd much more directly. As shown above, James Bridges, although he had seen Pontypridd, is only known to have designed voids in three-arch bridges and following Langley's pattern; but

Fig 9. Interior of spandrel of Coldstream Bridge in 1960 (By courtesy of County Surveyor of Northumberland)

in south Wales itself the first known use of voids after Pontypridd is in two single-arch bridges of long-span, namely, Wychtree Bridge (29 m (95 ft) span) and Dol-au-Hirion Bridge (25·6 m (84 ft)), both over the River Towy and both by the Edwards family.[20] Malkin credits both to William Edwards, but Dol-au-Hirion has a stone in the parapet carved with the name of his son Thomas and the date 1773. In both cases the cylindrical voids were placed over the haunches of the arch, as at Pontypridd.

Voids in the same position are found in a dozen or more other single-arch bridges in south Wales of which the construction dates are unknown, but none longer than about 18 m (59 ft) in span and some less than 12 m (39 ft 4 in). Probably none of these voids are necessary to the stability of the arches, and it looks as though circular voids became a fashion. It is clear, however, that the third motive named by Gautier, that of extra passages for floodwater, was sometimes present, because there are also a number of bridges of similar span with cylindrical openings through the abutments or approaches; these are lower than the voids over the haunches and were obviously made to allow extra discharge of floodwater. It is, indeed, possible that the designers intended the voids over the haunches to fulfil the same function and that William Edwards's motive had been forgotten. In at least one instance, Pont St. Clears in Carmarthenshire, the voids over the haunches pass tidal water twice a day.

Two known dates suggest a period for the construction of this group of bridges. One bridge with voids over the haunches, namely Pont ar Darrell in the outskirts of Brecon, was built in 1829, as recorded in an inscription on the parapet; and one with cylindrical openings through the approaches, Rhyd-y-Ceir Bridge near Llandovery, was built in 1813.[56] But as evidence that not all need be of the same period, Pont ar Ithon in Radnorshire was rebuilt with open cylinders through the abutments as late as 1934.[57]

The presence of voids over the haunches has not prevented severe distortion of some of these arches and indeed may have facilitated it. Considering that the body of the arch, even at Pontypridd, was built of unhewn masonry, long-term distortions could only be prevented by first-class workmanship, materials and maintenance. Wider bridges than Pontypridd gain less stiffening from their spandrel walls and Pont ar Darrell, for instance, is badly distorted. The most interesting distortion is that of Llanglydwen Bridge in Carmarthenshire. This is a segmental arch of 15·3 m (50 ft) span, 3·66 m (12 ft) rise and 6·1 m (20 ft) wide, with a 0·92 m (3 ft) diameter void over each haunch. The soffit of the arch can be seen to sag in transverse section between the (exterior) spandrel walls, and also, while the line of the arch on the upstream side remains a segment, on the downstream side it has bent to a shape distinctly pointed at the crown (Fig 10)—a distortion of the same kind as that of the old Ouse Bridge at York and of the third Pontypridd. It has been distorted in this way at least since 1936.[58]

A little out of Wales, there is a serious use of open cylinders for structural benefit in New Bridge over the Avon near Bath, a single arch of 30·5 m (100 ft) span built before 1788[59] with a cylinder of 2·75 m (9 ft) diameter over each haunch. This bridge can scarcely date from before 1762, as it does not figure in the controversy about long-span arches in Bristol, which is only ten miles away. By 1768 John Gwynn was at work on the first of his three multi-span bridges over the Severn at Shrewsbury (1768–74)[60], Atcham (1772–76)[60] and Worcester (completed 1781),[55] and he seems to have made no hollow spandrels, though in the Shrewsbury bridge there was a massive wall in each spandrel from back to back of the arches at least 1·22 m (4 ft) thick, and a cross wall intersecting it over the pier, the spaces between being gravel-filled.[61]

Another Shrewsbury architect, Thomas F. Pritchard, who is credited with the first design of an arch bridge of iron (an early

Fig 10. Downstream side of Llanglydwen Bridge in 1972

design for the Coalbrookdale bridge[62]), designed a single-arch bridge of brick to cross the Severn at Stourport, apparently in 1773.[63] The span was 41·5 m (136 ft) and the spandrel and haunch at each end were pierced by four circular voids, the largest of which was entirely over the abutment, and so would not affect the stability of the arch but would act as a flood passage. The bridge actually built at Stourport was a three-arch stone bridge and had a large masonry ring with four keystones over each pier, presumably round a circular void, and triple keystones on the main arches[64]; both of these features exactly similar to those on the elevations of Smeaton's bridges at Coldstream and Perth. It was finished in 1775[65] and is assumed to have been designed by Pritchard.[66]

William Hayward worked under both Pritchard and Gwynn on the Shrewsbury bridge.[60] In 1777 he made nine alternative designs for a bridge over the Tern at Atcham.[67] Three designs, of two and three arches, had open cylindrical voids over the piers and abutments, placed too high to pass floodwater; while all three of his single-arch designs had two or three cylindrical voids over each haunch. In the bridge as built (1778–80) there was 'a series of barrel arches of varying dimensions',[55] which presumably means cylindrical voids, in each spandrel of the 27·5 m (90 ft) span arch, but all hidden behind smooth spandrel walls.

A number of later single-arch bridges of moderate span in the west Midlands have open cylinders over the haunches and there are also several with cylindrical flood passages through the abutments. Of the latter, those of Barford Bridge may have been the first in the region if, as seems likely, they date from the partial reconstruction of the bridge in 1784 by the mason William Eborall,[68] who did a lot of bridge work in Warwickshire. Eborall has been credited with the design of Castle Bridge in Warwick,[69] for which he was certainly the masonry contractor.[70] It was built in 1789–93,[70] a single segmental arch of 32 m (105 ft) span, and in the interior of each spandrel are five longitudinal voids between parallel walls, the middle three under the carriageway and covered by brick arches, the outer two under the footways and covered by flat slabs.[71] The elevation resembles very closely that of Leafield Bridge, a mile upstream, by Robert Mylne, and the design has also been attributed to Mylne.[70] It appears to be the only bridge attributed to Mylne which has longitudinal voids in the spandrels, and is one of the early uses of this method in the west Midlands, though not necessarily the first.

Another significant use of the parallel-wall construction occurred in 1795–97 at Ashford, in the extreme south of Shropshire. A certain William Atkins contracted with the Justices of the Peace for Salop in 1795 to build a single-arch bridge of 24·4 m (80 ft) span over the Teme at Ashford, the 'internal parts of the spandrels to be secured by culverts or with brick walls with proper covering stones'.[72] There is little doubt that 'culverts' means cylindrical voids and 'brick walls with proper covering stones' refers to the longitudinal voids method with flat stone slabs over the voids. The significance of this contract is that it must have been approved by Thomas Telford, then Surveyor to the County; there is a manuscript specification which he probably drafted.[72] The bridge was built by 1797 with longitudinal voids, which were seen during recent repairs.[73]

Telford's use of parallel walls

It would seem that the first proposal of hollow spandrels at Ashford came from Atkins and the wording of the contract suggests that both types of hollow spandrel were well known. They would doubtless have been known to Telford, but he appears not to have used them before this date. He had built his first large stone bridge over the Severn at Montford in 1790–92[74] without hollow spandrels[75] and he did the same at Bewdley Bridge in 1798.[75] But the spandrels of Chirk Aqueduct, which was finished before 1801, were built 'with longitudinal walls instead of being filled with earth',[75] though the arches were only of 12·2 m (40 ft) span and he had thought earlier that 'in case of brick arches, that the spandrels must be brought up solid',[74] probably meaning solid brickwork and not earth-fill. In 1805–6 he built Tongland Bridge in Kirkcudbrightshire, of a single arch 34·2 m (112 ft) span, with parallel walls and voids which he described in his autobiography many years later and then added:

> 'I have ever since practised this mode, in order to lessen the weight incumbent on large arches, and the pressure outward against high wing-walls and spandrels.'[75]

He quoted the failure at Edinburgh in 1769 as evidence of the danger from the lateral pressure of earth-fill.

It was probably the occurrence of this remark in a printed book that prompted the assertion, oft repeated, that hollowing of spandrels by the parallel wall method was invented and first used by Telford. In fact he knew that he was adopting a method already in common use, for in the *Edinburgh Encyclopaedia* article on bridges, which he wrote with Alexander Nimmo in 1812, only six years after Tongland Bridge was built, Nimmo described the parallel-wall method as 'the customary practice of our most experienced bridge builders'.[21]

Theory applied to hollow spandrels

This remark is made towards the end of a lengthy treatment of arch theory in which Nimmo first repeats Emerson's derivation by fluxions of the distribution of weight required for equilibrium but then substitutes for it a purely geometric treatment leading to similar results. Unlike Emerson 50 years before, he does not reject variation of density in the spandrels as impractical and works out the density of the filling required for a segmental arch of 120° arc and thickness 1/10 of the radius, given a horizontal roadway; it increases from a value of one quarter of the density of the arch-stones, at the crown, to a value equal to that of the arch-stones, over the point about 60° from the crown, i.e. over the springings of the 120° arch. If the arch segment is more than 120° the required density becomes greater than that of stone and equilibration is not possible.

Nimmo proposes that the variation of density should be obtained by building parallel longitudinal walls of masonry which occupy one-quarter of the width of the bridge at the crown and thicken towards the haunch until they fill the spandrel at 60° from the crown. This is an entirely new suggestion with regard to the parallel-wall method. With reference to circular voids, he describes Pontypridd with some enthusiasm but observes that the cylindrical arches

> '. . . seem to press unequally on the arch, and only at a number of detached points; and though the widest tunnel may be placed just where the greatest evacuation may be necessary, yet this ill agrees with the gradual approximation to solidity, which we should find in passing down the back of the arch.'[21]

This statement is ambiguous. If Nimmo had used the theory as has been done in Fig 5 he would have seen that *all* the voids were placed correctly to obtain 'the evacuation' required and that, even with the voids increasing in size, the load, or 'solidity', also increases towards the haunches.

His article appears to be the only discussion of arch theory which ever gave a theoretical method for the design of hollow spandrels in masonry bridges. It also made the very first new suggestion about the proper form of the parallel walls since their introduction at Dumbarton (possibly) and Perth in the 1760s. The refinement of tapering the walls in plan seems, however, never to have been practised. At the date of writing it is probable that Nimmo himself had not yet designed a bridge; and the only drawings we have of a bridge built to his design, namely those of the Wellesley Bridge at Limerick (1824–34) show parallel walls which can only have been of constant thickness.[76] Amongst the many surviving drawings of Telford's bridge designs there is one which shows walls tapered in plan, namely the engraving of the design of his Avon Bridge near Hamilton in Lanarkshire,[75] built in or near 1825. The walls in a sectional plan taper from 0·46 m (1 ft 6 in) at the crown of the 24·4 m (80 ft) span segmental arch to 0·61 m (2 ft) over the springing, but a small note labels it as a 'plan of the wing and spandrel walls at 16 feet [5·24 m] above the springing' (Fig 11). The specification is also printed and requires the internal spandrel walls

> '. . . to be two feet [0·61 m] in thickness at ten feet [3·05 m] above the springing [below which was solid rubble] and to diminish to eighteen inches [0·46 m] at three feet six inches [1·07 m] below the level of the roadway,'[75]

but does not mention a longitudinal taper. Since recent inspection has shown that the bridge was built with internal walls of constant thickness both longitudinally and vertically, it must be concluded that the draftsman preparing the engraving either mistook the vertical taper in the specification for a horizontal one or drew the thickness of the walls as intended at their base on the extrados of the arch and not strictly 5·24 m (16 ft) above the springing.

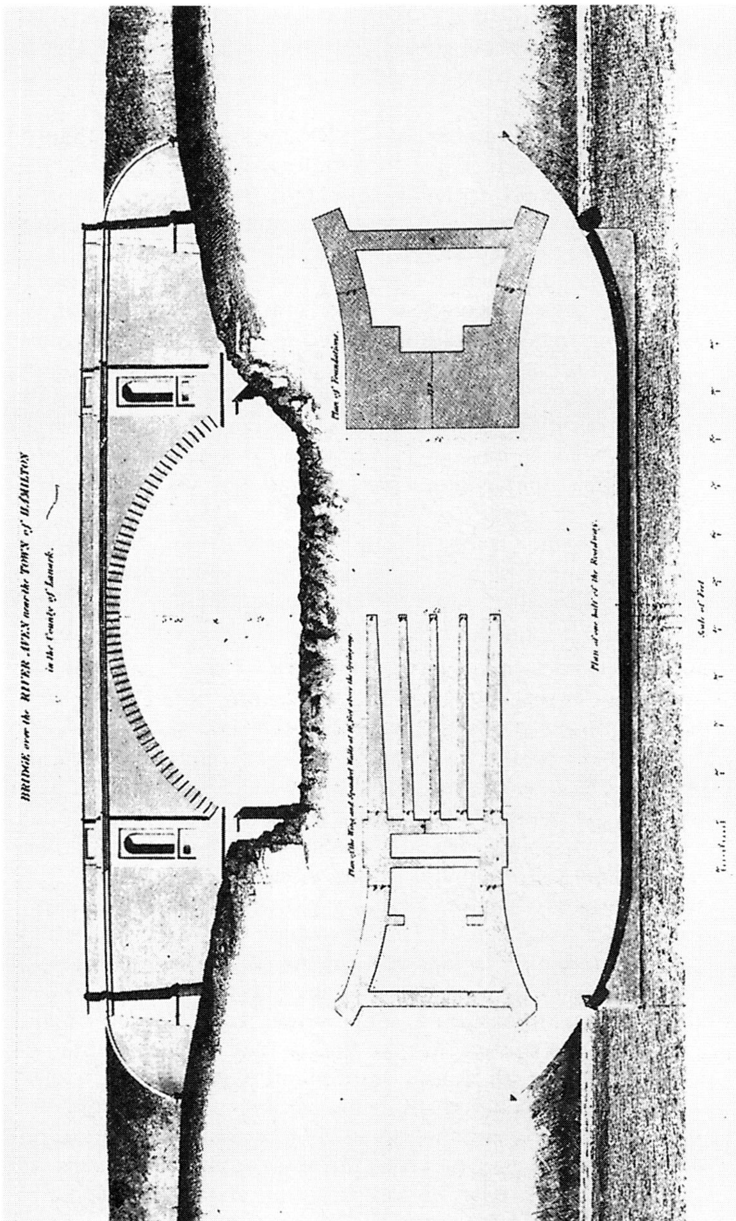

Fig 11. Engraving of Telford's design for Avon Bridge[75]

Conclusion

Nimmo's proposal to combine the stiffening effect of parallel walls with mathematical equilibration by varying the wall thickness was attractively inventive, but in fact the stiffening provided by the walls made equilibration unnecessary. The parallel wall method of construction had solved the problem of instability of arches together with those of excessive weight and lateral pressure in the spandrels. Both it and the cylindrical void method had been developed by practical designers without a single application of mathematical calculation, either to arches or spandrels, for which full documentary evidence can be found. Even Labelye, who claimed to have made calculations, adopted with success a design for the Westminster Bridge repair which was in flat contradiction of his stated theoretical principles.

Though the methods of mathematicians gave them so little help, there is no evidence either that the designers invented the hollow spandrel methods for themselves. On to their experience and their knowledge of Gautier they grafted the ideas of men who were really amateurs, such as Stukeley, Langley and Halfpenny; and the famous bridges of Edwards, Smeaton and their successors seem to owe a serious debt to this mixed company of original thinkers.

With respect to the application of theory the difference between the two methods was, that the parallel wall structure was too difficult for any theoretician to analyse at all—even Nimmo's late treatment making no quantitative assessment of the stiffening effect of the walls—but the great contribution of the walls to the stability of the arches was intuitively obvious; whereas with cylindrical voids the arch-spandrel structure was less stiff than without the voids and so there was increased need for the correct distribution of loading on the arch which mathematical equilibration could ensure. As very few designers were capable of doing the equilibration it was fortunate that the better method of construction made it unnecessary.

Amongst the common sources of designers' decisions, the simple statements which Emerson made about arches of 120° and 90° may sometimes have found a place, but much less often than the rules which Gautier had derived from his experience and observation. They also learned from their own and their contemporaries' work, both in success and failure. One example of this is Smeaton's list of bridge dimensions and another may be William Mylne's use of cylindrical voids in his Glasgow design a year before the collapse in Edinburgh, but at a time when the mistakes in the Edinburgh design must have already been clear to him.

Maturity in design is best evidenced, however, by some decisions not to use voids when they could be safely omitted, and this is most noticeable in the work of the best designers. After John Smeaton had used voids with parallel walls at Perth where the longest arch span was 22·9 m (75 ft) and at Dumbarton where a maximum reduction of weight was necesssary, he designed Banff Bridge with shorter arches, smaller spandrels and simple gravel filling[34]; but with similar short spans in his Hexham Bridge some years later he again made longitudinal voids,[10] clearly because the very bad ground made lightness again a prime requirement.[34]

A late design by Telford provides another example. At the same time as he designed the Hamilton bridge, of 24·4 m (80 ft) span and 6·1 m (20 ft) rise, with hollow spandrels he also designed Elvanfoot Bridge with gravel-filled spandrels though the span was 27·5 m (90 ft) and the rise 7·6 m (25 ft).[76] As this information comes from the specification only, the drawing being lost, the reasons are not clear, but the fact that the gradient of the roadway downward from the crown of the arch at Elvanfoot was to be 1 in 20 on one end and 1 in 27 on the other, while the roadway over the Hamilton bridge was virtually horizontal (Fig 11), may have meant that the greatest height of the spandrels at Elvanfoot was slightly less than at Hamilton. It would seem that only a very wide personal experience such as Telford's could give confidence to such a precise decision made without the help of theoretical analysis.

Acknowledgement

The author is indebted to librarians, archivists, engineers and other friends throughout the country for their help in finding and checking the information contained in the paper.

References

1. Gautier, H. *Traite des Ponts*, Paris, 1714.
2. Langley, B. *A Design for the Bridge at New Palace Yard, Westminster*, 1736.
3. Langley, B. *A Reply to Mr. John James' Review of the Pamphlets and Schemes . . . of a Bridge at Westminster*, 1737.
4. James, J. *Short Review of the Several Pamphlets and Schemes . . . of a Bridge at Westminster*, 1737.
5. Institution of Civil Engineers, drawings of T. Gayfere.
6. Labelye, C. *Description of Westminster Bridge*, 1751.

7. *The Gentleman's Magazine*, XXX, 1760, p. 364.
8. Halfpenny, W. *A Perspective View of the Sunken Pier . . . at Westminster* (engraving), 1748.
9. Langley, B. *Observations on a Pamphlet . . . Entitled, Remarks on the Different Constructions of Bridges . . .*, 1749.
10. Royal Society of London, 'Designs by the late John Smeaton, FRS', IV.
11. Emerson, W. *Doctrine of Fluxions*, 1743.
12. Muller, J. *Mathematical Treatise*, 1736.
13. Emerson, W. *Principles of Mechanics* 2nd ed., 1758.
14. Emerson, W. *Miscellanies*, 1776.
15. Muller, J. *Treatise on the Practical Part of Fortification*, 1755.
16. *The Gentleman's Magazine*, XXXIV, 1764, p. 534.
17. Williams, E. I. 'Pont-y-ty-pridd: a critical examination of its history', *Transactions of the Newcomen Society*, XXIV, 1945, p. 121-30.
18. *The Gentleman's Magazine*, XXIV, 1754, p. 528.
19. National Library of Wales, 'Plasybrain Manuscript', quoted by Williams *op cit*.
20. Malkin, B. H. *Scenery, etc., of South Wales*, 1804.
21. *The Edinburgh Encylopaedia*, Edinburgh, 1830, IV Pt. 2, 'Bridge'.
22. Ruddock, E. C. 'Details of the structure of Pontypridd 1756: a survey of early documents', *Industrial Archaeology*, August 1974.
23. Smeaton, J. *Mr. Smeaton's Answer to the Misrepresentations of his Plan for Black-Friars Bridge*, 1760.
24. *The York Courant*, 5 March 1810.
25. Sykes, J. *Local Records, 1833*.
26. Glamorganshire County Archive, unsigned drawing (assumed by T. M. Smith, see reference 22).
27. *Dictionary of National Biography*, XI, p. 365-6.
28. Dance, G. and Taylor, R. *Description of the Centre for the Great Arch of London Bridge* (engraving), 1760.
29. Guildhall Library, London, MS drawings by Edward Cooke.
30. Noble, J. *Miscellanea Invernessiana*, 1902.
31. Bridges, J. *Four Designs for Rebuilding Bristol Bridge*, Bristol, 1760.
32. Wood, J. *A Letter to the Commissioners for Rebuilding the Bridge at Bristol*, Bath, 1762.
33. Stratford, F. *Observations on the Letter by a By-Stander*, Gloucester, 1762.
34. Smeaton, J. *Reports . . . Made on Various Occasions*, 1812.
35. Corporation of London Records Office, 'Case of Robert Mylne, Surveyor', for payment of commission for works of Blackfriars Bridge, 1776 (MS).
36. Corporation of London Records Office, Surveyor's Bridges Plan No. 195.
37. [Mylne, R.] *Observations on Bridge-Building and the Several Plans Offered for a New Bridge*, 1760.
38. Riou, S. *Short Principles of . . . Stone Bridges*, 1760.
39. House of Commons Journal, XXIX Pt. 1, 1762, 153.
40. Northumberland County Record Office, 'Minutes of the Trustees for Coldstream Bridge, 1762–72'.
41. Institution of Civil Engineers, MS Reports of Sir John Rennie, V, 11–20, 1828.

42. Meade, M. K. 'Plans of the New Town of Edinburgh', *Architectural History*, XIV, 1971, p. 40–52.
43. Ruddock, E. C. 'The building of North Bridge, Edinburgh, 1765–75', *Transactions of the Newcomen Society*, 1974–75.
44. Town Clerk's Office, Perth, 'Minutes of the Commissioners for Perth Bridge', and letters concerning same.
45. MS in private hands, 'Blackfriars Bridge, London: Entries of the several orders drawn out by Mr. Robert Mylne.'
46. Mylne, R. S. *Masters Masons to the Crown of Scotland*, Edinburgh, 1893.
47. Glasgow City Archive, design for a new bridge at Glasgow, signed 1.8.1768.
48. Scottish Record Office, MS E732/18.
49. Historic Monuments Commission, Edinburgh, copy of design for Aray Bridge.
50. Graham, A. 'Archaeology on a Great Post Road', *Proceedings of the Society of Antiquaries of Scotland*, XCVI, 1962–3, p. 318–47.
51. Scottish Record Office, contract for Drygrange Bridge, 1778, MS GD244, Box 19.
52. Scottish Record Office, contract for Bridge of Dun, 1785, MS GD244, Box 19.
53. Duke of Roxburgh's Muniments at Floors Castle, letter in Section 3, Box 3/18, Bundle 46.
54. *Statistical Account of Scotland*, III, 609, 1792.
55. Public Works, Roads and Transport Congress, *British Bridges*, 1933.
56. Carmarthenshire Record Office, Quarter Sessions Order Books, 1813.
57. Private communication from Col. J. A. Paterson, former County Surveyor of Radnorshire.
58. Jervoise, E. *Ancient Bridges of Wales and Western England*, 1936.
59. Collinson, J. *History of Somersetshire*, 1791.
60. Ward, A. W. *The Bridges of Shrewsbury*, Shrewsbury, 1935.
61. Ward, A. W. 'The Reconstruction of the English Bridge, Shrewsbury', Selected Engineering Paper No. 66, Institution of Civil Engineers, 1928.
62. Maguire, R. A. and Matthews, P. 'The Ironbridge at Coalbrookdale: a reassessment', *Architectural Association Journal*, LXXIV, 1958, p. 30–45.
63. White, J. 'Notice of Mr. Pritchard's gradual progress in the application of iron to the erection of bridges', *Philosophical Magazine and Annals of Science*, XI, 1832, p. 81.
64. Hadfield, C. *Canals of the West Midlands*, 2nd ed., 1969.
65. *Berrow's Worcester Journal*, 15 December, 1773.
66. Colvin, H. M. *Biographical Dictionary of English Architects 1660–1840*, 1954.
67. Salop County Archive, copies of designs for Tern Bridge.
68. Warwickshire County Archive, minutes and accounts of Quarter Sessions.
69. Field, W. *Historical and Descriptive Account of the Town and Castle of Warwick*, 1815.
70 Private communication from Mr. M. W. Farr, Warwickshire County Archivist.
71. Extract from County Bridge Register, supplied by the County Surveyor of Warwickshire.

72. Salop County Archive, contracts, specification and letters concerning Ashford Bridge.
73. Private communication from County Surveyor of Salop, per Mr. Blackwall.
74. Gibb, Sir. A. *The Story of Telford*, 1935.
75. Rickman, J. ed. *Life of Thomas Telford*, 1838.
76. Weale, J. ed. *The Architecture of Bridges*, 1843.

11
Pont-y-ty-pridd: a critical examination of its history

E.I. Williams

The Name.

Most Welsh place-names are descriptive. They have been called " Odes in themselves," their terse aptness topographically being a recurring delight to those who understand them. Their origin is largely due to the synthetic character of the Welsh language, and while they retain their original forms, their meanings are apparent.

At some time or other a bridge was built near an earthen (cob) house and its portmanteau name followed; Pont, bridge; y Ty, the house; Pridd, earth,—Cobhouse Bridge. For many years it has been assumed,with the approval of histories of the town and of the bridge, that William Edwards's bridge on the Taff was the first and the only one to bear that name, and that the town Pontypridd owes its names to its proximity to this bridge. It is not so. " Pontytypridd " occurs four times in a manuscript written before 1700.[1] Again, some nine years before Edwards, in 1746, began his first bridge across the Taff, there was a Circulating School at Pontytypridd.[2] So the name was in use long before Edwards was born. Still, it is true, that in 1755, Edwards's edifice took over the name Pontytypridd from an earlier bridge in the hamlet of the same name, for in that year *A Plan and Prospect of Pontytypridd* was published. The modern spelling—Pontypridd—is a contraction evolved naturally, and was not uncommon in the eighteenth century. It followed an abbreviated pronunciation, and in this form, in 1856, was adopted officially.

Histories.

Much has been written about William Edwards and a few books have been devoted entirely to him and to his bridge at Pontypridd. He has found a place in many works of reference, including *The Dictionary of National Biography*. During the forty years centred around 1800 there was little travel on the Continent because of the unsettled state of Europe, so tours in Wales became fashionable. These led to the publication of many descriptions of the Country, in which Edwards's romantic story was generally included. Artists also, for the first time, discovered Wales, as the catalogues of the exhibitions of the works of British artists, during the same period, show. Nor has the technical study of the bridge been neglected,[3] and still continues. Since so much about Edwards is available, no biography of the clergyman—engineer—farmer is here attempted, but merely an examination of the accepted history of his endeavours at Pontypridd.

---

1 M. S. Ashmole, 1820[a], Bodleian Library.

2 *Welch Piety: or a Collection of the Several Accounts of the Circulating Welch Charity Schools, 1737.*

3 T. M. Smith, *Memoir of Pont-y-Tu-Prydd . . . , 1846.*
W. B. Coventry, *Pont-y-ty-pridd,* 1902.

PONT-Y-TY-PRIDD—A CRITICAL EXAMINATION OF ITS HISTORY

The chief sources of original information are:—

1. *T. M. A Remarkable Bridge at Pontytypridd in Glamorganshire.*[4]

This account will be dealt with at some length. There has been a tendency to slight it, by either ignoring it or cursorily noting it sceptically. There is no valid reason for this attitude. The identification of the writer should help to assess the value of his contribution, for in all accounts of the bridge he has been simply "T. M."

"T. M." was obviously a Welshman who thought in Welsh and sometimes translated his thoughts literally, many of his idioms being peculiarly Welsh. There are several of his articles in *The Gentleman's Magazine,* all pseudonymous. A rebus of his places him, in 1752[5], in Carmarthenshire, while in 1765[6] he wrote from the West Riding of Yorkshire. These facts alone could have led to the conjecture that he was Thomas Morgan, a Presbyterian Minister of Henllan, Carmarthenshire (1746-1760) and of Morley, Yorkshire (1763-1794), the intervening years being spent at Delph, Yorkshire. But positive identification is possible. In one of his two pocket-books (because of war conditions only one was available) at the National Library of Wales in this entry:—

"Letters and Short Essays of mine published in ye Gent. Magaz. - - - 1764 p. 564.
An Acct of Pontytypridd in Glamorgansh. wth a Copper Plate—T:M:"[7]

His list is incomplete. Thomas Morgan could write and even dared to cross quills with Joseph Priestley. He was interested in scientific matters, for his pocket-book mentions the lectures he attended and enumerates the scientific books he bought. In *The Gentleman's Magazine* he wrote of "Fireworks,"[8] "A Diving Ship to be rowed under water,"[9] "Tracts of Land Swallowed by the Sea,"[10] "Fossil Shells,"[11] and "An Extraordinary Alimentary Powder."[12] He was a real bookman, catholic enough in his reading to pay sixpence for *Rochester's Poems,* and threepence for the same author's *Letters,* strange companions for their owner's religious collection. But what particular qualifications had Morgan to write about Pontypridd bridges and their builder?

William Edwards was born in 1719, at Ty Canol (Halfway House), Groeswen, a farm in the parish of Eglwysilan, Glamorgan, but resided most of his life at Bryntail, another farm in the same parish. His father was Edward Dafydd (David),[13] the son, according to old Welsh custom, receiving as his surname, his father's baptismal name, later to have "S" added to it. On 7th January 1720, at Duffryn Uchaf, Groeswen, Thomas Morgan was born. Edwards and Morgan were neighbours until 1741, when Morgan went to Carmarthenshire to be trained for the Presbyterian ministry, much troubled during the two days journey there by his drunken companion—his father. These two young men had cultural and religious affinities. Hundreds of people in their neighbourhood had learned to read Welsh.[14] Both became nonconformist ministers. Edwards in 1745 and Morgan the following year. In 1760 Morgan left Wales for Yorkshire, where, he understood, the English did not object to pastors being shopkeepers as well, just as the Welsh approved of preaching and farming going together. Between 1743 and 1759, he visited his birthplace at least sixteen times, preached there on four occasions in Edwards's chapel, his visits varying from a few days to a few weeks.[15] So he witnessed most of Edwards's projects

[4] *Gentleman's Magazine,* 1764, pp. 564-565.

[5] *Gent. Mag.* p. 472. [6] p. 567. [7] N.L.W. MS. 5457.

[8] 1749, p. 217; [9] 1749, p. 249; [10] 1751, p. 60; [11] 1751, p. 588; [12] 1755, p. 7.

[13] "William the son of Edward David Bap: Feb. ye 8th 1719." Eglwysilan Register.
"Edward David buried the 6 of January 1726," Eglwysilan Register.
Malkin states that William Edwards was two years old when his father died.

[14] *Welch Piety, 1741.* Letter from David Williams, Eglwysilan, to Griffith Jones, Llanddowror.

[15] N.L.W. MS. 5457.

PONT-Y-TY-PRIDD—A CRITICAL EXAMINATION OF ITS HISTORY

at Pontypridd (the eastern end of the bridge is in the then Eglwysilan Parish about a mile from Bryntail) while his type of mind and life-long association with Edwards ensured his interest in the bridge-building.

But apart from these considerations Thomas Morgan's account is direct and factual and is accompanied by a plate that proves he had access to records now supposed lost, probably the *Plan and Prospect* previously mentioned. His is the most important contribution to the history of the bridge.

1a. Thomas Morgan's article was considered worthy of reproduction in *The Annual Register, 1764*,[16], where it appeared in slightly altered form, with two additions, one palpably wrong and the other probably so.

1b. THE PLASYBRAIN MANUSCRIPT.[17]

This is a MS. of special interest in a commonplace book at the National Library of Wales. It is a transcript of a document by Evan Evans, Ieuan Brydydd Hir (1731-1789)—Evan the Tall Poet—a renowned Welsh Antiquary, a great collector and transcriber of old Welsh manuscripts, but best known to Englishmen for his supplying the material that enabled Gray to write some of his famous poems.[18] Evans took *The Annual Register* account and corrected or added to it where necessary. For a few years about 1780 he lived in Monmouthshire and Glamorgan.[19] What is important is Evans's declaration " whatever was incorrect in this printed Account has been here carefully set aright from the Architect's own mouth, so that the whole may be depended upon as real matter of fact." The N. L. W.. transcript was made in 1789 by John Williams, Plasybrain (Crow Mansion). The original by Evans is not among the few of his many manuscripts examined by the writer, but the transcript appears to be genuine and reliable.

2. (Henry Penruddocke Wyndham). *A Gentleman's Tour through Monmouthshire and Wales, in the months of June and July, 1774*.

This was first published anonymously in octavo, in 1775, and reprinted in 1781 and 1794. The author made a second journey in the summer of 1777, and both tours were described in a handsome, illustrated quarto, published at Salisbury in 1781. Wyndham was a noted topographer and scholar and was well acquainted with Welsh antiquities, as some of his published works testify. His *Gentleman's Tour* was the first of many such works relating to Wales and was highly praised by Richard Gough.

With few exceptions,[20] every published account of Pontypridd Bridge for the past one hundred and seventy years has been Wyndham's, including that in 1804, of the oft-quoted authority, Dr. Malkin, who, though he recorded most of what is known about Edwards, contributed nothing new to the history of the bridge. Several authors had, before that date, told the story[21] largely in Wyndham's words; others, after 1804, in Malkin's, but the information of all is the same.

3. Benjamin Heath Malkin. *The Scenery, Antiquities and Biography of South Wales*.

Because of its unique biographical sketch of William Edwards, this work is mentioned. The author deserved to be better informed about the history of the bridge.

There are two editions, a quarto, with illustrations, in 1804, and in two volumes octavo, in 1807. It is an excellent work, stylishly written, and the best of its kind about South Wales. In

16 p. 147.

17 N.L.W. MS. 15c.

18 *Some Specimens of the Poetry of the Antient Welsh Bards* with *Dissertatio Bardis, 1764*.

19 Brit. Mus. Addit. MSS. 15025 and Llanover MSS. No. 31, p. 94.

20 G. W. Manby. *Historic and Picturesque Guide . . . Glamorgan . . . , 1802*, pp. 174-178. Based on *Gent. Mag.*, 1764.

21 Richard Gough. *British Topography, 1780*, I, pp. 503-504. Briefly summarises *Gent. Mag.*, 1764.

PONT-Y-TY-PRIDD—A CRITICAL EXAMINATION OF ITS HISTORY

1803 Dr. Malkin visited William Edwards's son David, at the latter's historic home, Beaupré, near Cowbridge, Glamorgan. There David Edwards related the story of his father's life, and all later writers have drawn upon it unquestioningly and almost exclusively for their accounts of the famous bridge-builder. Simple, significant incidents and experiences of William Edwards's youth, and his struggles against misfortunes at Pontypridd comprise much of the narrative. Few dates are given (some inaccurate) and the precise times of the disasters at Pontypridd are unfortunately wanting. How much David Edwards depended upon tradition for his descriptions of events that occurred before he was twelve years of age is not known.[22] However, little would be known about William Edwards but for his son David and Dr. Malkin.

William Edwards's Bridges at Pontypridd.

The existing bridge is said to be Edwards's third attempt. Thomas Morgan rightly gave full particulars of four attempts, only to be mistrusted and disregarded later.

The First Bridge. This bridge, begun in 1746, was finished, stood for two years and two months and was then swept away by a flood.[23] Thomas Morgan says it was in four arches; other accounts say three. No illustration of it is known. Edward Daniel Clarke toured South Wales in 1793, when he wrote of this bridge:—" The ground in which the foundation was laid, proved unfavourable and soon convinced the architect, that, even if the flood spared his work, the instability of its base would soon be the cause of its fall." It may be that this circumstance contributed to Edwards's decision to bridge the river in one span. He had agreed, for £500, to provide a bridge that would stand for seven years; so he proceeded with the " very whimsical scheme."[24]

The Second Bridge. Edwards's first of three attempts to " build a bridge in one arch." Thomas Morgan's account and that of the *Plasybrain* MS. agree in essentials.

Thomas Morgan.	Plasybrain MS.
" But when he had almost finished the arch, the center timber work gave way, and all fell to the bottom."	" Just after the first single arch was finished, and before the center was struck, a flood came and carried all away."

There is no indication that Edwards yet suspected the equilibrium of his great arch.

A MS. in the National Library of Wales[25] describes how, with sadness, William Edwards saw the timbers of one of his bridges floating down the flooded Taff. This gave rise, some years ago, to the foolish belief that he had built a wooden bridge. What timbers were they? They could not have been those of the first bridge, for it stood finished for over two years; nor those of the third for its centre was struck; nor those of the existing bridge. Thomas Morgan's chronicle completes the story of the floating timbers. They were those of the second bridge.

The Third Bridge. After the collapse of the second bridge, Edwards erected stronger timber work for his third attempt.[26] Accounts agree as to the cause of his failure this time. The arch was finished and the centering had been removed. About six weeks later,[27] because the arch was

21 (cont.) James Baker. *A Picturesque Guide to the Local Beauties of Wales,* 1792, 1, pp. 87-88.

(E. D. Clarke) *Tour through the South of England, Wales . . . 1793,* pp. 178-181.

Collection of Welch Tours, 1798, pp. 159-160.

(A. B. L. Maudet de Penhouë) *A Pedestrian Traveller. Letters describing a Tour through part of South Wales, 1797,* pp. 20-21.

T. Evans. *Cambrian Itinerary,* 1801, pp. 119-120.

22 " David the son of William Edwards baptiz'd March 1 1745," Eglwysilan Register. A few months before his father was ordained.

23 N.L.W. MS. 15c. 24 *Gent. Mag.,* 1764. 25 N.L.W. MS. 9927 D.

26 *Gent. Mag., 1764* and *Plasybrain MS.* 27 *Plasybrain* MS.

ill-balanced, its keystones were forced out and " it fell again to the bottom." This occurred on the 7th November 1754. A contemporary account mentions the disaster. " Thursday 7th Nov. 1754 . . . the floods have carried away an arch built over the Taaf 144 feet wide."[28] So all three disasters occurred during floods.

The Fourth and Existing Bridge. For long has this bridge been admired, and often has it been extolled in prose and verse.

Wyndham, in the first edition and the reprints of his work already noted, mentions appreciatively the feat of Thomas Williams, a local wheelwright, in building the timber support for the arch. Why he omitted this information from his revised edition is not clear. There is some reason for believing this Thomas Williams to have been Edwards's co-pastor at Groeswen.

There has been considerabe conjecture as to what, besides masonry, occupies the space in the abutments between the road surface and the arch. It is said that it may be filled with charcoal.[29] The *Plasybrain* account gives entirely new information, thus:—" in order to lessen the quantity of matter in the abutments pressing upon the arch and thereby bring it to an equipoise with that on the crown, he has contrived three cylindrical arches and one Semicircular in each abutment, the Semicircular are hidden by the side walls. It is also strengthened by very large pieces of timber laid in the work between the great Arch and the Cylinders."

Engineers must decide whether any final pronouncement upon the stability of the bridge is possible, until the dimensional particulars of the structural features of it given in the *Plasybrain* account are obtained.

Attempts have been made to discredit Thomas Morgan's account of four bridges. The date of the collapse—7th November, 1754—of the third (so called second) bridge disposes of the objections. They are based upon the erroneous statement by Malkin, that it fell in 1751. The three years error provide plenty of time for what was accomplished in the real second attempt. The existing bridge dates from 1756, so its construction was accomplished in not more than two years time. It could hardly have been finished by the end of 1755, for that would be less than fourteen months after the failure of the preceding bridge. " William Edwards, 1756 " was the inscription on it in 1802.[30] The first inscription, on a small stone and without date, was " W. Edwards fec."[31] The present one reads:—" William Edwards, 1750. Repaired by Edward David and Thomas Evan, 1798." The very bad condition of the parapet about 1800 makes the date of repairing suspect. Something worthy of the builder should replace so misleading a record.

Edwards and Smeaton.

It has been suggested that the planning of Pontypridd Bridge was partly Smeaton's idea, but no evidence has been produced that really supports the claim. Whether Smeaton ever met Edwards or corresponded with him, directly or indirectly, is not known. That Smeaton knew of him sometime after 1756 is certain. On February 18th, 1760 Edwards sent a letter describing his bridge, together with a plan of it, to Mr. Capel Hanbury, a wealthy ironmaster of Pontypool, Monmouthshire, who next year showed a drawing of the bridge (possibly the plan) to the *Society of Antiquaries*.[31a] Edwards had then completed a large bridge at Usk, near Pontypool. This letter and plan came into Smeaton's possession, probably not later than 1765, when Hanbury died. Afterwards they went into the collection of Sir Joseph Banks. And this appears to be the origin of the Smeaton story. Among the members of the first Smeatonian Society, 1771, was an Edwards,

[28] *Gent. Mag., Historical Chronicle, 1754*, p. 528. Also mentions severe flooding in Carmarthenshire at the same time. Was Thomas Morgan the reporter?

[29] T. M. Smith. *Memoir of Pont-y-Tu-Prydd . . . 1846.*

[30] G. W. Manby. *Historic . . . Guide . . . Glamorgan, 1802.*

[31] and [31a] Richard Gough. *British Topography, 1780.* II, pp. 503-504.

PONT-Y-TY-PRIDD—A CRITICAL EXAMINATION OF ITS HISTORY

but he may have been Lawley Edwards, who was credited with "being a good hand at the bottle."[32] Printed reports do not help. So there, for the time, the matter rests.

There is an oddly ambiguous statement in *Archæologia*[33]—"The Welch mason Edwards, in a wonderful bridge over the Taaf, in Glamorganshire, learnt from experience, what Perronnet's theory had taught him. Over each haunch of the Pont y ty Prydd there are three tunnels." What exactly does this mean? It may appear to say that Edwards got his idea of the tunnels from the famous French architect, who began his ten-year task of building Orleans Bridge, in 1750, when Edwards was in the middle of his ten-year struggle at Pantypridd. Presumably, what the writer intends to convey is that Edwards and Perronet solved similar problems, the one by experiment and the other by calculation.

WILLIAM EDWARDS AND LORD WINDSOR.

The Plan and Prospect of Pontytypridd, 1755 was dedicated to Lord Windsor, and it has been said that the builder was indebted to him for financial aid. Malkin states that Edwards's family did not know "that he (Edwards) was ever indebted for any emolument but to his own industry and abilities;" and that is the accepted view. Thomas Morgan's positive declaration upon the matter is:—"By this time the mason was greatly in debt and greatly discouraged.[34] But the Lords Talbot and Windsor (who have estates in the neighbourhood) pitied his case, and being willing to encourage such an enterprising genius, most generously promoted a subscription among the gentry in those parts." The tactful interpolation "estates in the neighbourhood" merits investigation. Besides a desire to help a courageous engineer, there was another reason why Herbert, Lord Windsor, wanted a bridge at Pontypridd. Glamorgan was being industrialised, and between 1723 and 1752, he and his father before him, had granted leases of vast areas of land on both sides of the Taff, for mining iron and coal. One, dated 1748, when Edwards was completing his first bridge, is a remarkable document, from which the following is a quotation: "with full power and free liberty to and from the sd. Thomas Morgan, his exors., ads., assns., his and their servants, Agents and workmen, to dig for, make, and sink pits, and to make trenches, ditches, gutters, ponds, and levels, and to erect limekilns for burning of lime, engines, banks, stanks, sluices, and all other conveniences for the better scouring for getting, having and obtaining all such iron mine, coal, and stone, and the earth and rubbish raised out of the said pits, trenches, ditches, gutters, ponds and levels, and the iron mine, coal, stone and tile raised out of the said Works, and the lime that shall be burnt to lay on the ground next the mines, pits, kilns and other places where the same shall be found, burnt or raised, or on any other place being waste or common within that part of the said Manor of Senghenith called Supra Cayach convenient for that purpose and to take, carry away with horses, carts or any other carriages, and to dispose of the same at his and their will and pleasure, and to his and their proper use and benefit."[35] Lord Windsor was the most influential person in the district and the chief promoter of the scheme by which the Hundreds of Miskin and Senghenydd were to provide a bridge, those hundreds with their buried riches being separated by the turbulent Taff. So there was nothing remarkable or exceptional in his action. For similar reasons, a few years later, he subscribed to a fund for paying for the next of Edwards's bridges—at Usk, Monmouthshire.

Wyndham corroborated Thomas Morgan's history; and he was of the gentry. In the second edition of his *Tour* he wrote—"It may be some satisfaction to the reader, as it was to me, to

32 *Fenland Notes and Queries*, IV, p. 324.

33 Samuel Ware. *Observations on Vaults*, XVII, p. 48. Communicated to the *Society of Antiquaries* on March 5th, 1812.

34 "distressed" *Plasybrain* MS.

35 John Lloyd. *The Early History of Old South Wales Ironworks, 1906*, p. 20.

hear that the county had nobly indemnified and even rewarded the heroic perseverance of the Cambrian architect."

Here again the *Plasybrain* MS. is unequivocal and gives also the cost of the bridge.

GEMT. MAG. AND ANNUAL REGISTER.	PLASYBRAIN MS.
"A subscription among the gentry by which a sum of money was raised that enabled him to complete the bridge in one arch for the last time."	"a subscription among the gentry by which the sum of £100 was raised which enabled him to complete the arch for the last time."

Thomas Morgan said nothing about the cost of the bridge, but *The Annual Register* mentioned it.

THE ANNUAL REGISTER.	THE PLASYBRAIN MS.
"This surprising piece of masonry, though built of stone, cost but £700."	"This surprising piece of masonry . . .cost from first to last £1,153 18. 2."

If £100 was the total gift Edwards received he must have lost heavily. That sum alone cannot be reconciled with what Morgan and Wyndham wrote about the subscription. Other gifts have been reported.[36] Gough states that Edwards lost £600 but does not give the source of his information, which was possibly the Capel Hanbury letter previously cited. It could be expected that Hanbury would show both the drawing and the letter to the *Society of Antiquaries.*

A SELECT LIST OF PAINTINGS, DRAWINGS AND ENGRAVINGS OF PONTYPRIDD BRIDGE.

PAINTING:

Richard Wilson (1714-1782), a Welshman and one of the foremost British landscape painters. Date, about 1760. The painting was engraved.

DRAWINGS:

Paul Sandby (1725-1809). The father of water-colour art. Friend and admirer of Wilson. This water-colour drawing was in the Burdett-Coutts Collection.

Michael Angelo Rooker (1743-1801). Pupil of Sandby and engraver of some of Wilson's paintings. This finished water-colour drawing was exhibited at the Royal Academy in 1798 and is now in the National Museum of Wales. 12½" × 15". (See Plate XII).

Samuel Hieronymus Grimm (1734-1794). His *Pontypridd in Glamorganshire* (probably a water-colour drawing) was exhibited at the Royal Academy in 1780. Grimm, after completing a commission for Gilbert White (Selborne),[37] toured South Wales with Wyndham in the summer of 1777 and fourteen of his drawings (Pontypridd not among them) were engraved for Wyndham's quarto edition of his *Tour through Monmouthshire and South Wales.*

ENGRAVINGS:

Plan and Prospect of Pontypridd, 1755. No copy known to exist. Recorded by Thomas Morgan, thus:—"This was in the year 1755, when a copper plate plan and prospect of this surprizing arch was published." There is some confusion in the account which cannot be elucidated until more evidence becomes available.

Plan of a Remarkable Bridge at Pontytypridd in *The Gentleman's Magazine, 1764.* Line, 4¾ in. × 7⅝ in. Probably a copy of the above *Plan,* reduced in size.

Peter Charles Canot after Wilson. Line 14 in. × 20¼ in. Exhibited in 1766 at the exhibition

[36] *Seren Gomer, 1834,* tud. 68.

[37] Rotha Mary Clay, *Samuel Hieronymus Grimm, 1941.* pp. 75-76.

PONT-Y-TY-PRIDD—A CRITICAL EXAMINATION OF ITS HISTORY

of *The Society of Artists of Great Britain.* Later the plate was dated *17th July 1775* and published in Boydell's *Collection of Prints, engraved from the most capital paintings in England.*

Godfrey. Line 5½ in. × 8 in. In Gough. *The Antiquarian Repertory, 1776.* Reprinted 1807.

S. Alken after J. Smith, 1794. Aquatint, 5 in. × 7⅜ in. Rarely found printed in colours. In W. Sotheby. *A Tour through Parts of Wales.*

John Hassell. Drawn and Engraved by, 1794. Aquatint, 7¼ in. × 10¼ in.

View of Pont-y-Praid, the new Bridge over the River Taaf. Line, 3¼in. × 8¾in. No date. Early and unique in that tht view is downstream and shows the hamlet through the arch. In Brit Mus.

(Attributed to Compte Armand B. L. Maudet de Penhouët). Two views respectively 3½ in. × 6⅛ in. and 3¾ in. × 6¼ in. Line. In *Letters describing a Tour through parts of South Wales. By a Pedestrian Traveller, 1797.*

George William Manby. Small mezzotint vignette in the artist's *Picturesque Guide from Clifton through the counties of Monmouthshire, Glamorgan and Brecknock, 1802.*

John Laporte, Drawn and Etched by, 1804, 7 in. × 9⅛ in. In Malkin, *Tour of South Wales.*

John George Wood, Drawn and Etched by, 1811, 5⅝ in. × 8¾ in. In the artists's *Rivers of Wales.*

Woolnoth after J. P. Neale, Line 4 in. × 6⅛ in. In *The Beauties of England and Wales (1815).* Another plate was wrongly lettered *Pontypridd* in the first issue of this work.

John Brown. Etched by, 3½ in. × 7¼ in. Vignette on the engraved Title-page of T. M. Smith, *Memoir of Pont-y-Tu-Prydd, 1846.* The engraved title-page is dated 1838.

Original steel engravings of the bridge, octavo size, appear in Henry Gastineau, *Wales Illustrated, 1829,* and Thomas Roscoe, *Wanderings and Excursions in South Wales (1836).*

In addition to the above, there exist many woodcuts and small copper-plates of little interest.

There is an attractive painting of the bridge on a *Nantgarw* dish, in the National Museum of Wales, c. 1818.

MINIATURE PORTRAIT.

Mr. Edwards Architect in Wales. By J. T. Hill. Exhibited at the Royal Academy in 1779. Said to be a copy made for John Morris (later Sir John), in 1806, from the original painted for the Empress of Russia.[38] Given by a descendant of Sir John Morris to Dr. W. T. Edwards, a descendant of the Architect, after whose death it was presented to the National Museum of Wales. (See Plate XII).

ENCOMIUMS.

" Virtutem incolumen odimus."

" The Cambrian architect emerged from his native mountains (and) left this monument of his exquisite taste and genius to vie with the finest works of antiquity."

Edward Daniel Clarke. Author of works on Greek Antiquities.
Presented marbles to Cambridge University.

" Had the remains of such an arch been discovered among the ruins of Greece and Rome, what pains would have been taken by the learned antiquaries to discover the architect."

Henry Penruddocke Wyndham.

[38] N.L.W. MS. 9927 D.

PONT-Y-TY-PRIDD—A CRITICAL EXAMINATION OF ITS HISTORY

" Great architect! illumed
By nature's light, thy daring genius scorned
An imitated grace: no sculptured form,
Triton, or nereid, or wrought river-god
With meretricious ornament disturb
The simple grandeur of thy bold design."
William Sotheby, 1794.

" Pont fawrglod hynod yw hi—yn rhywfodd
Ni lwyddodd ail iddi;
Bwa mawr uwch llawr y lli'
Coeliwch, o waith gwas Celi."
Thomas ab Ioan neu Eos Glan Rhymni,
Nantgarw, 1834.

William Edwards and Groeswen Chapel.

Edwards's progress as an architect was rapid, for he was but twenty-seven when he started his Pontypridd undertaking. Equally rapid was the progress of his religious activities. His son David described him as very wild until about eighteen years of age. At twenty-two he was preaching, partly perhaps, through the influence of his own labourer at Bryntail, Harri Smith, a godly eloquent young man. On the Second of May, 1741, Edwards and four others were signatories to a letter of importance in the development of nonconformity in Wales; so at twenty-five he was a person of standing in the religious movements of his country.

In 1743 a small meeting-house was built near the site of the present Independent Chapel, Groeswen, beside a field named Waenfach. In 1745 Edwards became the first pastor,[39] jointly with Thomas Williams, who died twenty years later, and Edwards became sole minister. He was recommended to the Church by Edmund Jones, " The Old Prophet," one of the quaintest Welshmen of the eighteenth century and author of published and unpublished works on Welsh folklore. For forty-four years Edwards served his congregation.[40] He died on 7th August, 1789 and was buried two days later in the graveyard of the Parish Church. *The Gentleman's Magazine*[41] and *The Annual Register*[42] printed obituaries. Quotations from the former follow.

" 7th August, 1789. At his home near Caerphilly, William Edward, architect and bridge-builder, or the Rev. William Edward, for he sustained both characters with assiduity and ability. The celebrated bridge on the river Taaf called Pont-y-tu-Pridd, by the English, New Bridge, was constructed by this extraordinary man . . . His fame was diffused through the kingdom . . . He retained his passion for religious exercises . . . He conducted a very large and mingled congregation . . . to the age of 71 (sic), at which he died, after sustaining a long illness with exemplary patience. The reputation of his piety is very respectable, even among contending sects, that of his morals was unblemished; and his inventive genius and talents will entitle him to a perpetual rank among the real benefactors of mankind."

Postscript.

In 1944 the Pontypridd Urban District Council set aside a sum of money for the care of William Edwards's grave. There remains an opportunity for attending to his greater monument,

[39] About 1750 according to Malkin.
[40] T. Rees and J. Thomas. *Hanes Eglwysi Annibynol Cymru, 1872,* II, tud. 372-377; 280-382.
[41] 1789, p. 767. [42] 1789, p. 218.

his bridge, and by so doing, rid the town of a hideous piece of vandalism. Nothing calculated and planned could better demonstrate the lack of appreciation of the treasure inherited by the town, than the thoughtless action of its Council of about forty years ago, in suspending a public lavatory from the large eastern void of the bridge; and there it remains, much of its interior visible from the bridge. Since that day, a fine park and more stand to the credit of the Council, including the clearing away of several old houses near the bridge, so that now there is no excuse for retaining what has long offended the many who prize the majestic arch and cherish the memory of its creator.

The paper was illustrated by valuable engravings and MS. material from the author's own collection which he was about to present to the National Museum of Wales.

The discussion on this paper was taken with that of the succeeding one.

NOTE. In the Smeaton Drawings, 6 vols., in the possession of the Royal Society, there is a sketch (Vol. IV, Fol. 90v) entitled: "This plan being the South Prospect of New Bridge over the river Taaff in Glamorgan-Shire. Erected by William Edward, Mason, Native of ye S[d] County." This may possibly be the drawing referred to on p. 125.—ED.

PLATE XII.

WILLIAM EDWARDS, AET 60, from the miniature by J. T. Hill.

WATER COLOUR DRAWING BY M. A. ROOKER, 1798.
Both illustrations by Courtesy of the National Museum of Wales.

12
William Edwards's bridge at Pontypridd

E.C. Ruddock

The arch bridge which has spanned the Taaf at Pontypridd since 1756 is remarkable, first for the story of its construction by a heroic mason, William Edwards, and secondly for its very unusual form. The span, at the date of construction, was longer by at least 30 ft than that of any other arch in Britain, the arch was recklessly slender and the spandrels pierced by open cylindrical voids never seen in Britain before. The details of the history of its construction were established in an excellent survey of early documents by E.I. Williams in 1945 [1]; the present paper attempts to establish the details of the structure as it stood when first finished, mainly from early scale drawings and a model. As it is complementary to Williams's paper, opportunity is taken to reconsider one or two matters concerning which he missed vital references, and a little is said about alterations before 1846.

Documents

No bridge can stand for 200 years without repeated maintenance and repairs, and so the surviving structure, although very important evidence, is far from conclusive as to the original details. The documents considered here are three manuscript scale drawings, one of them engraved, another early engraving, and a model, together with written evidence relating to them all:

Drawing 1.[2] An unsigned elevation among the 'Designs by the late John Smeaton, FRS' in the Library of the Royal Society in London.
Drawing 2.[3] An elevation signed 'J. Essex' and dated 1762, in the Drawings Collection of the Royal Institute of British Architects in London.
Drawing 3. A larger sheet of plans, elevations and sections to scale, with freehand views of details and the surroundings of the bridge, unsigned, in the Archive of Glamorgan County Council in Cardiff. All were engraved and published in *Minutes of Proceedings of the Institution of Civil Engineers*, Vol 5 (1846).[4]
Model in possession of W.C. Mylne about 1830,[5] now lost.
Gentleman's Magazine Engraving. Unsigned elevation, published in the magazine, December 1764,[6] and reproduced in the *Scots Magazine*, February 1765.[7]

Drawing 1

The drawing is reproduced as Fig. 1. It is drawn in ink and the elevation washed dark grey, both rather roughly, and is obviously not by Smeaton's hand. The dimensions marked are the span (140ft) and rise (35ft) of the arch, and the figures '9', '6' and '4' (apparently diameters in feet) in the three circular holes at each end. The holes are left unwashed. A scale is given, and the segment of the arch soffit is extended in a light line to a full semicircle, the diameter being dimensioned 175ft. The arch ring is shown of constant thickness 2ft 6in, composed of thin stones of equal thickness, about 12in,

WILLIAM EDWARDS'S BRIDGE AT PONTYPRIDD

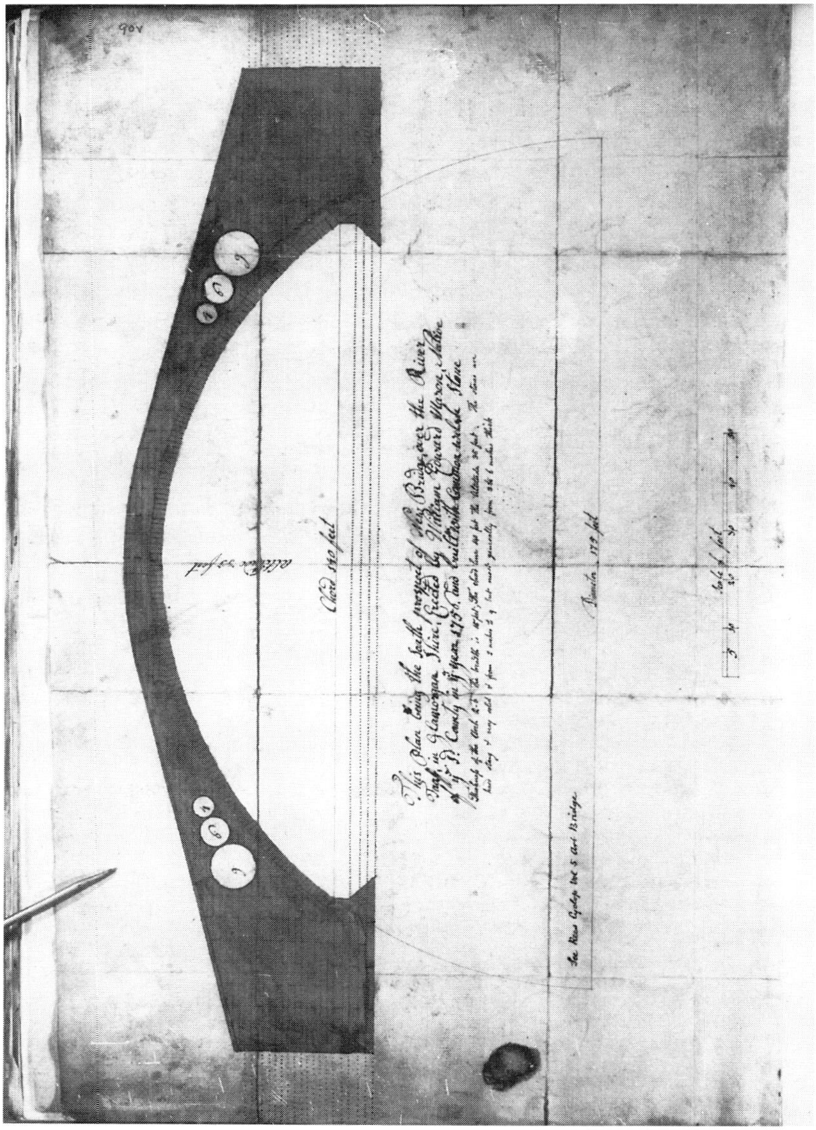

Fig. 1 (by courtesy of the Royal Society of London)

and all of the full 2ft 6in height. All the masonry of the spandrels and parapet is shown as horizontal courses of equal thickness and regular broken jointing – clearly a conventional rather than a precise portrayal. At the top of the parapet is a coping of long, thin stones, laid to a curve which cuts across the horizontal bedding of the courses below. The thickness through the whole bridge at the crown scales about 8ft. There is no break or line between the abutment wingwalls and the spandrel and parapet walls, but the foundations of the abutments are stepped out under the arch soffit and presumably also laterally from the base of the wingwalls. The circular holes are not surrounded by arch rings – another evidence of rough or hurried drawing – and are not even separated by sufficient thickness of masonry to allow such rings.

The title, in the hand of the draughtsman, is:

> This Plan, being the South prospect of New Bridge over the River Taaf, in Glamorgan Shire, Erected by William Edward Mason, Native of ye Sd. County in ye Year 1756, and built with Common Welsh Stone.

To this is added in a neater and smaller script:

> Thickness of the Arch 2–6 The breadth is 15 feet, The chord line 140 feet the Altitude 35 feet. The stones are hard strong and very solid and from 2 inches to 9 but most generally from 3 to 6 inches thick.

Elsewhere on the drawing in the same script is the note:

> See Rees Cyclop Vol 5 Art Bridge.

These added notes are in the hand of John Farey junior (1791–1851), who was personally connected with the production of Rees's *Cyclopaedia*[8] from 1805, when he began making drawings for some of the articles.[9] He continued to make drawings for the *Cyclopaedia* for many years and later wrote several articles himself, but it is more likely that the 'Bridge' article, which was first published in Part 1 of Vol 5 in June 1805,[10] was written by his father, John Farey senior (1766–1826). Neither the 'Bridge' article nor the longer one on 'Canals' in Vol 6 Part 1 (1806), which was also by Farey senior, shows any special knowledge of Smeaton's works, though Farey senior had been known to Smeaton in the later years of his life[11]; and the account of Pontypridd in the 'Bridge' article is almost all quoted from Malkin's *Scenery, etc., of South Wales* (1804).[12] There is also positive evidence that the Smeaton drawings first came into the hands of the Fareys only in 1809 or 1810. This was for the purpose of copying and reducing many of them for the engravings printed with the *Reports of the Late Mr John Smeaton, made on Various Occasions* in 1812[13]; and when Farey junior started this work in 1810 his father described the collection at his disposal as

> the miscellaneous bundles of papers, which belonged to the late ingenious Mr John Smeaton, the civil engineer, now in Sir Joseph Banks's possession.[14]

Banks had bought all Smeaton's remaining professional papers within a year of his death in 1792[15] and the reports published in 1812, which filled three large volumes, were all from these manuscripts.[16] As the reports had been copied out for the printer much earlier,[17] it is likely that by 1809 the originals were back with the drawings and therefore that the 'miscellaneous bundles' comprised the whole collection bought by Banks.

The drawings are now bound in six volumes,[18] each with an ornately inscribed title page. In the first three volumes this inscription concludes: 'collected and arranged by John Farey, 1821'; and after the title page of Vol 1 is a note by Farey junior about Smeaton's manner of making and copying drawings. Many of the drawings bear notes in the same hand, often referring to books of Smeaton's letters both sent and received, as well as the printed reports and other papers; and one of these notes in Vol 1 was made in or after 1835.[19] It can therefore be concluded that the drawings were in Farey's keeping at least until 1821 and probably long after that; and that he held the other papers for at least long enough to make a thorough study of them with the associated drawings.

Of the notes on Drawing 1, the short reference to the *Cyclopaedia* might have been added at any time after Farey received the drawings and is unrelated to Smeaton's other papers; but there is good independent evidence that Farey copied the longer note from a letter he had found amongst the 'miscellaneous bundles'. Thomas Rees, FSA, who was a contributor to the *Cyclopaedia,* and also assisted his namesake the editor in arranging the plates,[20] is quite certain to have known Farey junior who drew so many of them; and Rees was certainly quoting either from Farey's note on Drawing 1 or from its source, which he named, when he wrote in 1815:

> The following particulars are given by the builder himself, in a letter to Mr Hanbury of Pont y Pool, in Monmouthshire, dated the 18th February, 1760 which accompanied a plan and elevation of the bridge, now among Mr Smeaton's papers in the possession of Sir Joseph Bankes:—'the thickness of the arch is two feet six inches; the breadth is fifteen feet; as to the quality of the stone, they are hard, strong, and very solid, and from two inches to nine, but most generally from three to six inches thick.[21]

There is a small difficulty about identifying the drawing referred to here as Drawing 1, in that Drawing 1 is only an elevation of the bridge, but the mistake of calling it a 'plan and elevation' would be easily made. It is confirmed by Richard Gough in *British Topography* Vol 2 (1780) that Mr Capel Hanbury of Pontypool not only had a drawing of Pontypridd but showed it to the Society of Antiquaries in London in 1761;[22] and as Hanbury died in 1765,[23] it might have been easy for Smeaton to acquire both letter and drawing at that date.

Farey and Rees may have known positively that Drawing 1 was made by William Edwards himself, having his letter to compare the scripts. This is not certain now, but all the known facts support the proposition that Drawing 1 was provided by Edwards, if not drawn by him, no more than four years after he finished the bridge.

Drawing 2 (Fig. 2)

This is the work of a much better draughtsman. It is in ink lines with the masonry washed grey and the circular holes black. It has a neat border, is signed in one bottom corner 'J. Essex delinr' and dated in the other '1762', and bears a rough pencil note in another hand 'Plan of the new Bridge over the River Taaf, 12 miles above Cardiff 1762'. Nothing is known of the drawing before its purchase by the Royal Institute of British Architects in 1961.

The span and rise are dimensioned 140ft and 35ft respectively; the arch ring scales 2ft 6in, the thickness right through the bridge at the crown scales 8ft; the parapet is topped by a line of long, thin stones; the masonry of the spandrels and wings is shown with no break or line between the two; the abutment foundations are stepped; there are no arch rings round the circular holes. In all these respects Drawing 2 agrees exactly with Drawing 1, and might even be copied from it, but there are three distinct, if small, differences. First, the circular holes at each end scale 8ft, 5ft and 3ft in diameter. Secondly, there is a projecting keystone at the crown of the arch – a feature

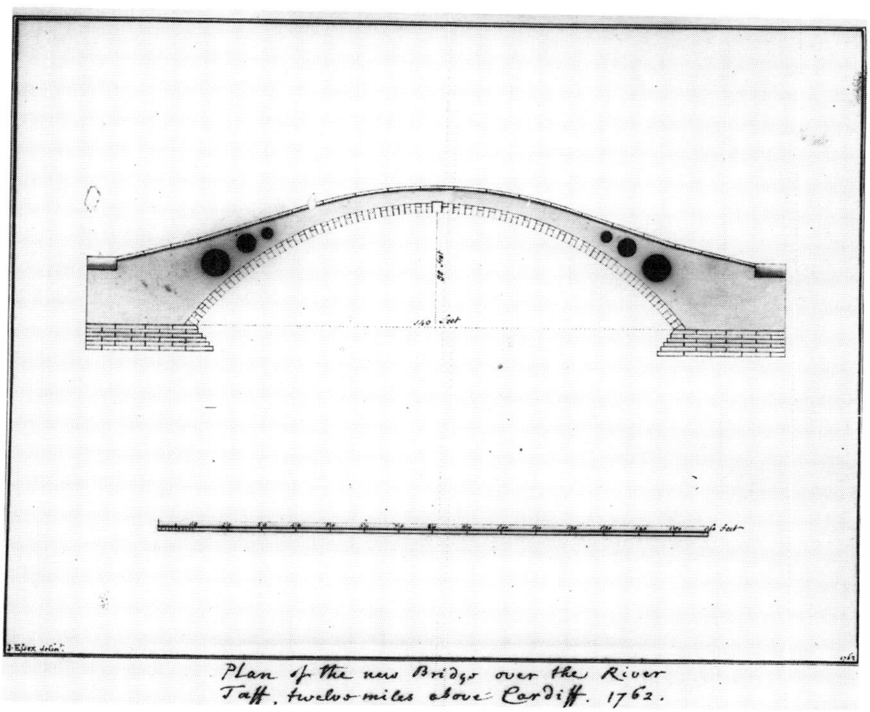

Fig. 2 (by courtesy of the Royal Institute of British Architects)

WILLIAM EDWARDS'S BRIDGE AT PONTYPRIDD

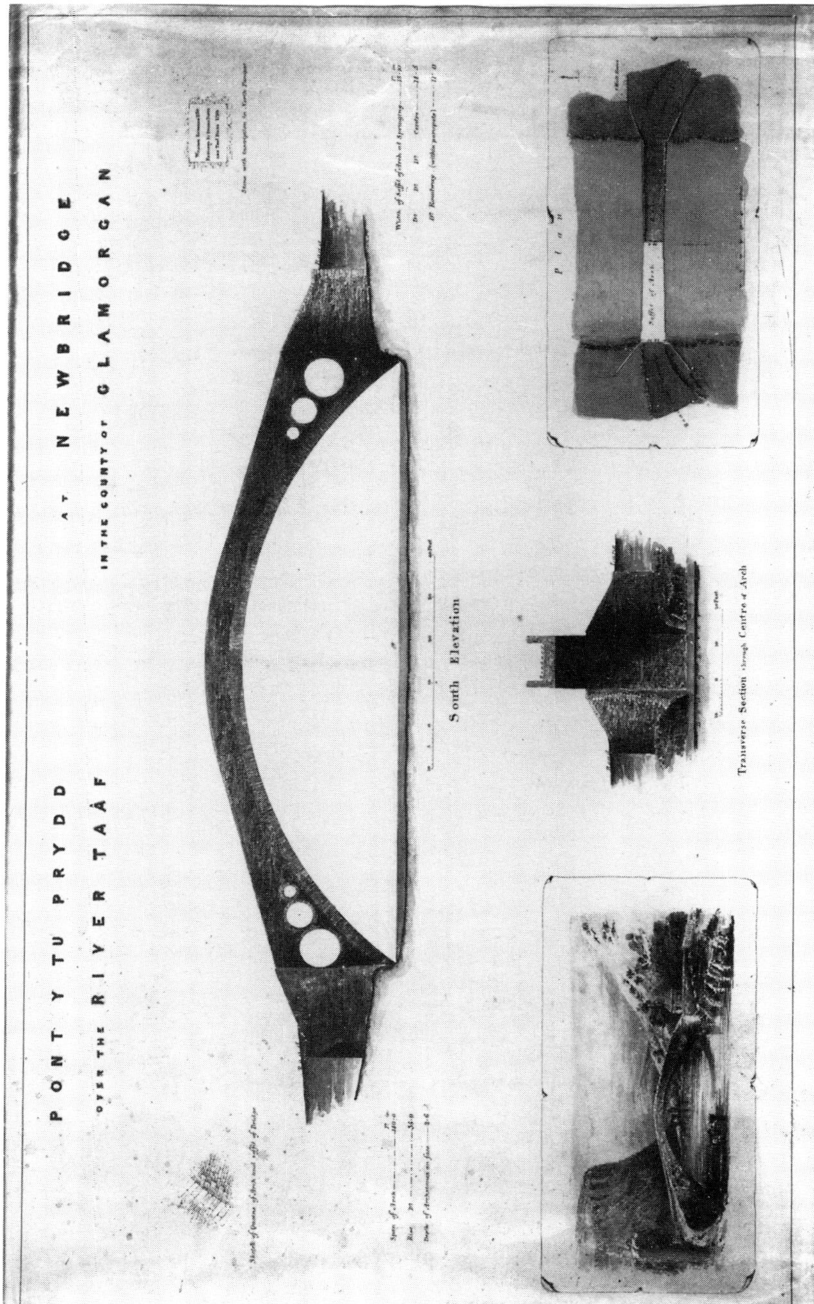

Fig. 3 (by courtesy of Glamorgan County Record Office)

which might be added by any architectural draughtsman as something proper to all drawings of arches, if not of the arches themselves. Thirdly, every second voussoir of the arch ring is shown, not as a single stone 2ft 6in high, but as two stones one over the other making up the full thickness of the ring. The last is an important difference for several reasons. It suggests that Essex either visited the bridge or copied an early accurate drawing other than Drawing 1, and the agreement in the main dimensions and other details is therefore doubly valuable. It also reveals as original these double arch-stones, which are shown also in Drawing 3 some seventy years later (Figs. 3 and 4) and are unchanged today, although they must detract, at least to some extent, from the strength and stability of the arch. The continuance of this detail on the face of the arch suggests strongly that in the arch, as distinct from the spandrels, wings and parapets, nothing more than minor replacement of stone has taken place since the bridge was completed in 1756.

Drawing 3 (Fig. 3) *and Model*

This is a beautiful drawing, finished in ink and water colours. It shows a large elevation and cross-section at the crown, a plan half at road level and half at arch soffit, various notes, freehand details of the masonry at the quoin of the arch and of the inscription stone, and a perspective sketch of the bridge and adjacent landscape. There is neither date nor signature, but the information and probably the drawing itself can be attributed with confidence to T.M. Smith, because all the views and notes are reproduced accurately in the four plates which were published as illustration to his *Account of Ponty-y-tu-Prydd* in 1846. He first read this paper to the Institution of Civil Engineers in 1838,[24] and quoted local people's memories of high floods in 1816 and 1829. He must therefore have visited the bridge between 1829 and 1838. According to his *Account* he had measured the span and rise as 140ft and 35ft respectively:

> These dimensions give a diameter of 175 feet; and in the accompanying drawing the arch is shown a segment of that circle. At the time the author made his survey, he was unable to determine the form of the arch more precisely. The width of the soffit is 15 feet 10 inches at the springing, and diminishing to 14 feet 5 inches at the crown, by six offsets on each face, varying from 2 inches to 3 inches in depth. The roadway is only 11 feet wide over the centre of the arch. . . . The arch-stones, throughout each face, are of parallel width; they are 2 feet 6 inches deep, and vary from 6 inches to 9 inches in thickness, and from 12 inches to 18 inches in the depth of the bed. The soffit is all rubble masonry, two or three courses making up the dimensions of each arch-stone [ie each of the large stones at the vertical edge of the arch seen on the elevation (see Fig. 4, Smith's detail of 'Quoins of Arch and Soffit')]; and, from the slaty description of the material, little, or no, labour appears to have been necessary in forming the joints. The masonry of the spandrils and the parapet is of a similar description, being irregularly laid, the stones varying from 1 to 9 inches in thickness.

This description has been quoted at length because it is in almost perfect agreement with the statements in Edwards's letter to Hanbury seventy years before, but provides extra details, all of which may be assumed to have obtained from the date of first construction. Measurements by the present author in 1970 of the width of the road and parapet at several points agree approximately with Smith's figures for the width of the bridge, but there is less regularity than he infers, the widths at the two ends differing by 5 inches, and the six offsets being only on the arch and not carried up the spandrel as shown on Drawing 3. The 2ft 6in depth of the arch-stones on the face is well maintained, and their lengths on the soffit are about 12–18in as Smith said, but their thicknesses along the arch are less regular than he reported, and many less than 6in, which agrees better with Edwards's letter to Hanbury than with Smith's statement. A series of accurate measurements has been made very recently by Mr B.J. Collins of Glamorgan Polytechnic to determine the present shape of the arch curve. They show it to be still within an inch or two of a circular arc, but an arc of longer radius than the $87\frac{1}{2}$ft quoted by Smith and therefore of lower rise than the 35ft given by all the early drawings. This alteration could be the result of either a shortening of the arch masonry caused by the compressive stress in it or a slight outward movement of the abutments. In fact both are to be expected in any arch which has stood for two hundred years.

Several other important features are shown in Drawing 3. First, the circular voids are shown surrounded by arch rings, which were omitted from Drawings 1 and 2, and the voids scale 9ft, 6ft and 3ft in diameter (to the nearest foot in each case). Secondly, the spandrel walls are parallel only over the arch; landward from the springings the wingwalls splay out at a wide angle. This change from the parallel wings shown in Drawings 1 and 2 might possibly have been part of the alterations to which Smith referred in the following passage:

> The narrowness of the bridge and the steepness of the ascent were found inconvenient and dangerous, and, about twenty years ago, the roadway was, in consequence, raised at each end, and the surface paved.

As another date in the printed *Account* shows that he revised it in 1846, he was dating the alterations about 1825, but Drawing 3 shows the inscription stone in the parapet reading, as it still does:

> William Edwards 1750
> Repaired by Edward David
> and Thos. Evans 1798

The splayed wingwalls also show clearly in a water-colour drawing exhibited at the Royal Academy in 1798 (and printed with Williams's paper in 1945),[25] so either Smith's 'twenty years ago' was a wild guess or the raising of the roadway was an improvement made long after the splaying of the wingwalls. The splaying of the wings seems to be the biggest 'repair' after the first construction and the most likely event to merit a new inscription; this would date it at 1798. The earlier inscription, according to Gough,[26] read simply:

> W. Edwards fec.

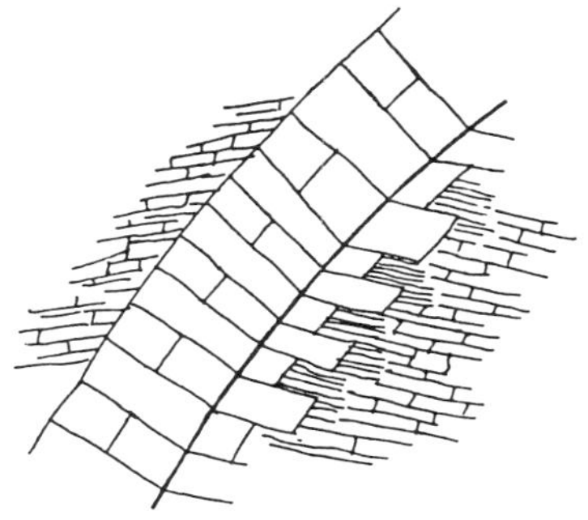

Fig. 4 'Quoins of Arch and Soffit' (after Smith)

Fig. 5 Cross-sections at Crown of Arch in Smith's Third (left) and Fourth (right) Plates

The drawing shown at the Royal Academy in 1798 showed the copings of both parapets broken and missing in several places, and Drawing 3 shows a coping of small stones, roughly square in section but rounded at the top, laid transverse to the top of the parapet. The coping is of similar form today and it must be concluded that it was first put on sometime between 1798 and 1829.

The shape of the coping is quite unimportant structurally, but it is linked in the plates accompanying Smith's *Account* with a feature of the greatest importance. Both Drawing 3 and the cross-section of the arch in the third of Smith's plates show only the stones forming the face of the arch on each elevation of the bridge as 2ft 6in deep. The remainder of the archstones are shown little more than a foot deep (Fig. 5). There is no mention of this in the text but both Drawing 3 and the second plate bear a note:

'Depth of Arch stones *on face* 2–6' [present author's italics]

Smith makes no claim to have met anyone who had special knowledge of the internal construction of the bridge, but a fairly likely source for his drawing of the cross-section is suggested by the drawings in his fourth plate. In this plate are shown a 'sketch of the centre' in elevation, with the arch ring laid on it, and a cross-section through the centring and the bridge itself at the crown of the arch (Fig. 5). In this cross-section the parapets are slightly lower than those in the section in the third plate and the coping is of a quite different shape, slightly overhanging each side of the parapet and with a pointed top. The cross-section of the road-metal, fill, and the arch proper is exactly the same as in the third plate.

In the text of the *Account* Smith explains that his two views of the centring were drawn from a model owned by W.C. Mylne MInstCE. The inclusion of masonry in both views, and masonry which was in some ways untrue to the state of the bridge at the date of Smith's visit, suggest that the model was not just of the centring but of the bridge also, and showed details both of the parapet and of the cross-section of the arch. The suggestion is supported by the supposition that the model owned by W.C. Mylne in the 1820s or 1830s was the same as that acquired by his father, Robert Mylne, the architect of Blackfriars Bridge, in 1764. Robert wrote in his diary at the end of that year –

> *Memorandum for end of year* – Received about 3 months ago a modell of the Welch Bridge from Mr Southwell. £2 2s 0d paid to the maker and the expenses of carriage.[27]

There can be little doubt that the model he received was of Pontypridd, which was just becoming well known in 1764. The price paid would cover a substantial model, and, since he calls it a model of the bridge, it would certainly show the masonry and possibly the centring as well. A model of the centring alone would not have been referred to as a model of the bridge.

The likely inference from these facts is that the cross-section of the coping shown in Smith's fourth plate is that of the original coping of long thin stones which appears in elevation in Drawings 1 and 2, and that both it and the cross-section of the arch which is common to the third and fourth plates and Drawing 3 were copied from the model bought by Robert Mylne in 1764. The early date of the model makes it very credible that its maker should have been able to ascertain the details of the internal construction of the arch, but while we do not know his identity it is impossible to ascribe great authority to his model or the cross-section drawings which may derive from it. What can be said is that if these drawings of the cross-section are true, the survival of Pontypridd has not yet been explained satisfactorily by structural theory.

Gentleman's Magazine Engraving

This shows the elevation only, to a smaller scale than that of Drawing 1 or Drawing 2 (Fig. 6). The span, rise and arch thickness agree with the dimensions in those drawings and the masonry is all as in Drawing 1, except that the circular voids are surrounded by arch rings, the voids each scale about 1ft less in diameter than those in Drawing 1, and the coping, which

Fig. 6 from the *Gentleman's Magazine*

extends only over a short length of the parapet near the crown of the arch, is shown without joints. It is inscribed:

A remarkable Bridge at Pentytypridd in Glamorganshire

and:

Newbridge. This Magnificent Piece of Work was constructed and erected by William Edward a Welsh Mason, in the Year 1756.

Diameter	175	1st Cylinder
Chord	140	2nd Cylinder
Altitude	35	3rd Cylinder
Abutments	32	Height of the Abutments 20

It would appear that the cylinder diameters were omitted by mistake.

This was probably the first accurate engraving. It is so similar to Drawing 1 that it could well have been copied from it, but, even if it was, it is important confirming evidence because it accompanied an article in the *Gentleman's Magazine* written by Edwards's personal friend Thomas Morgan [28] and so supports the proposition that Drawing 1 originated from Edwards himself.

Materials and Filling of Spandrels

Williams [29] and, more particularly, Hamilton [30] in a parallel paper, criticised Smith's *Account* too severely. As regards the story of the failure of Edwards's earlier attempts to bridge the Taaf, their criticism was justified, for Smith repeated Malkin's story [31] which Williams showed to be wrong; but Smith's details of the masonry were from his own observation and perhaps Mylne's model and deserve respect. Regarding the sources of the materials for the masonry, he was probably quoting from local hearsay during his visit:

> The bridge is built of hard and durable grey sandstone, from the quarries in the neighbourhood; the lime used was from the well-known Aberthaw cliff, on the north side of the Bristol Channel.[32]

The statement about the stone agrees with Edwards's letter to Hanbury and the source of the lime is confirmed by the very first architect who is known to have visited the bridge after its completion. This was James Bridges of Bristol, who went there in 1759 or 1760:

> I . . . made a journey into Wales to enquire after the famous Limestone, produced at Pennorth and Aberthaw; also, to see the single Arch Bridge, over the Taaff, in Glamorganshire, built by Mr Edwards, and Mr Williams. . . . Here the above Lime-Stone performs wonders; the Arch-Stones being only 2 feet 6 inches long.[33]

With regard to the filling of the spandrels, Smith wrote: '. . . it is said the spaces . . . were filled up with charcoal . . .'[34] Williams quoted from the Plasybrain Manuscript, an account said by its author to be 'carefully set aright from the Architect's [ie Edwards] own mouth':

> In order to lessen the quantity of matter in the abutments pressing upon the arch and thereby bring it to an equipose with that in the crown, he has contrived three cylindrical arches and one Semicircular in each abutment, the Semicircular are hidden by the side walls. It is also strengthened by very large pieces of timber laid in the work between the great Arch and the Cylinders.[35]

About Smith's statement that the spandrel filling was charcoal, Williams was non-committal, but neither he nor Hamilton seemed to be aware of Smith's source for this information. Smith was quoting, but not in full, from the following passage in Thomas Telford's article on the history of bridges in the *Edinburgh Encyclopaedia*, written in 1812:

> We have been informed, by that excellent and respectable engineer Mr Jessop, who was then a clerk with Mr Smeaton, that Mr Edwards consulted that eminent man with respect to rebuilding this large arch. Mr Jessop does not recollect the advice which Mr Smeaton gave, but the bridge was rebuilt in 1755 (sic); . . . and Mr Evans, a native of Wales, who was afterwards chief engineer to the Royal Canal of Ireland, informed Mr Jessop, that, in addition to the cylinders being left hollow, the spaces between them were filled up with charcoal.[36]

Telford is not a reliable historian, as will be shown later, but as Evans is not known to have been in Ireland until about 1774,[37] he was probably in Wales when verbal tradition about the bridge was still quite fresh.

Smeaton and the Design of Pontypridd

The reference by James Bridges to the bridge as 'built by Mr Edwards, and Mr Williams' need cause no question since Wyndham, in 1775, identified the contribution of Thomas Williams, a local wheelwright, as the building of the centring.[38] But the suggestion that John Smeaton was concerned in the design, dismissed by Hamilton[39] and doubted by Williams,[40] needs to be reconsidered.

Telford's statement cannot be entirely true, for Smeaton did not meet Josias Jessop, William's father, until the Spring of 1756,[41] the year when Pontypridd was finished. William was then only ten years old, and did not become Smeaton's clerk or pupil until at least three years later.[42] As it was almost thirty-five years after that before William first met Telford,[43] and possibly some years more before he related the story, the fault of memory is likely to have been Jessop's. He must certainly have told Telford something, and probably much as Telford wrote it. What he was recalling might have been a visit by Edwards to Smeaton after 1759 when Jessop was present, possibly responding to a request by Smeaton for information about Pontypridd. Alternatively, Smeaton might have been consulted by Edwards late in 1754 or in 1755 before he rebuilt the bridge and have recounted the fact to Jessop some years later. This is a very reasonable supposition; for if, as Thomas Morgan records,[44] an extra sum of money was raised by the promoters of the bridge to assist Edwards after the collapse of his third attempt to bridge the river, it is most unlikely that they would have allowed him to spend it without first consulting some architect or engineer of repute about the design. We know that Robert Weston was directed to Smeaton as the right man to rebuild the Eddystone Lighthouse at the end of 1755 by no less a person than the President of the Royal Society,[45] and it is very easy to imagine Edwards receiving similar direction only about a year earlier.

If Smeaton was consulted about the design, it would be natural for Edwards to send him a drawing of the bridge then or later, and Drawing 1 might have come into his hands in this way. Happily, its authority as evidence of the original form of the bridge would not be reduced. But the evidence that Drawing 1 is the drawing sent by Edwards to Hanbury is as strong as the evidence that Smeaton learned of Pontypridd directly from Edwards, and it is best to accept both as well-attested, but quite independent, occurrences. Smeaton's acquisition of the drawing, with Edwards's letter to Hanbury, was a sequel which must be considered largely accidental.

WILLIAM EDWARDS'S BRIDGE AT PONTYPRIDD

References

1 E.I. Williams, 'Ponty-y-ty-pridd: A Critical Examination of its History', *Transactions of the Newcomen Society*, XXIV (1945), 121–30.
2 Royal Society of London, 'Designs by the late John Smeaton, F.R.S.', IV, Fol 90v.
3 Royal Institute of British Architects Drawings Collection, G3/11.
4 T.M. Smith, 'Account of Ponty-y-tu-Prydd', *Minutes of Proceedings of the Institution of Civil Engineers*, V (1846), 474–7.
5 See Smith, op cit, 476.
6 *The Gentleman's Magazine*, XXXIV (1764), 564–5.
7 *The Scots Magazine*, XXVII (1765), 75–6.
8 *The Cyclopaedia: or Universal Dictionary of Arts, Sciences and Literature*, ed A. Rees (1819). I am indebted to N.B. Harte of University College, London, for clarifying the dates of first issue of separate volumes and roles of contributors to the Cyclopaedia. For some references see N.B. Harte, 'Rees's Clocks Watches Chronometers and Naval Architecture: A Note', *Maritime History* III (1973), 92–5, and 'On Rees's Cyclopaedia as a Source for the History of the Textile Industries in the Early Nineteenth Century', to be published in *Textile History*.
9 See *Dictionary of National Biography*, IV, 1066–7.
10 *The Philosophical Magazine*, LVI (1820), 218–24.
11 *DNB* IV, 1066.
12 B.H. Malkin, *Scenery, etc., of South Wales*. (1804), 86–8.
13 The circumstances of publication are fully described by A.W. Skempton, 'The Publication of Smeaton's Reports', *Notes and Records of the Royal Society of London*, XXVI (1971), 135–55.
14 *Phil Mag*, XXXVI (1810), 102–5.
15 Skempton, op cit, 136, 140.
16 Ibid, 136–8.
17 Ibid.
18 RSL, 'Designs by the late John Smeaton, F.R.S.' Also published on microfilm by the E.P. Group in 1972. An earlier detailed catalogue of the collection is *Newcomen Society Extra Publication* No 5 (1950), with a valuable preface by H.W. Dickinson and A.A. Gomme.
19 RSL, 'Designs . . .' I, fol 104v. This was first noted by Dickinson & Gomme, op cit, xii.
20 *Phil Mag*, LVI, 221, and *Cyclopaedia*, I, Preface, v.
21 T. Rees, *The Beauties of England & Wales, XVIII: South Wales* (1815).
22 R. Gough, *British Topography*, II (1780), 503–4.
23 Williams, op cit, 125.
24 Smith, op cit, 474.
25 Williams, op cit, Plate XII.
26 loc cit.
27 A.E. Richardson, *Robert Mylne, Architect and Engineer* (1955), 65.
28 See note 6. For identification of the writer as Edwards's friend, see Williams, op cit, 122–3.
29 op cit, 125.
30 S.B. Hamilton, 'Pont-y-ty-prydd: Notes on the Technical Significance of a Remarkable Bridge', *Trans Newcomen Soc* XXIV (1945), 131–7.
31 loc cit.
32 op cit, 476.
33 J. Bridges, *Four Designs for Rebuilding Bristol Bridge*, Bristol (1760), 57.
34 op cit, 475.
35 Williams, op cit, 123.
36 *The Edinburgh Encyclopaedia*, Edinburgh (1830). The article 'Bridge' is in vol IV Part 2. The date of writing (1812) is given in the text and internal evidence shows the section 'History of Bridges' to have been written by Telford.
37 V.T.H. & D.R. Delany, *The Canals of the South of Ireland*, Newton Abbot (1966), 46.
38 H.P. Wyndham, *A Gentleman's Tour through Monmouthshire and Wales*, (1st edn 1775) quoted by Williams, op cit, 125.
39 op cit, 134.
40 op cit, 125.
41 J. Smeaton, *Narrative of the Building of the Eddystone Lighthouse* (1791), 45.
42 A.W. Skempton & E.C. Wright, 'Early Members of the Smeatonian Society of Civil Engineers', paper to be published in *Trans Newcomen Soc*.
43 See, for instance, A Gibb, *The Story of Telford* (1935), 29.
44 *GM*, XXXIV, 564.
45 Smeaton, op cit, 38.

Appendix	*Glossary of terms*
Abutment	Block of masonry which resists the oblique thrust from the end of an arch.
Arch	A curved structural member, supporting vertical loads across an opening.
Arch ring	The vertical face or edge of an arch showing on the elevation of a bridge.
Centre or centring	Temporary structure (usually of timber) on which an arch is built.
Coping	Last course of stones at top of a wall, usually shaped to throw off rainwater.
Crown (*of arch*)	Highest part of an arch, at top of the curve.
Fill	Material – usually gravel – filling space between spandrel walls and/or wingwalls below road surface.
Keystone	The highest and last stone placed in an arch. In a bridge arch, there is a line of keystones across the width of the arch, but a projecting or carved keystone was often placed on each elevation.
Parapet (*wall*)	Side wall above road level for protection of road users.
Quoin	An angle in masonry, as where the vertical face of a bridge spandrel meets the arch soffit.
Rise (*of arch*)	The vertical height from the lowest to the highest point of the soffit.
Soffit	Bottom surface or underside.
Spandrel (*or spandril*)	The material (or void space) overlying one half of a bridge arch under the road surface.
Spandrel walls	Side walls enclosing this material or space.
Voussoirs	The stones forming an arch, wedge-shaped in hewn masonry, but flat or irregular in rubble.
Wings (*or wingwalls*)	Side walls extending landward from the spandrel walls to contain the fill under the road.

13
The foundations of Hexham Bridge

E.C. Ruddock

The story of the construction and collapse of John Smeaton's bridge in 1777–1782 is told with liberal reference to contemporary practice, his own previous experience, and other attempts to bridge the Tyne at Hexham. The safety of the foundations against scour and against bearing capacity failure is assessed by modern analyses with the aid of a recent soil survey, and the contributions of skirts and girdles to the stability of the foundations is also investigated. The collapse is shown to have followed the classical pattern of failure caused by scour.

On présente l'histoire de l'établissement et de l'écroulement du pont de John Smeaton entre 1777 et 1782 en faisant souvent mention aux techniques contemporaines, à sa propre expérience, et aux autres essais de jeter un pont sur le Tyne à Hexham. La sécurité contre les affouillements et la capacité portante des fondations sont évaluées par des analyses modernes, en se servant d'une reconnaissance récente des sous-sols; on examine en plus les effets des 'jupes' et des 'ceintures' sur la stabilité des fondations. On montre que l'écroulement du pont se passa selon la manière classique d'écroulements par affouillements.

The collapse of John Smeaton's bridge over the Tyne at Hexham in March 1782 was the only important failure in his career. Although it has been mentioned in many books on engineering history, the fullest narrative of its design, construction and failure is still the *Memorial on Hexham Bridge* (*Reports* **3**, pp. 299–320) written by Smeaton himself in May or June 1783 and printed with other letters and reports concerning the bridge in his collected *Reports* in 1812. Twenty-two original drawings, or fair copies, are also extant in Smeaton's own collection of drawings, now in the library of the Royal Society of London (*Designs* **4**, pp. 131–148). The *Memorial* is the primary source of the historical description which follows and only facts derived from other sources are individually referenced. The dimensions used in the second part of the paper are derived from either the drawings or the *Memorial* unless otherwise noted.

HISTORY

Current practice and Smeaton's common methods

Several methods of making good foundations in water were in common use in Britain in the 1770s. The method most generally favoured was to form a watertight cofferdam, pump out the water and excavate about 1 m into the river bed, lay a timber 'grating' or platform at that level and the first course of masonry on it. Piles might be driven under the grating, and this was Smeaton's common practice, used for the bridges of Perth and Coldstream on bottoms where the gravel was deep and had no soft material underneath. He added what he called a 'fence' or 'casing' of timber sheet piles about 2·5 m long, driven just under the edge of the pier to ensure that the gravel under it would be retained whatever scour might happen elsewhere (*Reports* **1**, pp. 188–190 and **3**, pp. 236–239; *Designs* **4**, pp. 125, 125v, 157). It corresponds to what is nowadays called a 'skirt'. For Banff Bridge, which he designed in 1772 to replace one swept away by a flood after only three years' use (Cramond, 1891, pp. 289, 321), he contrived to make the casing serve as cofferdam as well (*Reports* **3**, pp. 349–352; *Designs* **4**,

p. 120v); the bottom was a similar gravel. In these, as in many other works in running water, he required the foundations to be surrounded by slopes or mounds of loosely dumped rubble stones, and that similar stones be dumped in any hollows which the flow of the rivers might make in their beds, because he had 'abundantly experienced that good quarry rubble would resist the action of a current to a greater degree than any kind of gravel' (*Reports* **3**, p. 309). He expected this rubble to fill up with gravel and finer particles and become, with such adjustment of shape as the flow itself caused from time to time, a stable and permanent bottom, needing only occasional inspection and minor replacement of material.

Other bridge designers often ordered unstable bottoms to be 'penned' or 'framed and sett'. These terms meant paving with heavy stone blocks or setts held within a frame of thick timbers laid flat on the bottom and tied down by short piles. This work might extend round the piers themselves or, alternatively, over the whole bottom between the piers and a short distance up- and downstream.[1] By preventing scour near the piers it removed the danger of the pier foundations themselves being underwashed.

The other method of founding was in caissons and it was most used where the ground was so permeable or the tides were so high that cofferdams could not be dewatered. The Tyne at Hexham, in a span of thirty years, was to call for all these methods.

'Gravel crust' bottoms

Every gravel river-bottom was subject to scour and Smeaton's defence had always been rubble stone; but a gravel crust over very weak material was especially difficult. Before his design for Hexham in 1777, Smeaton had recognized such conditions at least twice. In 1760 he bored the bed of the Clyde at seven different crossings by the town of Glasgow and found at all of them 'mud or sleech' underlying a gravel which varied from 0·3 to 1·6 m thick (*Reports* **1**, pp. 333–7). At the worst site of the seven, where the gravel was loose and 1 m thick, he thought it possible to found a bridge if the bed between the piers was protected by 'piling, setting and framing'. But he recommended that the bridge be built at another site where the gravel was more compact, and he was content to design his bridge there with a combined width of waterway through all the arches equal to the existing breadth of the river, so as to cause no increase of mean velocity of the current. The design shows piers only 4 m wide founded on timber platforms no more than 0·3 m below the surface of the river bed (*Designs* **4**, p. 114). There is a timber or stone 'girdle' laid on the projection of this platform tight against the side of the first course of masonry.

This design was submitted along with his proposals for development of the river, and these proposals involved considerable control works, including damming out the tides (Renwick, 1912, pp. 47–49) so he probably expected no violent currents to attack his bridge. The bridge and control works were both abandoned, but when another bridge was designed in 1768 by William Mylne, not for the site Smeaton chose but for the one he thought worst, Mylne drew just the same detail for the foundations of his piers (Mylne, 1768). They were built, however, not like that but with bearing piles and a surrounding sheet-pile skirt, an acknowledged copy of Smeaton's method at Perth Bridge then nearing completion (Ramsay, 1768). They gave some trouble just after they were finished and had to be protected by a 'dam' downstream, to raise and steady the water under the bridge.

The framing and setting which Smeaton mentioned here he never suggested again for a large bridge and he criticized the method at least once because decay of the timber made it impermanent (*Reports* **3**, pp. 345–348). What foundations he would have proposed, had there been no intention to control the river, we cannot tell.

[1] See Fig. 12.

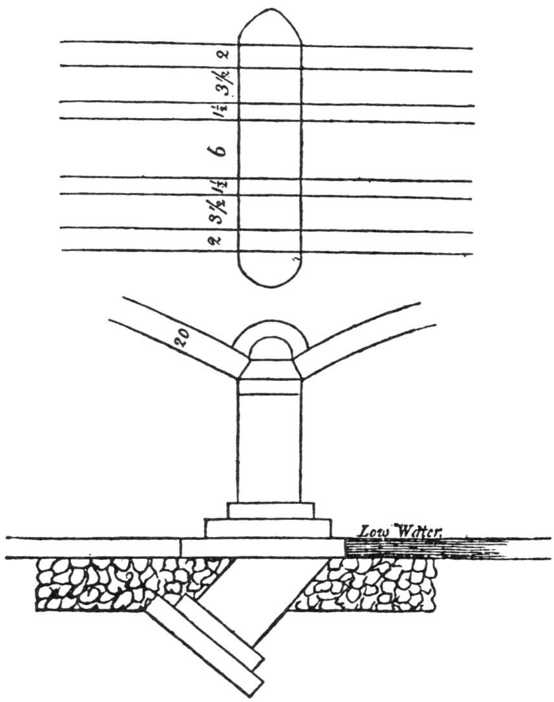

Fig. 1. Sketch of Dumbarton Bridge failure (Smeaton: *Reports* 1, p. 339)

At Dumbarton in 1768 he saw the failure of two arches of the newly-built bridge, caused by a gross tilt and settlement of the pier between them (*Reports* **1**, p. 339)—a movement which today would be called a bearing capacity failure, if his sketch is at all accurate (Fig. 1). He found the river bottom to be a crust of gravel 0·6 m thick over mud so soft that an iron bar sank in by its own weight to a depth of 12 m. That the failure was caused by simple overloading of the ground is confirmed by the fact that it happened 'without any flood, external violence, or previous notice' (*Reports* **3**, p. 294). Smeaton recommended that the sunken pier be left as it was, the gravel crust made good, apparently with rubble stones (Fig. 1), and a new pier be built on top of the old one, making it and the spandrel over it much lighter than before. He believed this was done with success and recalled it when Hexham Bridge failed fourteen years later (*Reports* **3**, pp. 293–294). Recent works confirm that the bridge had hollow spandrels (F. A. MacDonald and Partners, 1934; Babtie, Shaw and Morton, 1960).

Earlier bridge work at Hexham

In making his design in 1777 (Fig. 2), Smeaton had several other precedents to consider. He himself had made the first known design for a bridge at Hexham in 1756, at a site between Tyne Green on the South and Hermitage on the North bank, some distance downstream of the West Boat (Fig. 3). What he knew of the form of the river bed and the nature of the material neither his drawing nor his estimate shows, but the estimate does show that he intended to lay the foundations, of 'rough block stones', several feet below the bottom. He allowed for

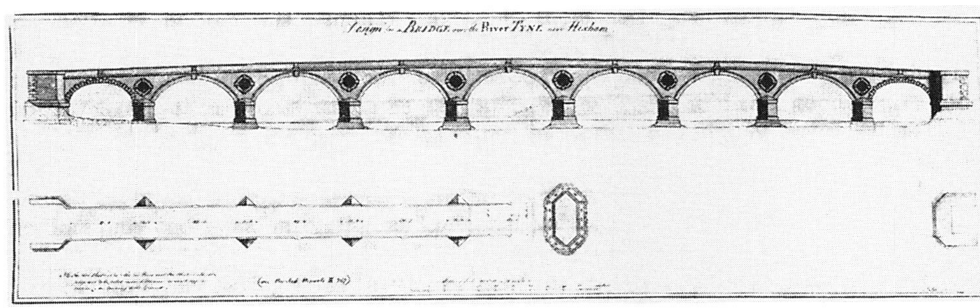

Fig. 2. Smeaton's design drawing, 1777 *Courtesy of the Royal Society*

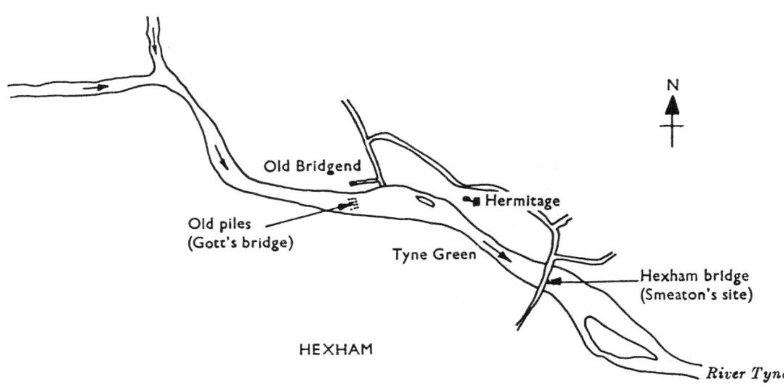

Fig. 3. Map of Hexham and the Tyne

cofferdams and pumping but no bearing piles. The bridge was to be of four long arches each 23 m span, with piers 3 m wide, the total width of waterway under the arches being 92 m (Smeaton, 1756).

He had a second invitation to design a bridge for Hexham, apparently in 1764, but declined to undertake it due to pressure of other work (Smeaton, 1764). The request came from Sir Walter Blacket who owned the estate of Hexham Abbey, and Sir Walter later engaged John Gott to design a bridge. This was built a little upstream of the West Boat, where Old Bridgend now stands on the North Bank (Fig. 3). Gott was well known to Smeaton as a bridge contractor in the West Riding and superintendent of works on the Aire and Calder Navigation, and Smeaton saw some of the building of the bridge when passing and heard more, certainly from Blacket and probably also from Gott. The piers were founded in 'broad cofferdams of earth' kept dry by chain pumps. There were tapered timber piles driven 3 to 4 m and supporting platforms of 0·3 m timbers 1 m below the river bed. On these platforms the masonry was built.

In the hours of darkness of 16–17 November, 1771, less than two years after its completion, all the superstructure of this bridge except one abutment was destroyed by the 'great inundation' of the Tyne which tore down six other bridges and left only one intact (Sykes, 1833). Because it happened at night the sequence of failure was not observed, but at least some people believed the bridge to have failed 'for lack of elevation and not for want of proper foundation'

Fig. 4. Platform for one pier of John Gott's bridge, photographed 1974

(*Errington Papers*, pp. 41–46). The likeliest meaning of this phrase would seem to be that, the whole bridge being low, the flood first filled all the arches to the crowns and then accomplished the destruction of the superstructure without disturbing the foundations.

Smeaton wrote in his *Memorial* in 1783 that he found this failure impossible to explain until he heard from Jonathan Pickernell in 1776–7 of later investigations of the river bed (described below), when he concluded that 'the violence of the water [had] taken off the crust of gravel, wounded also by the excavation for the piers, so as to let loose the quicksand' which he believed must extend under all the foundations. He was encouraged in this view by the knowledge that Blacket had paid a bond for £3000 to the County Justices rather than take the risk of rebuilding the bridge but absolved his undertakers from a similar bond, so sure was he that they could not have foreseen the violence of such a flood; and Smeaton was further encouraged by the sight of the platform which formed the base of one of the piers standing for several years after the collapse tilted sideways out of the water at about 45° to the horizontal.

Three years later, however, in his answers to a series of 'interrogatories' which were put to all the witnesses in the lawsuit between Errington and the County, he offered a different explanation in which he considered the tilted platform as possibly the only one disturbed by scour, and that at one side only, thus causing the tilt (*Errington Papers*, pp. 129–130). He referred also to the bridge's 'lack of elevation', but in the sense, not of a low profile of the whole bridge, but of a high ratio of span to rise in each arch with consequent large horizontal thrust which would cause a progressive collapse of all the arches once a single pier had moved.

Examination of the site today supports this later explanation. It is clear that the crust of gravel was not generally 'taken off', since most of the piles of two of the piers still stand erect and little displaced, protruding from 0·3 m to 1·0 m out of the river bed, while the platform of a third pier lies in its place, complete and only slightly tilted, and presumably with its piles intact under it (Fig. 4). There is a part of another platform tilted just like the one Smeaton described and, if it is the same one, it argues that the gravel at this site is in fact uncommonly difficult to 'take off'.

Smeaton was called to report on the destruction by this flood, not of the Hexham Bridge but the bridge at Newcastle (*Reports* **3**, pp. 260–266). This he did in conjunction with the local engineer John Wooler, who was thereafter responsible for the reconstruction of the northern half of Newcastle Bridge. Wooler was also engaged by the Justices of the County in 1774 with regard to Hexham Bridge; first to approve a design by a certain John Fryer which local builders offered to build for £5000 (plus the materials of the former bridge) and later, when they withdrew and there was no answer to oft-repeated advertisements for another undertaker, to make his own design of a bridge to be built by direct labour under his supervision (*Quarter Sessions*, 1774–1775). Borings made by a 'surveyor' had shown, at a site about 50 m upstream[2] from the former bridge, 'a bed of clay laid at no more than four feet under the bed of the river', and so Wooler designed timber platforms to be placed on the top of the clay with piles driven below, the same sort of foundations as Gott had used. But when the resident supervisor Jonathan Pickernell dug for the first foundation in 1775 he found at 4 ft (1·2 m) depth not clay but 'quicksand full of bubbly springs, and of so loose a texture, that by hand only, a bar of iron entered into it 46 ft (14 m) without meeting any resistance; and that a trial pile of whole timber entered 26 feet (8 m), at 2 inches and a half (62 mm) per stroke of the ram without stopping'. In a letter dated July 1775 (*Errington Papers*, p. 3), Wooler pronounced the bridge impossible unless the whole river bed were paved with a 'wall' 2 m high and 13 m wide founded 1·6 m below the bed on timber 'sleepers' bedded in the quicksand at a close spacing of 0·3–0·45 m and protected by lines of closely-spaced timber piles driven 8 m into the quicksand along the whole of the upstream and downstream edges of the wall. This was much more than ordinary penning and would have made a true raft foundation for the bridge. It was obviously very expensive and never seriously considered.

The Justices, already committed by more than £3500 spent on preliminary works and materials (*Errington Papers*, pp. 164–165), continued to advertise for undertakers and continued to employ Pickernell. He had already made borings up and down the river, finding everywhere loose sand or quicksand under only a small thickness of gravel. One site he bored was a mile downstream from the former ones, a little below the East Boat, and it was an interest in this site, which adjoined his estate, that brought a suggestion from Henry Errington, Esq., that he would contract to build a bridge if John Smeaton should think it feasible, design it and direct the construction.

First design, 1777

Smeaton chose a line across the river near the downstream end of a still pool below the East Boat (Fig. 3) and made soundings himself from a boat, but only to a depth of 2·7 to 3·0 m. He used a pointed iron bar, counting the blows of a hammer required to drive it, which he thought to give a better estimate of the compactness of gravel than a boring tool.[3] He found that the bar was 'very considerably less resisted, and particularly in the main current, after it was

[2] In his 1786 evidence Smeaton said 50 m downstream (*Errington Papers*, pp. 129–130), but as this is the only reference to a downstream site it is considered a mistake.

[3] As modern engineers use penetration tests instead of borehole samples in coarse-grained soils.

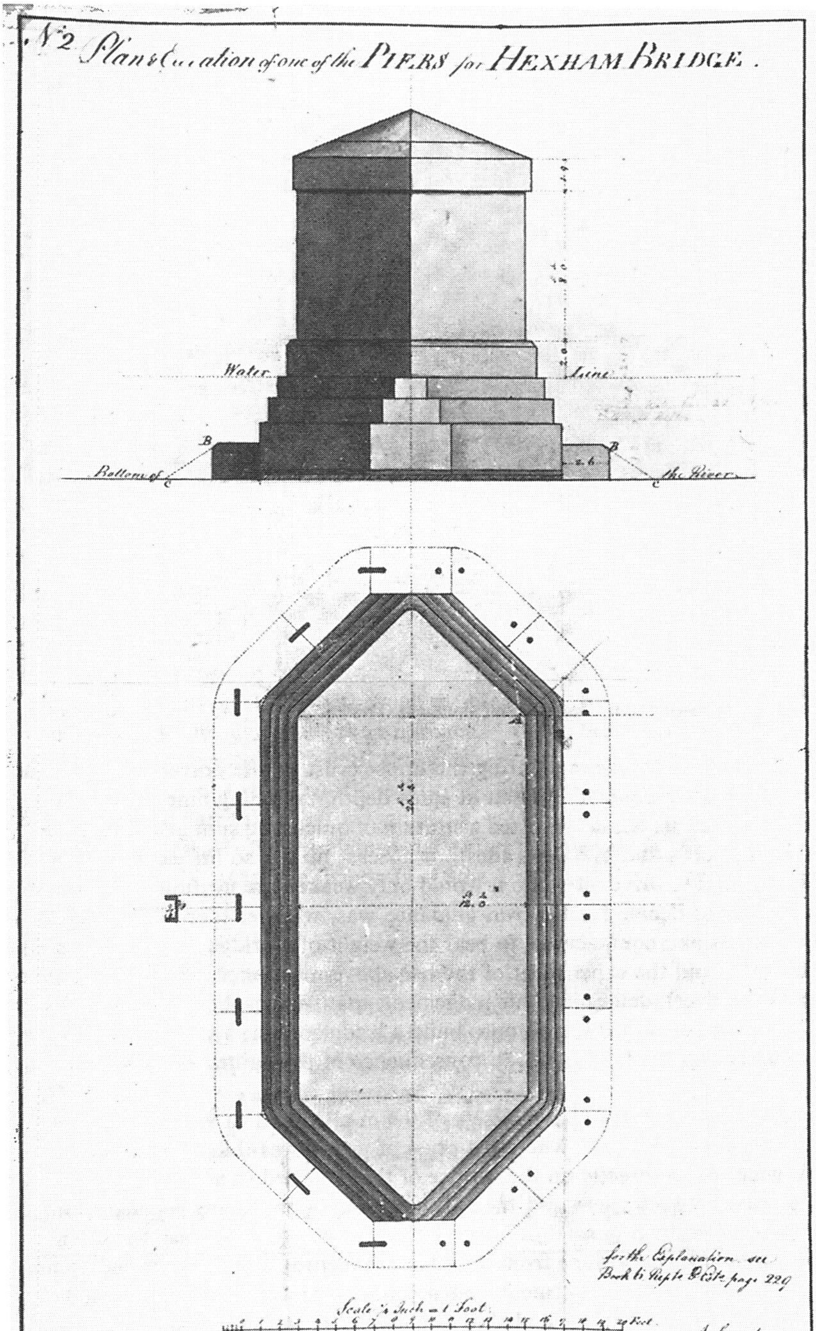

Fig. 5. Smeaton's design for a caisson pier 1777 (Design 1) *Courtesy of the Royal Society*

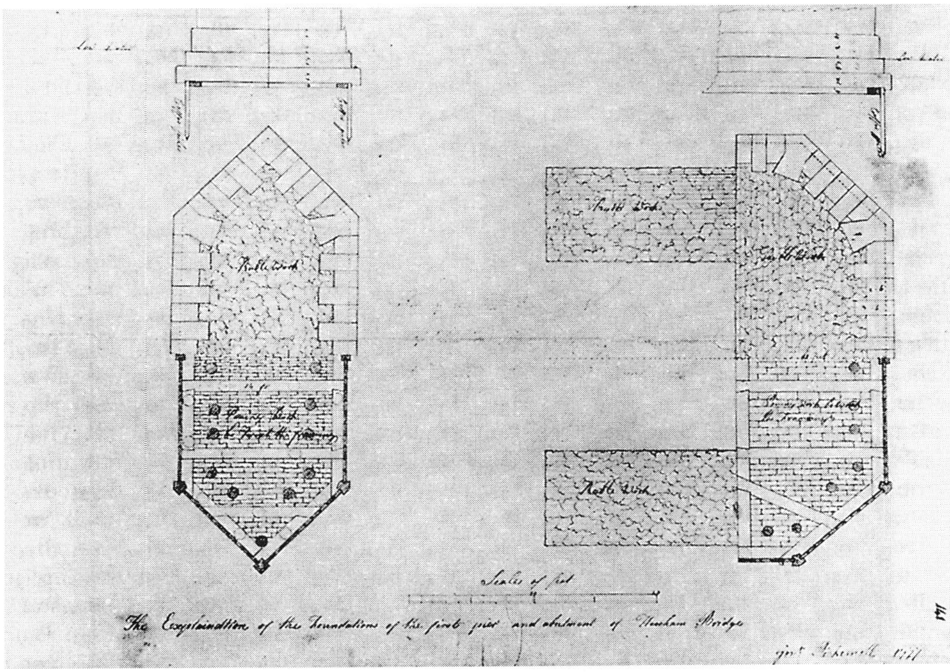

Fig. 6. Pickernell's drawing of a piled pier and abutment (Design 2) *Courtesy of the Royal Society*

driven some feet, than it was in entering the upper crust of the gravel bed [and] he thought himself well justified in concluding, that at some depth, exceeding nine or ten feet (2·7–3·0 m), at this place, there either actually existed a stratum of quicksand similar to that at the West end of Tyne Green [Gott's and Wooler's sites], or at least matter so little compact or capable of bearing weight, that to drive piles into it would only weaken the stratum. The question therefore, that he had to decide for his own guidance was, whether there was a bed of gravel of sufficient thickness and compactness to bear the weight of a bridge, in case it was unwounded and unbroken? And the experiment of the bar above-mentioned (which was tried in several places across the river), determined his judgement, that what he had felt and experienced was sufficient. . . . He concluded therefore to build a bridge of nine arches instead of seven that it might have more legs to stand upon, in consequence of the natural weakness of the stratum' (Fig. 2).

The 'legs', or piers, were to have offsets of 0·6 m all round to widen the foundations, and towards the middle of the river where the crust of gravel was thinnest and weakest he would place these wider bases directly on the surface of the river bed so as to maintain the full thickness of the crust between them and the soft stratum. For defence against scour each base would be surrounded by a 'girdle' of heavy shaped stones, each at least one tonne weight and all cramped together, and a slope from the edge of the girdle down to the bed composed of 'a deposition of rubble, stone and cement, called beton'. Since a cofferdam would 'wound' the gravel crust, the piers would be founded in caissons (*Reports* **3**, pp. 270–273). His drawing of this construction (*Designs* **4**, p. 134v) is given here as Fig. 5 and in subsequent analysis the proposed method will be called Design 1.

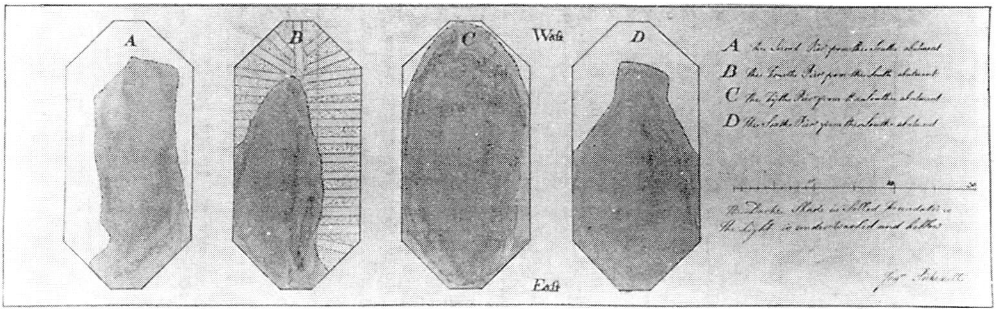

Courtesy of the Royal Society

Fig. 7. Pickernell's plans of the underwashed piers 1778. The dark areas are those remaining in contact with the ground, the white areas are underwashed

For the piers next to the abutments and the abutments themselves, where the gravel seemed harder and possibly deeper, Smeaton proposed founding by his common method in cofferdams, with bearing piles, grating, skirt of sheet piles about 1·5 m long, and rubble defending slope. This is shown in Fig. 6 (*Designs* **4**, p. 141). The gratings are shown by this and other drawings to have been laid at from 0·85 to 1·40 m below the assumed low-water level. The depths below the river bed are not recorded. This method of construction will be called Design 2.

The contract drawing is given in Fig. 2 (*Designs* **4**, p. 131). The middle arch had a span of 15·5 m and the total width of waterway through the nine main arches was 129 m, with a further 16·5 m of 'land arches' through the south approach. The width obstructed by the piers was 29 m. On the back of a note of these dimensions amongst Smeaton's drawings (*Designs* **4**, p. 130v) is a second list of figures which seem to be the dimensions of the arches of Gott's bridge and these would show that that bridge had a waterway of only 113 m; the middle arch was certainly 21·4 m in span. Two drawings made after the collapse in 1782 (*Designs* **4**, p. 148; *Errington Papers*, p. 7) show that the spandrels of Smeaton's bridge were built hollow to lighten the superstructure.

Changes in design during construction (1778–80)

Use of the cofferdam method (Design 2) was extended to the second pier from the north end as well as the first from each end. The founding of the other five piers began in the spring of 1778 and Smeaton was present to see the first one founded. By the end of July three more were founded and built up above water level. The placing of the girdle and beton slope, however, had begun only at one pier when a rapid flood scoured the bed from under all four at their upstream ends and along the sides to an average width of 0·4 m and they tilted and settled by up to 0·45 m (Fig. 7).

Smeaton decided that the girdle and beton were not enough protection and ordered sheet piling to be driven at about 1 m distance from all the faces of each pier, with a 'screw clamp' or waling to tie it just below low-water (*Reports* **3**, pp. 275–278; *Designs* **4**, p. 139). He hoped it could be used as a cofferdam and pumped dry while small stones, gravel and sand were packed by a hand ram into the voids to underpin the pier, but this proved impossible and a small diving bell was used for the underpinning. The piles were cut off just above the waling at about low-water level and the space within the casing filled up with rubble and sand, paved on

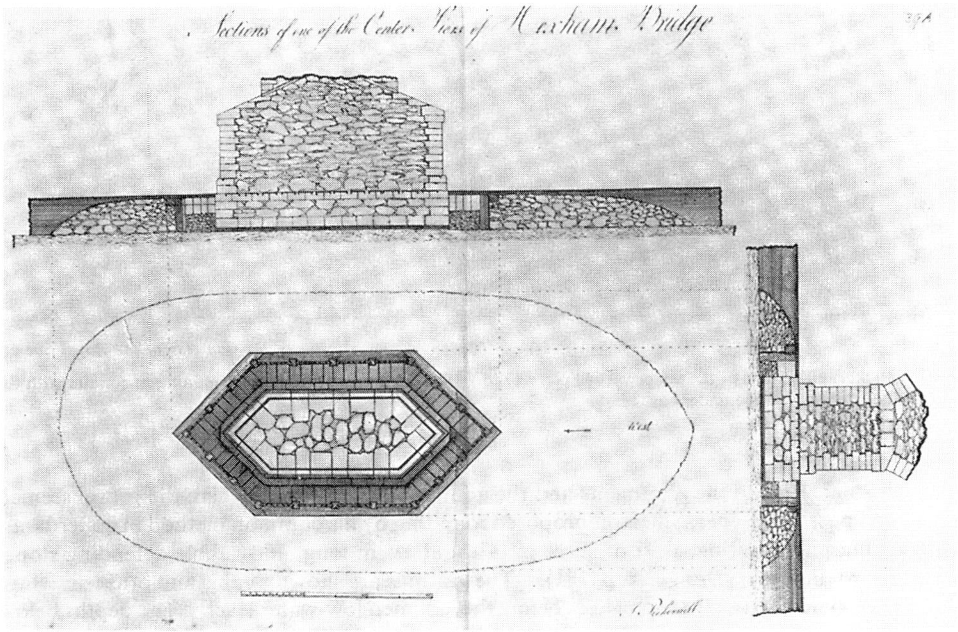

Fig. 8. Pickernell's drawing of a pier as built 1779 (Design 3) *Courtesy of the Royal Society*

top with large squared stones. Outside the piling a slope of coarse rubble was dumped, extending about 3 m out at the sides of the pier, 10 m upstream and 7 m downstream. Pickernell's drawing of this (*Designs* **4**, p. 139A) is reproduced as Fig. 8. The masonry was completed with adjustments for the tilting and extra cramps and tie courses. He ordered the sheet piling to be driven 3 m into the gravel at the upstream end of each pier, reducing to 2·1 m at the downstream end, but the piles at one of the middle piers would drive only 1·5–2·1 m and at another, the seventh from the north, they were badly deflected by large stones in the gravel. He assumed that the remainder were driven as ordered. In subsequent analyses this construction, with the piles penetrating 1·83 m into the existing river bed, will be called Design 3.

On 12 December, 1778, when the piling and most of the underpinning had been completed, but not the rubble slopes, a flood rose to within 0·22 m of the top of the impost (top course of the piers) and Pickernell reported that there was a fall of 0·69 m from the upstream to the downstream end of the piers. In 1779, when most if not all of the rubble had been placed, a fall of 1·15 m was reported during another flood. Both these floods were 'sufficient to tear up and remove the natural bed of gravel . . . wherever there was a particular set upon it', but caused no distress to the piers, and the rubble slopes were found to have survived, being usefully filled up and covered with gravel in some places.

In 1780, as the bridge was being finished, it was reported that the bottom was wearing deeper in the channels between the piers and Smeaton advised Pickernell that the rubble slopes might be allowed to settle until their tops were 0·92 m down the face of the sheet piling if their surfaces were at a gradient of no more than 1 in 2 and the bottom of the channels no more than 2·44 m below low-water (*Reports* **3**, pp. 291–293). This settling of the surface of the slopes would be

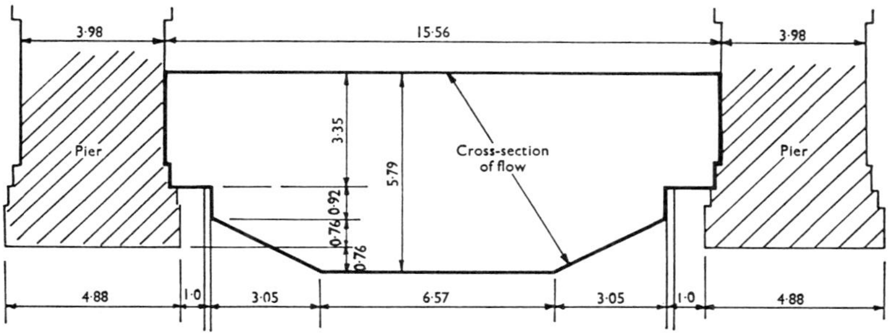

Fig. 9. Waterway and piers of middle arch. Limiting safe conditions envisaged by Smeaton (Design 4)

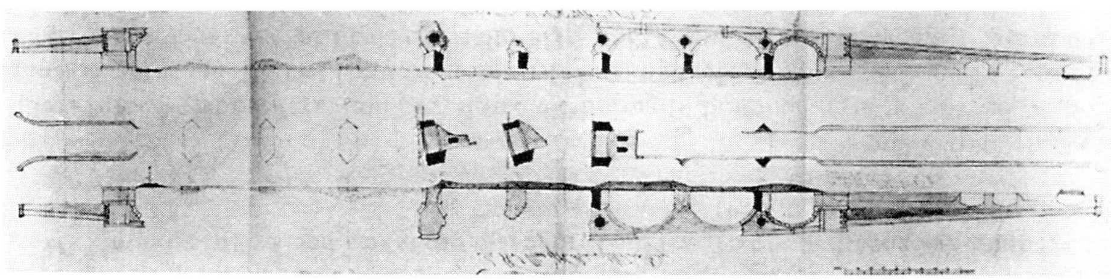

Fig. 10. Smeaton's drawing of the ruins 1782 *Courtesy of the Royal Society*

caused not by the current removing the rubble but by scour of the natural gravel bed near and under the toe of a slope, with the rubble rolling down the slope as well as settling and compacting. Fig. 9 shows the limiting condition which Smeaton was describing and this will be called Design 4 in the analyses given below.

The collapse, 1782

The facts of the bridge's collapse on the morning of 11 March, 1782 were briefly told by Errington's agent John Donkin who watched it (*Errington Papers*, pp. 24–25). He noted the levels of the water at both ends of the piers, which differed by between 1·22 and 1·53 m, and probably about 1·37 m. Particles of lime first began to fall from the fourth arch from the north end; this increased for about a minute, then cracks spread through the arch and spandrels for about another minute and then the third pier and the third and fourth arches collapsed together. Five other arches fell within the next half-hour, leaving only the first three at the south end, but large parts of two of these fell shortly afterwards. The two middle piers stood tilted at 10° or more, their west (upstream) ends sunk in the river bed and their east ends lifted clear above it (*Designs* 4, p. 148). The piers founded on piles and gratings fared no better, two at the northern end being reduced to mounds of stones (Fig. 10).

Smeaton's explanations of the failure

In several explanations during the ensuing years Smeaton gave three reasons for the failure. The first and most important was that the floods in the River Tyne had been shown capable of a degree of violence far greater than he had thought possible (*Errington Papers*, p. 26). He calculated from the heights of fall reported that the velocity of the water through the bridge during the flood of 1779 was 4·73 m/s and that in the fatal flood of 1782 upwards of 5·10 m/s. He insisted that if he had believed it possible that there should ever be a fall of as much as 0·76 m (with corresponding velocity of 3·86 m/s) through the bridge he 'never should have recommended to Mr. Errington to have undertaken to erect a bridge upon that bed of gravel'.

Secondly, he emphasized the weakness and susceptibility to scour of the river bed, implying that it was at least as bad in these respects, and perhaps significantly worse, than he had anticipated (*Errington Papers*, pp. 26 and 129–130). He also deduced, quite precisely, that though the rubble defences could withstand the scouring effects of water moving at 4·57 m/s (an approximate statement of the velocity in the 1779 flood) they could not withstand a current of 5·10 m/s. It is clear that he had previously expected the rubble to withstand any current which might happen in the Tyne.

Thirdly, he expressed twice in the year after the failure the idea that 'the incumbent weight of a body of water of 5 ft (1·53 m) in thickness above the bridge superior to what was to counterbalance it below had actually forced down the whole stratum of the gravel into the soft matter that lay beneath it and caused the Bridge itself to take a heel up the stream' (*Errington Papers*, p. 26; *Reports* **3**, p. 294).

Litigation and rebuilding

Through six years of argument and litigation Smeaton resolutely refused to 'risk his credit' on the design of another bridge at Hexham and Errington declined to fulfil his contract to uphold the bridge for seven years (Anon, 1788). He offered instead to pay the Justices £3000, which was Smeaton's estimate (*Errington Papers*, pp. 107–108) of the cost of reinstatement according to the original design, to which the contract referred; a jury eventually awarded the Justices £4000 damages in 1788 (Anon., 1788). The Justices engaged Robert Mylne, the designer of Blackfriars Bridge, as their technical witness and as part of his survey of the ruins in 1783 he had a single boring made near the second pier from the north end, which was one of the piers built on piles because Smeaton thought the gravel deep enough there. Mylne's 'borer' drilled to a depth of 7 m and reported the ground to be 'uniformly a composition or congeries of roundish and flat stones, gravel and sand, of equal quality and consistence in the whole of that depth' (*Reports* **3**, pp. 296–298).

Mylne considered that a bridge could be built safely on this bottom and when the case was concluded the Justices duly built one on the same, or very nearly the same, site by direct labour under the direction of their Bridge Surveyors, William Johnson and Robert Thompson (*Quarter Sessions*, 1788–1796). How much help these men received from Mylne is not clear, but it was almost certainly limited to the foundation design, for the superstructure (Fig. 11) is a replica of Smeaton's bridge, and must include much of the stone of it, salvaged from the river after the collapse.

The type of foundations on which the bridge now stands is shown by an isometric drawing at the County Surveyor's office, which is reproduced as Fig. 12, and there is a small-scale plan and elevation drawn and signed by William Johnson in 1796, now in the Drawings Collection of the Royal Institute of British Architects, which confirms that these are the original foundations. The piers stand on timber piles and platforms and the whole river bed is framed and sett both under the arches, for 1 or 2 m upstream and for almost 8 m downstream from the ends of the piers. The framing is tied down by no less than five lines of closely-spaced piles extending right across the river. The cost to the county was just over £8000 (*Quarter Sessions*, 1788–1796), compared with a contract price of £5700 for Smeaton's bridge.[4]

[4] Evidence by John Donkin during the legal proceedings shows that Errington's actual expenditure was about £6060, including an allowance for the services of his agent, etc.

Fig. 11. Hexham bridge today. The cantilevered footways were added in 1964

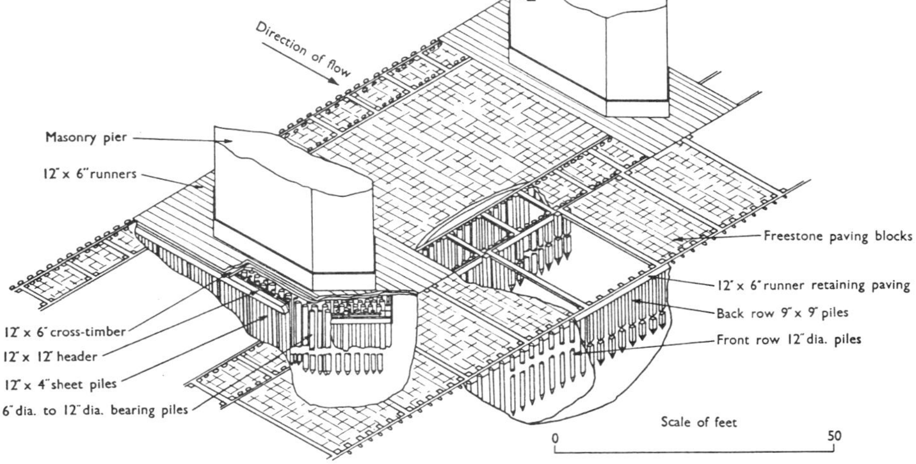

Courtesy of Northumberland County Surveyor

Fig. 12. Drawing of foundations, twentieth century

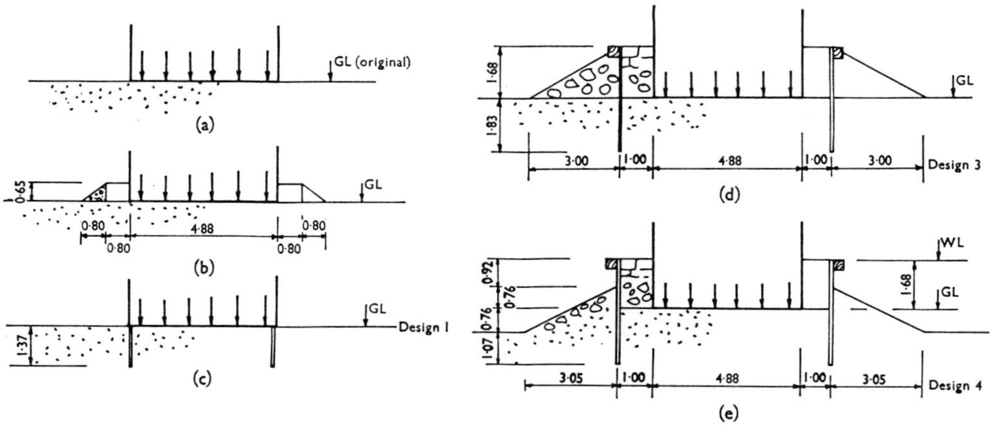

Fig. 13. Sections of pier foundations for bearing capacity calculations

Later history and recent soil survey

In 1831 the County Surveyor, in a review of bridges under his care, commented that 'the foundation masonry of the abutments and piers has not been laid at a proper depth below the river, by which ill-judged economy, frequent and expensive repairs will be indispensable' (*Surveyors Bridge Book*, 1831). Major repairs, mostly to the 'penning' cost £644 in 1831–1832, £355 in 1867, and £92 in 1883–1885 (*Expenditure*, 1831–1885). In this century the penning has been concreted over and the concrete extended to form an apron stretching at least 13 m downstream over the whole width of the river.

An extensive series of borings has been made recently along the line of the Hexham and Corbridge Bypass and this shows the general character of the ground in the area (North Eastern RCU, 1973). All the deposits overlying the rock appear to be alluvial in origin and their depth is generally about 20 m. Gravel/sand mixtures containing boulders and rock fragments, and generally dense or medium-dense, are the most common materials and where the line crosses the Tyne about 900 m upstream from Gott's site, this material extends from the river bed right down to rock. Along a section of the line which runs parallel with the river abreast of Gott's site but some 300 m to the North, layers of silty sand and sandy silt were found at depths varying from 1 m to 6 m in six successive boreholes. Further east the rock level was shallow and these layers absent. Two boreholes were made in the north bank on the downstream side of the existing bridge (i.e. Smeaton's site) and within 25 m of the abutment. Under 3–4 m of made ground there was in one hole a stratum of dense to medium-dense sandy and clayey gravel extending to a depth of 16 m, and in the other hole similar material but covered on top with a layer of 'medium-dense clay' between the depths 3·8 and 5·2 m. The lowest standard penetration value recorded in these two holes was 16 in this clay layer. Slight 'blowing' of the gravel was recorded at 12 m in the first hole, but this is the only suggestion of potential quicksand conditions.

These findings must throw doubt on Smeaton's impression that there was a general layer of quicksand or other silty material only a metre or two below the bed along at least a mile of the river; but with regard to the actual site on which he was building the two borehole logs do not conflict with the findings of his probing survey, and there is no check on his findings about the river bed at midstream.

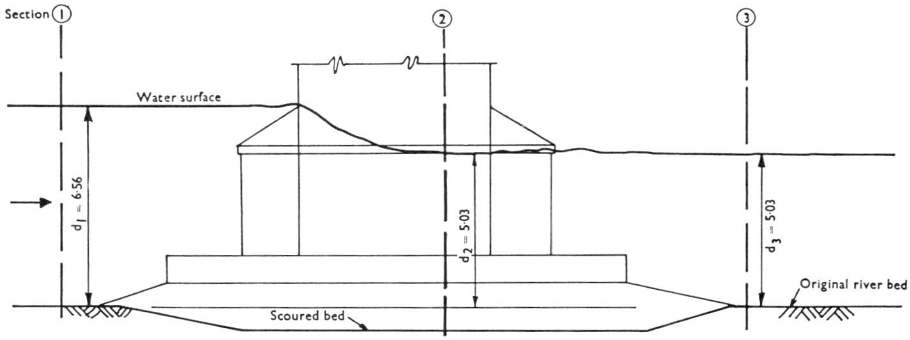

Fig. 14. Profile of flow through middle arch (Design 4)

ANALYSIS

There are several small discrepancies between dimensions given in Smeaton's design drawings and what are apparently record drawings made by Pickernell at the site, and where these occur Pickernell's values are preferred. The normal or low-water level is taken as 1·68 m above the (original) natural surface of the river bed at the caisson piers. This conforms to the level of water shown in Pickernell's drawing of one of the middle piers (Fig. 8). His drawing of an end pier and abutment (Fig. 6) shows that both the river bed and the base of the masonry were higher there, being only 0·4 m and 0·7 m respectively below low-water.

Hydraulic calculations and assessment

Smeaton's calculations of velocities of flow from the observed fall are considered in relation to Design 4, since they were made after he had advised that Design 4 would be safe. Design 4, as it applies to the middle arch and its piers, is illustrated by Figs 9 and 13(e).

His results all correspond to the simple relationship $h = v^2/2g$, where h is the loss of potential energy head in passing the bridge—or the 'fall',—and v is the velocity gained. This, as he said, was a 'rule' stated in the hydraulic textbooks of the time, but it was proved for, and properly only applied to, the discharge from a static tank through a small orifice. It could not be applied with any accuracy to the discharge through the arches of a bridge because there would be a considerable velocity of approach. The amount of error is shown by calculating the velocities through the middle arch by a modern textbook method (Webber, 1968, pp. 202–204).

The longitudinal profile of the water is known to be of the general form shown in Fig. 14. Webber finds that

$$\frac{v_2^2}{2g} - \frac{v_1^2}{2g} = d_1 - d_2 = 1{\cdot}53 \text{ m} \qquad (1)$$

For continuity of flow,

$$A_1 v_1 = A_2 v_2 = A_3 v_3 \qquad (2)$$

The appropriate width of flow area at cross-sections (1) and (3) is that of the middle arch plus the half-widths of the adjacent piers, i.e. 19·54 m. A_2 is the area of flow shown in Fig. 9. Inserting values, the equations yield $v_1 = 4{\cdot}05$ m/s; $v_2 = 6{\cdot}84$ m/s; and $v_3 = 5{\cdot}40$ m/s.

Smeaton's calculation gave a single velocity, applying correctly to section (2), of about 5·10 m/s. Although the difference between this and the modern result is not large it could be quite disastrous in view of the fact that the transporting power of a current, and therefore the onset of scour, is found to be proportional to the fifth or sixth power of the velocity (Webber, 1968, pp. 164–6). Smeaton's statement that the river bed was stable under a current of 4·57 m/s and yet completely torn up by a current of 5·10 m/s is much more reasonable in the light of this modern knowledge than it must have seemed in his own day.

However, even the velocity of 3·86 m/s which he said would have dissuaded him, could he have foreseen it, from ever undertaking the bridge, is very much higher than the velocity which would nowadays be considered safe over even the coarsest of gravel (Neill, 1973, pp. 90–94; Webber, 1968).

The location of maximum scour agrees well with modern experience. The worst scour during the flood of 1778 took place near the shoulder angles of the upstream ends of piers (Fig. 7) and on the day of the collapse two piers in the middle of the river developed a steep tilt upstream (Fig. 10), clearly due to concentration of scour at their upstream ends. From the evidence of observations at full scale and in hydraulic models (for instance, Neill, 1973, pp. 76, 94, 98; Terzaghi and Peck, 1948, pp. 409–11) this is now considered the 'classical' pattern of scour round bridge piers. The sharp 135° shoulder angles used by Smeaton in the plan of the piers cause more eddying and therefore greater scour at the upstream ends than would rounded surfaces. He was well aware (*Reports* **1**, pp. 99–102) of the use of smoothly-curved shoulders by the French engineers and of the hydraulic arguments for them, but in most if not all of his bridges he adopted shoulder angles of 135°, with the point of the cutwater forming a right-angle.

Foundation methods and bearing capacity

A precise analysis of the final mechanism of collapse is not attempted because it would require accurate knowledge of the properties of the soil and the shape of the river bed surface just before the collapse. Such data are not available. The following verbal arguments and calculations attempt, firstly, to assess the effects on the stability of the foundations of the two expedients which Smeaton used to protect the piers from the effect of scour. There is little doubt that he expected these devices, here as well as in other bridges, to increase the bearing capacity and reduce the settlement as well as fulfilling their primary function. The two devices were the girdle of masonry and/or loose rubble and the skirt of sheet piling. The second purpose of the analysis is to determine the approximate values of the factors of safety against a failure of bearing capacity at various stages of the design and construction.

The five designs considered are shown in Fig. 13 (a)–(e). Fig. 13(a), which Smeaton did not propose at all, is included as a control case with no girdle or skirt. Fig. 13(c) differs from Smeaton's Design 2 in omitting both the bearing piles and the rubble slope; this is done to isolate the effect of the skirt. The length of the skirt, 1·37 m, corresponds to Pickernell's drawing (Fig. 6).

Table 1. Footing on river bed (Fig. 13(a))

Assumed ϕ'	30°	35°
N_γ	22·4	48·0
Ultimate bearing capacity q_0, kN/m^2	437	937
F_b (load factor for pressure 240 kN/m^2)	1·82	3·90
N_γ for failure under 240 kN/m^2	12·3	12·3
Hence, ϕ' mobilized by 240 kN/m^2	25·9°	25·9°
and $F_S = \left(\frac{\tan\phi'}{\tan\phi'_{\text{mobilized}}}\right)$	1·19	1·44

Effect of skirts. The effect of a skirt can be considered in relation to each of the recognized modes of failure of shallow footings (Vesić, 1973). In the case of general shear, a skirt which extended through the zones of plastic equilibrium and some distance into the unsheared soil below would offer some resistance to the formation of the shear zones and so increase the bearing capacity. But since each of the skirts in Fig. 13(c)–(e) would lie entirely within the zones of plastic equilibrium they would not add anything to the soil's resistance to general shear failure. Because they are aligned more or less with planes of shear in the (theoretical) zones of radial shear and the friction between timber and soil is usually less than the internal friction of the soil itself, they would be more likely to reduce the resistance to general shear.

If the failure were by punching shear, the primary surfaces of shear (Vesić, 1973) would be more or less vertical and close to the planes of the skirts; and so the skirts would be likely again to decrease rather than increase the resistance to failure.

The third mode of failure, by local shear, involves considerable settlement of the footing with the initial shear flow and lateral expansion of the soil confined to zones close under the edges of the footing. There is no doubt that such shear strains and the resulting settlement would be reduced by the presence of even a short skirt, particularly if it were restrained from lateral movement at its top as Smeaton's skirts were. The effect of the skirt would be to prevent lateral expansion of the soil in the critical zones and also to reduce the vertical pressure on these zones by carrying some of the pier load directly down to lower levels.

Factors of safety and the effects of girdles. The properties of the soil are assumed to be: $\gamma'=10$ kN/m^3, and $\phi'=30°$ or 35°. The actual friction angle of the gravel at and near the surface of the river bed would almost certainly lie within these limits. Weaker material may have been present at greater depth, as Smeaton believed, but the recent site investigations make it virtually certain that such material would be, at worst, a silt or sandy clay in which pore-pressures would dissipate as quickly as the load was applied (two years for the whole dead load); and the assumption of a cohesionless material with $\phi'=30$–$35°$ at all levels is therefore justified.

From Smeaton's initial drawings the volume of solid work carried by one of the middle piers is found to be 578 m^3. Since this includes ashlar, rubble masonry and gravel fill, an average density of 2000 kg/m^3 is assumed. For the bearing area of 45·1 m^2 given by Fig. 5, the bearing pressure is found to be 254 kN/m^2. With the water at the assumed low level, buoyancy reduces the effective pressure to 240 kN/m^2. This pressure, and the dimensions of one of the middle piers, are used in all the calculations to facilitate comparisons, although Design 2 was applied only to end piers where the pressure and dimensions were somewhat different.

For a footing on the river bed (Fig. 13(a)) of width 4·88 m and length/width ratio about 2, calculations using Vesić's (1973) tables of bearing capacity factors give the results shown in Table 1.

As bearing capacity theory can take account only of uniform surcharge loads outside the footing, the effect of the girdle in Smeaton's Design 1 (Fig. 13(b)) cannot be quantified. And to obtain an approximation to the effect of the girdle in Design 3 (Fig. 13(d)) it is necessary to assume that its actual weight (assuming $\gamma'=10$ kN/m^3) is distributed uniformly over the half-width of the opening between the piers, namely 7·33 m (from Fig. 9).

Calculations following Vesić then give the results shown in Table 2.

Hence the girdle in Design 3 gives a useful increase in the factor of safety. The effect of the girdle in Design 1 would clearly be much smaller. The change from Design 3 to Design 4 is a lowering of the river bed by 0·76 m and of the surfaces of the slopes by 0·92 m, while the surfaces inside the skirts are unchanged. Considering this to be equivalent to a reduction of the uniform surcharge of 0·84 m, the calculations can be extended to give the values shown in

Table 3. Since the footing base is now higher than the river bed, these calculations are not strictly justified but they provide a rough measure of the difference between Designs 3 and 4. Design 4 clearly represents a condition approaching failure.

Stability of skirts as anchored bulkheads. The simple stability calculations commonly applied to sheet pile bulkheads would show that the skirts in Designs 2, 3 and 4 are all unstable. This is because the calculations ignore friction between the soil and the bulkhead and between the soil and the surcharge (in this case, the bottom of the pier). In practice, the skirt of Design 2 would be perfectly stable. The skirt of Design 4 comes nearest to the conditions of a common anchored bulkhead, and a simple 'free earth support' calculation gives a factor of safety of 0·82 when the passive resistance of the slope is ignored but a uniform surcharge of gravel is assumed, of the full height of the slope (1·52 m), resting on the gravel of the passive zone below. This suggests that there was a serious risk of outward rotation of the skirt if the river bed and slopes were scoured to the shape drawn as Design 4.

Discussion

Skirts and girdles. A girdle as big as that of Design 3 gave a considerable increase in the bearing capacity but it also reduced the area of flow and so increased the intensity of scour. A skirt under the edge of the pier, as in Design 2, could have no such bad effect but it, too, proved an inadequate defence against underwashing; it is unlikely to have increased the bearing capacity.

Scour and the sequence of failure. Whether the condition of Design 4 had been reached, or had actually been passed, at any pier before the fatal flood is not known; it exists only as a drawing of what Smeaton thought to be safe. The calculations, both for bearing capacity and bulkhead stability, suggest that the margin of safety was, at best, small. Taking account of the ease with which the river bed was known to scour (after the flood of 1778) it was not right to call Design 4 safe at all.

Table 2. Design 3 (Fig. 13(d))

Assumed ϕ'	30°	35°
Surcharge pressure, kN/m^2	5·7	5·7
N_q	18·4	33·3
Shape factor (for $L/B = 2$)	1·29	1·35
Depth factor (for $D/B = 0{\cdot}12$)	1·03	1·04
Additional bearing capacity, kN/m^2	139	264
Total bearing capacity	576	1201
Revised F_b	2·40	5·00
Revised F_S	1·33	1·61

Table 3. Design 4 (Fig. 13(e))

Assumed ϕ'	30°	35°
Change of surcharge pressure, kN/m^2	−8·4	−8·4
Reduction of bearing capacity, kN/m^2	−204	−389
Residual bearing capacity, kN/m^2	372	812
F_b	1·55	3·38
F_S	1·13	1·37

The fact that scour is never uniform makes it almost certain that the collapse of each pier began with a yield of the ground at the toe of the sheet piling wherever the scour was deepest (usually near the upstream end); the release of lateral restraint would then allow a sudden settlement of the pier with consequent cracking of the arches and spandrels and a rapid progress towards collapse. The fact that five or six piers gave way in half an hour is surprising, however, and can only be explained by supposing either that the velocity of the current was still increasing rapidly or that the obstruction of the fallen arches caused such an increase of velocity round the remaining piers.

A bearing capacity failure. The conditions at Dumbarton Bridge present an important contrast. Babtie, Shaw and Morton (1960), from borehole surveys for the new bridge upstream, reported soft river alluvium of great depth and they had experience of similar material through the whole town area east of the old bridge. Allowable bearing pressures are generally assessed at no more than 67 kN/m^2, suggesting ultimate bearing capacity of little more than 160 kN/m^2. This applies only near ground level, the allowable pressure reducing with depth. Smeaton's report of a stronger 'crust' is therefore confirmed. If there were no offsets at the feet of the piers the actual pressure under them would be about 160 kN/m^2. As there are offsets, the actual pressure is less, but the width of the offsets is not known and so neither is the value of the pressure. There is also extensive protection of the river bed and the piers by piling and setting. Babtie, Shaw and Morton found serious settlement only at one pier in 1960, the evidence being a downstream tilt of 125 mm in the length of the pier.

The bearing pressures at Dumbarton have therefore always been much higher than modern design would permit but there has been no shear failure since the partial collapse in 1768. Unfortunately there are few records from 1768 and the amount of load applied before the collapse and its rate of application are not known. The small uniform surcharge provided afterwards by setting on the river bed appears to have been carefully maintained. The foundations at Hexham almost certainly had higher factors of safety against shear failure, but shear failure occurred after the sudden scouring of the river bed and rock slopes, which destroyed the surcharge of the girdles and probably also the toe support of the skirts.

Science and design. The only quantitative criterion of safety Smeaton quoted was a height of 'fall' through the bridge and the only parameters he calculated were velocities of flow. He made these calculations after the failure by a rough rule and, as there were better methods available which, with his undoubted skill in fluid mechanics, he could have used, we must conclude that he did not think bridge design worthy of, or amenable to, accurate quantitative predictions.

For assessing the strength of foundations there was no quantitative theory. Like other engineers of his day, Smeaton judged the type and condition of the soils with simple tools and then designed on the basis of his own previous experience. He took full advantage of the opportunity, which the contemporary speed of construction afforded, to change his design in response to observations of foundation behaviour during the construction period. At Hexham this was not enough to compensate for the error in his initial judgement that the 'gravel crust' could resist the erosive power of the river.

In his idea that the difference of water level upstream and downstream would tend to overturn the bridge there is an interesting hint of the slip-circle mode of collapse, now so important in soil mechanics theory; but on permeable soil there could be no overturning moment because the effective stresses in the ground would be independent of the height of water over it. The reason the bridge 'took a heel upstream' is to be found solely in the well-known concentration of scour at that end of the piers.

ACKNOWLEDGEMENTS

The Paper is part of a programme of research supported by grants from the Nuffield Foundation and the Edinburgh University Travel & Research Fund. Dr P. R. Vaughan and Professor A. W. Skempton gave advice on the analytical methods and Mr T. Turner gave useful comment on an early draft. For access to documents and permission to quote them, sincere thanks are due to the organizations and persons named in the reference list and also the Burgh Engineer of Dumbarton.

REFERENCES

Most of the references are in three collections, of which two have been published. Location of individual letters, drawings, etc. is indicated by the short title of the collection, with page or folio numbers.

Printed books and papers

Anon. (1788). *Narrative of the proceedings relative to Hexham Bridge.* London: J. P. Coghlan.
Cramond, W. (1891). *Annals of Banff* **1**, pp. 289, 321. Aberdeen: Spalding Club.
Neill, C. R. (Ed.) (1973). *Guide to bridge hydraulics.* Toronto: University Press.
Sykes, J. (1833). *Local records.* Newcastle-upon-Tyne.
Terzaghi, K. & Peck, R. B. (1948). *Soil mechanics in engineering practice.* London: Chapman & Hall.
Webber, N. B. (1968). *Fluid mechanics for civil engineers.* London: Spon.
Renwick, R. (1912). *Extracts from the records of Glasgow* **7**, pp. 47–49.
Vesić, A. S. (1973). Analysis of ultimate loads of shallow foundations. *Jnl Soil Mech. Fdn Div. Am. Soc. Civ. Engrs* **99**, SM1, 45–73.

Collections

Designs by John Smeaton, vol. 4, in the library of the Royal Society of London. Published on microfilm by E. P. Wakefield, 1972.
Errington Papers. Manuscript legal papers at Northumberland County Record Office (ref. ZAN M17/57). (By permission of the Society of Antiquaries of Newcastle-upon-Tyne.)
Reports of the late John Smeaton, FRS (1812). 3 vols. London.

Other manuscripts, drawings and unpublished reports

Babtie, Shaw & Morton (1960). *Report on Dumbarton Bridge.*
Expenditure, Records of (1831–85). At office of Northumberland County Surveyor.
Johnson, W. (1796). Plan and elevation of Hexham Bridge. In Drawings Collection of Royal Institute of British Architects (ref. H8/5).
Macdonald, F. A. & Partners (1934). *Dumbarton Bridge Reconstruction,* **1934.**
Mylne, W. (1768). Contract drawing for bridge at Jamaica Street, Glasgow. In Glasgow City Archives.
North-Eastern Road Construction Unit (1973). *Contract plans for Hexham and Corbridge Bypass.*
Quarter Sessions order books. At Northumberland County Record Office.
Ramsay, A. (1768). Letter to Lord Loudoun, 15 September. In 1768 Bundle 6 of the Marquess of Bute's 'Loudoun Papers'.
Smeaton, J. (1756). Plan and estimate for a bridge at Hexham. In *Trevelyan papers* at Newcastle-upon-Tyne University Library (No. 69).
Smeaton, J. (1764). Letter to Duke of Portland. In *Portland papers* at Nottingham University Library (ref. PwF 8385).
Surveyor's Bridge Book (*c.* 1831–80). At Northumberland County Record Office.

14
Pulteney Bridge
Jean Manco

Bridges lined with shops have always been rarities, but Robert Adam's creation in Bath has more than novelty value. His restrained composition of curves and rectangles was one of the visual delights of English Neo-classicism. The uncluttered lines of the river façades were designed to be effective at a distance, while the elevations to the road provided an interesting play of light and shadow for passers-by.

Paradoxically, the success of the concept destroyed the architecture. Within twenty years of its building, commercial pressures on this prime shop site dictated an expansion upwards, with a butchery of those graceful elevations which fortunately Adam was not alive to witness. Then in the nineteenth century, individual leaseholders enlarged their shop windows or cantilevered out over the river as the fancy took them. Storms and road widening also took their toll. By 1948 the buildings had 'become pathetic travesties of the original design.'[1] In that very year, Bath City Council removed a wooden structure from the back of one shop; the tide had turned. From then on the story has been one of gradual restoration.

BATHWICK ESTATE AND THE PULTENEYS

Bathwick was a rural parish when William Pulteney acquired it in 1726.[2] The 600-acre estate lay just across the Avon from the city John Wood the elder was even then contemplating with a developer's eye (Fig. 1). As Bath expanded, the potential of Bathwick must have become obvious. Wood wrote in 1742:

> Many of the Citizens of Bath have already beautified [Bathwick] Village by their little Places of Retirement in it, so nothing is more probable than that the higher Lands of the whole Lordship will soon be inriched with nobler Villas, and the lower Grounds be covered with such Structures as will augment the Body of the City, and become the chief Beauty of Bath in respect to her Works of Architecture.[3]

Two hurdles had to be overcome. Firstly most of the land was held by tenants on lifehold leases; these William Pulteney refused to renew. Gradually, he transferred to short-term leases, thus freeing the land for eventual development.[4] He did not live to see the outcome. Pulteney, by then Earl of Bath, died in 1764 leaving Bathwick still a village. His immense fortune was inherited by his brother General Henry Pulteney. The childless general, dying just three years later, willed his whole estate to Frances, daughter of his cousin Daniel Pulteney. Frances was married to an Edinburgh lawyer, William Johnstone, who had adopted the Pulteney name. This energetic and frugal Scot immediately began to make plans to develop the Bathwick Estate. The remaining problem was that the only direct route from Bath to Bathwick was by ferry. The

General died in October 1767.[5] By the following February, Pulteney was conferring with Bath City Council about his proposals for a new bridge.[6]

The Corporation's own plans fitted in with Pulteney's. The City Improvement Act of 1766 had made provision for a new guildhall and market, involving a major reorganization of the east side of the High Street. The time was ripe for Pulteney to purchase the land north of this to cut an access route through to a new bridge across the Avon. To his wife's chief trustee, the Earl of Darlington, Pulteney explained:

> The pump Room must always be considered the center of Bath, and therefore the nearer we can bring our ground to that center, the more valuable it will be for building . . . it is proposed by the Corporation to carry a parade along the side of the River, from the Grove to this New Bridge which will give a short & commodious passage for chairs & persons on foot . . . By means of which a good dale [sic] of the Bathwick ground will be within a sixpenny fare of the Pump room, playhouse, coffee houses & assembly rooms.[7]

BUILDING THE NEW BRIDGE

Once Pulteney had convinced both the Corporation and his wife's trustees of the soundness of his plans, the next step was an Act of Parliament. Since Bathwick was held in trust, lands on the estate could not be sold or mortgaged without involving Parliament, a lengthy and expensive business which had to be twice repeated. The first Act, passed in 1769, empowered the trustees to raise £3,000 through mortgage to finance the scheme for a toll-free bridge and access street and to purchase and exchange land for the street.[8]

However, it was not long before Pulteney grew more ambitious. In August 1769 he sent Lord Darlington an elevation of a comparatively simple three-arch bridge, designed by Thomas Paty of Bristol.[9] Paty's own estimate for the work was £4,569, but Pulteney had obtained a second estimate from local builders John Lowther and Richard Reed for only £2,389. 11*s*. It looked as though £3,000 would at least cover the cost of the bridge, but already Pulteney was toying with the idea of adding shops. The second estimate included the cost of four shops, two at each end of the bridge. Pulteney was clearly anxious to press ahead with building. The masons estimated that they could complete the bridge by September the following year, if the contract was made out in time to get the pillars above water before the winter.[10] It seems, however, that work did not start that year and by the summer of 1770 the brothers Adam were directing the building.[11]

Probably Pulteney approached the Adams initially for plans that would put flesh on his proposals for a developed Bathwick. Certainly Robert Adam was paid for such plans while Pulteney Bridge was being built, although surviving plans are dated 1777 and 1782.[12] The Adams' ambitious Adelphi scheme on the bank of the Thames may well have led Pulteney to entrust them with his river-side development in Bath. They were also fellow Scots, though there is no evidence that any of the family was personally known to Pulteney before 1770.

We shall never know for certain who first suggested putting shops actually on the bridge, but presumably it was Robert or James Adam, who were undoubtedly familiar with both Italian and English examples. The Adams had visited Florence and Venice,

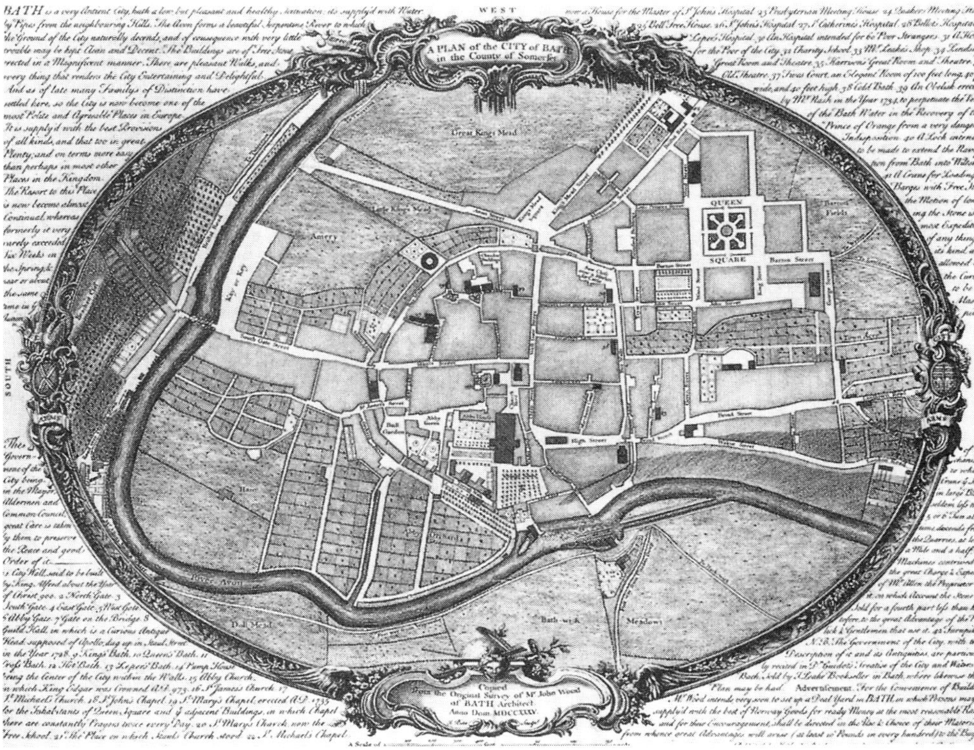

Fig. 1. *John Wood's map of Bath in 1735 (Bath Central Library)*

Fig. 2. *Design for a bridge from* I quattro libri dell'architettura di Andrea Palladio, *re-drawn by Bertotti Scamozzi*

where they could scarcely have escaped noticing the Ponte Vecchio and the Ponte di Rialto.

While the Ponte Vecchio is overlayed with the accretions of centuries, the Rialto is the product of a single mind. In 1588 Antonio dal Ponte won the commission to design a stone replacement for a wooden bridge. His was an audacious construction, ignoring classical principles to provide a startlingly elegant solution to the problems of the site. A single arch carries two arcades of shops sloping up to porticoes at the apex which allow a view of the river.[13]

One of the rejected classical designs for the Rialto was that by Andrea Palladio. This must have been known to Robert Adam, since the plans for it appeared in Palladio's enormously influential *I quattro libri dell'architettura* (Fig. 2). Palladio's design was clearly the inspiration though not the blueprint for Adam's Pulteney Bridge. Stripped of its heavier ornamentation and with the references to temple architecture muted, this homage to ancient Rome emerged from Adam's hands as the coolest of English understatements. There may have been practical reasons for certain changes. It is doubtful whether Pulteney would have approved the cost of a mass of statuary, and central porticoes overlooking the river would bring him no shop rents. Adam chose a different focus for his design. A lofty Venetian window formed the centrepiece on each side, set amidst an array of tripartite arches along the street façades. There are distant echoes here of Diocletian's Palace, measured and drawn by Adam on his Grand Tour (Fig. 3).

Palladio's design included tabernacles for statues above the piers of the bridge, inspired by the bridge of Augustus at Rimini.[14] Adam abandoned these in favour of the elegant simplity of circular windows, which had the practical purpose of lighting the cellars below the shops. Whether or not we have Pulteney's frugality to thank, Adam's penchant for gilding the lily was here held in check; there are none of the swags or medallions that we expect of him. In 1768 he had decorated his proposed bridge at Syon House with engaged caryatids on the piers, linked by swags and rosettes.[15]

English influences can be discounted. On the London-to-Edinburgh road that the Adams must have travelled, York and Durham still had houses on their bridges. The irregular clusters of buildings shown in contemporary views have charm to the modern eye, but would scarcely have been admired by Robert Adam.[16] (The only survivor among England's medieval housed bridges is the High Bridge in Lincoln, dating in its present form from 1540.)[17]

Clearly bridges with shops were not by any means a new idea, but Adam's designs caused consternation in Bath. In July 1770, Lord Darlington had been informed of the new scheme before the foundations were laid. His approval was sought at this early stage because Adam's plans required a width of 50 ft rather than the 30 ft of the Paty design. It would be more expensive from the foundations upwards.[18] His reply was sufficiently encouraging for a start to be made immediately, but Bath Corporation, who had not been consulted, wrote to Pulteney in protest. They claimed that the circulation of air would be prevented and smoke from houses on the bridge would prove a nuisance. A more cogent argument was that for the shops to be profitable, the bridge would have to be so crowded as to impede crossing traffic. 'It has been for some years past an uniform practice throughout the Kingdom to avoid and condemn

Fig. 3. *From Robert Adam,* The Ruins of the Palace of the Emperor Diocletian at Spalato in Dalmatia *(Venice, 1764)*

incumbrances of this kind.' Pointing out that London and Bristol had cleared their bridges of houses, they clearly thought it perverse that Pulteney was proposing to bring this outdated phenomenon to Bath. Their most withering comment was that 'A street on a Bridge of no greater breadth than thirty feet will never be thought, according to the modern Ideas of Building, to have an elegant Appearance'.[19]

The Council had not actually seen Robert Adam's plans at that point and so did not realize that he intended to increase the width by 20 feet. Adam's own copies were purchased in 1833 by Sir John Soane and are now in the Soane Museum, so we are fortunately in no doubt of his intentions (Figs 4–9). The admired southern river façade is sometimes referred to as the 'front' of the bridge. It is certainly easy to think of the present dilapidated jumble on the north side as the 'back', but this is misleading. Given the passion for symmetry that dominated this period in architecture, it would be very startling indeed if Adam had intended his bridge to have different façades to the north and south rows of shops. Not surprisingly, all the evidence shows that they were originally identical. Adam's elevations are not labelled to indicate to which row they refer; clearly they were meant for both. His design for the 'Lodges at the end of Mr Pulteneys Bridge at Bath' shows identical pavilions (Fig. 5). His plan is headed 'Plan of the shops on the New Bridge at Bath: This side to be exactly the same as the other' (Fig. 9). Adam planned twenty-two shops, each with a staircase to an attic above. Twelve of them also had a cellar, reached by a ladder beneath the stair. The puzzle of these plans is the small mystery of the concealed door; where is the entrance to the central shop of each row? The answer must be that what looks all window in his elevation to the road is actually a Venetian door. The views by Thomas Malton and Thomas Howell show that Adam's design was carried out in every detail, with both sides of the bridge identical (Figs 10–12).

Costs mounted during building. In 1771, the trustees were cajoled into petitioning for a second Act of Parliament to raise a further £3,500. This was passed in 1772, when 'great progress' had been made in buildings.[20] By March 1773, accounts could be

Fig. 4. *'Bridge for William Pulteney Esqr at Bath' by Robert Adam (Sir John Soane's Museum)*

Fig. 5. *'Lodges at the end of Mr Pulteney's Bridge at Bath' by Robert Adam (Sir John Soane's Museum)*

prepared. The figure for building alone was £8,183. 5*s*. 2*d*., while the total cost of the project had risen to nearly £11,000.[21] Whatever they thought of this, the trustees had been presented with a fait accompli. What could they could do but agree to a third Act of Parliament to reimburse Pulteney?[22] The accounts included an estimate of £576 to finish the work, so it would seem that completion was only months away. The first print of the New Bridge was dated 30 November 1773, by which time some of the shops 'fit for any genteel business' were probably in use. Tenants for the remainder were sought from January 1774.[23]

The builders were local men. Reed and Lowther were the masons on the bridge itself, but the masonry of the shops was entrusted to Singers and Lankeshere. A much repeated story is that Reed went bankrupt as a result of his speculation in the bridge,[24] but this is clearly apocryphal. He was paid in exactly the same way as the other craftsmen. The plumber and glazier was Thomas Warr Atwood, who, as a Councillor, was among the signatories of the Corporation's protest to Pulteney. Possibly he resented a London architect being consulted. Like several other Bath craftsmen of the period, he was also an architect and developer and had used his influence to obtain the appointment as City Surveyor. Since the old city prison had to be demolished to clear the approach to the new bridge, Pulteney provided a site for a new prison across the river in Bathwick. Adam supplied designs for this, but they were set aside by the Corporation in favour of plans by Atwood.[25]

THE PATRON AND HIS ARCHITECT

The 1773 accounts for the bridge include a payment to Adam for 'a Plan of the Streets Buildings and Squares of a New Town on the Bathwick side of the River.'[26] They were to come to nothing. In December that year the Boston Tea Party signalled the start of the American War of Independence and the Bath building boom was halted. For many years, Adam's elegant and urbane bridge led out onto meadows rather than a Palladian townscape. The great expense of Pulteney Bridge had not yielded immediate returns for the Bathwick Estate. In fact, 'before Pulteney-street was built, the population of Bathwick was so small, and the intercourse with the city so trivial, that tenants could not be found to occupy the shops on the bridge.'[27] But there is nothing to suggest that

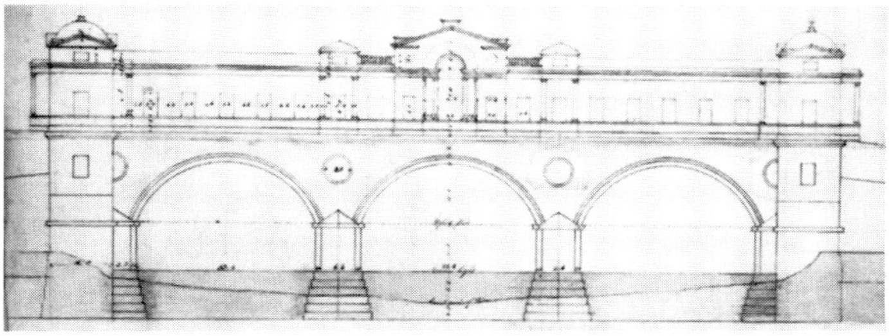

Fig. 6. *'Front of the New Bridge & Shops at Bath for Sir William Pulteney Bart' by Robert Adam (Sir John Soane's Museum)*

Fig. 7. *'Front of Shops next the Road for the new Bridge at Bath For Sir William Pulteney Bart' by Robert Adam (Sir John Soane's Museum)*

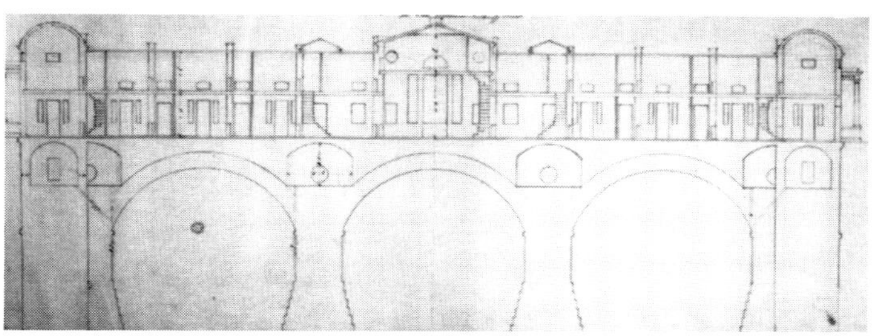

Fig. 8. *'Section of the Bridge & Shops over it at Bath for Sir William Pulteney Bart' by Robert Adam (Sir John Soane's Museum)*

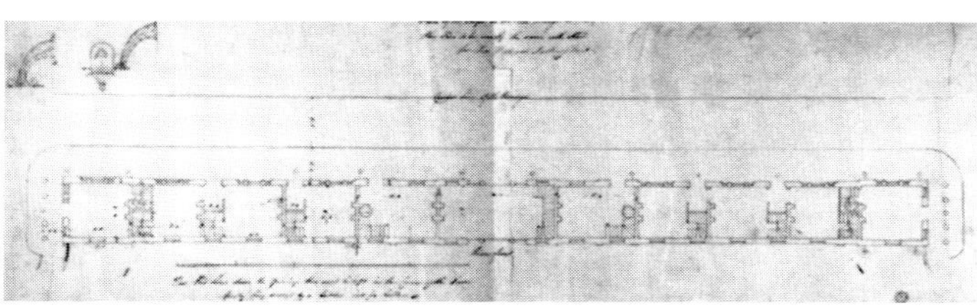

Fig. 9. *'Plan of the Shops on the New Bridge at Bath this side to be exactly the same as the other for Sir William Pulteney Bart' by Robert Adam 'The Red lines [circles] shew the openings that must be left in the Crown of the Arches whereby they descend by a Ladder into the Cellars' (Sir John Soane's Museum)*

Pulteney regretted his choice of architect. He continued to involve Robert Adam in his plans for Bathwick. He probably also recommended Adam to his brother, Sir James Johnstone of Westerhall, co. Dumfries. In 1790, Adam designed the Johnstone family mausoleum in Westerkirk Churchyard for Sir James. Pulteney was among the pall-bearers at Robert Adam's funeral in March 1792. This has led some writers to describe him as a close friend of Adam's. However, the other pall-bearers were a duke, two earls, a viscount and a lord,[28] so one suspects that rank had taken precedence over

Fig. 10. *'The New Bridge at Bath' Aquatint 3 March 1779 after a drawing by Thomas Malton (Victoria Art Gallery, Bath)*

Fig. 11. *Watercolour by Thomas Malton of the south side of Pulteney Bridge dated 1785 (Victoria Art Gallery, Bath)*

Fig. 12. *Pen and wash drawing by Thomas Howell of the north side of Pulteney Bridge, with Bath Abbey visible in the background (Collection Bath Preservation Trust, 1 Royal Crescent, Bath.) The only other known view of the north side before it was rebuilt is an aquatint by A. Robertson and R. Faulder dated 1 October 1792.*

intimacy with the deceased. William Pulteney might be the sole commoner among them, but he was 'supposed to be the richest commoner in the kingdom'.[29]

ALTERATIONS BY BALDWIN

Whether he was a friend or a prominent client, Pulteney had the tact to see Adam to his grave before desecrating his handiwork. Thomas Baldwin, Bath's City Surveyor after Atwood's death in 1775, was the architect of the long-delayed development of the Bathwick estate, which began in March 1788.[30] Samuel Hieronymous Grimm drew Pulteney Bridge in July 1788, with its long, high-arched causeway stretching into rural Bathwick. When Grimm returned the next June it was to quite a different panorama. 'The New or Pulteney Town' was springing up, with Argyle Buildings nearing completion.[31]

That lovely unity of design was shortly to be lost. On 26 March 1792, less than a month after Adam's death, a building lease of the major part of the New Bridge was issued to William Henry Marks with Baldwin's plans for alterations (Fig. 13). Excluded were the two end houses on the east, previously four shops. It seems that the end pavilions had been amalgamated with the shops next door at an early stage. However, the demand for more spacious units had clearly not been satisfied. Balwin's plans provided for the conversion of sixteen shops into six large ones, with a raised roof permitting a higher ceiling and larger windows on the first floor. Adam's recessed, columned windows were converted to bays, with larger bay windows replacing the west end porticoes. No doubt it all made sound commercial sense, but Adam's street elevations were ruined. Window heads pushed through the string course; the cornice had become split-level and the integrating pattern of arches destroyed. Marks covenanted to finish the work by 25 March 1794 and it seems that he completed on schedule, for no. 1 was leased by him to a coachman in April 1794. Ironically, the lease to Marks provided that, after these alterations, the façades should never be altered or varied.[32] In fact, Baldwin's was just the first of many distortions of Adam's original vision.

FLOOD, STORM AND REBUILDING

On 26 September 1799, the *Bath Chronicle* reported 'This morning the inhabitants of the houses situate on the New-bridge were very much alarmed by the pier of the centre arch giving way, by which two or three of the houses are endangered.' High floods were blamed for the damage.[33] An engineer 'experienced in building bridges and aqueducts' was consulted and within days the houses over the fracture had been taken down. The *Bath Journal* immediately expressed the hope that all the houses would come down and stay down. They appealed to Henrietta Laura, Lady Bath, to consider 'the beautiful access which laying open the bridge . . . would make to that much and justly admired part of town.'[34] Clearly some Bathonians had never been reconciled to this strange hybrid structure in their midst. By December, estimates were being sought for reinstating Pulteney Bridge. This work was directed by John Pinch the elder,[35] who had replaced Baldwin as architect to the Pulteneys. 'After great labour and perseverance' the workmen finally achieved a firm foundation for the damaged pier by

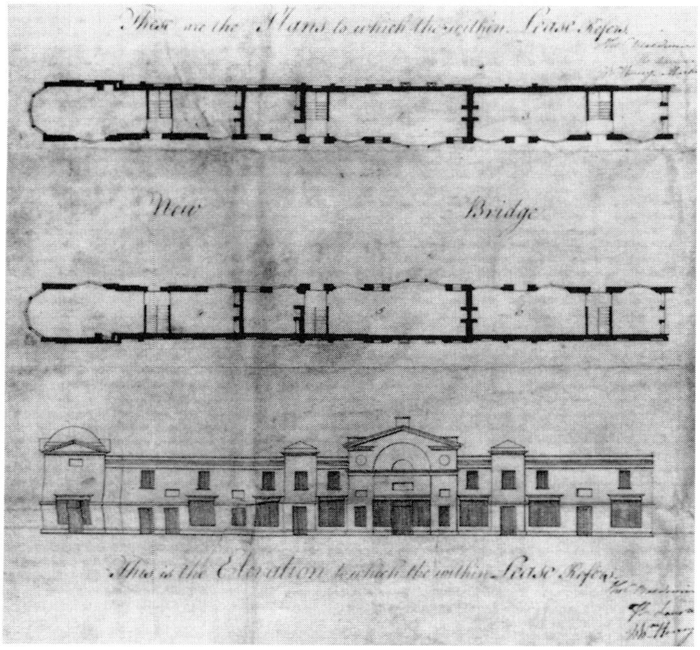

Fig. 13. *Plans for the alteration of Pulteney Bridge by Thomas Baldwin March 1792 (Bath Record Office)*

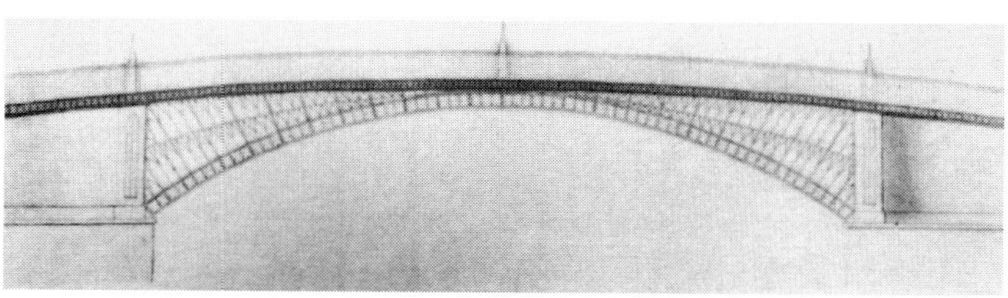

Fig. 14. *'Plan and Elevation of a Design for PULTENEY BRIDGE in the CITY OF BATH' by Thomas Telford (Collection Lord Barnard)*

5 November 1800. Just four days later a 'heavy flood . . . so violently strained the remaining pier of Pulteney Bridge which had hitherto withstood the force of the water that . . . it gave way carrying with it the house of a stay-maker erected on it with the whole of his property, the family having but just time to save their lives.' Only the north side of the bridge was affected.[36]

The rapid decision was taken to erect a temporary bridge upstream, while Pulteney Bridge was rebuilt 'on a most elegant plan, and to consist of only one arch.'[37] The temporary timber bridge, built in three months under Pinch, was opened on 15 March 1801. It was still standing in 1860, linking the buildings of the Northgate brewery on either side of the Avon.[38]

The design for a new single-span bridge was by Thomas Telford (Fig. 14). Presumably Telford was the engineer consulted by Pulteney when the bridge first gave way. Telford was born in the same parish as Pulteney and Telford's cousin acted as land steward to the Johnstones of Westerhall. When Telford moved to London in 1782, Sir

Fig. 15. *Pulteney Bridge as it stands today (Bath City Council)*

Fig. 16. *Engraving by J. and H. Storer 1 December 1822 after a drawing by Joseph West. On the right is the bowed end to the north side of Pulteney Bridge (Bath Central Library)*

James Johnstone arranged for his brother William Johnstone Pulteney, who was in London, to consult Telford about improvements to Westerhall. Pulteney then commissioned Telford on his own account to make alterations to the vicarage of Sudborough, Northants. When Pulteney became MP for Shrewsbury, Telford renovated Shrewsbury Castle for his use. The two seem to have had a close and cordial relationship. In 1792/3, Telford visited Bath and was very critical of recent development in Bathwick and elsewhere. Though he does not name the guilty architect, the reference is clearly to Baldwin and may have persuaded Pulteney to transfer his patronage to John Pinch. In the 1790s Telford built bridges by the score, including Buildwas, his first iron bridge. Using iron, he evolved a single-span design less obstructive to river navigation and less vulnerable to flood damage.[39] These were the very considerations that moved Venetian officials to prefer Antonio dal Ponte's design for the Rialto to that of Palladio. Needless to say, Telford's plan for an iron Pulteney Bridge was never carried out, but the Palladian model came close to suffering a second rejection two and a half centuries after the first. Possibly Pulteney found that it would be cheaper to rebuild half a bridge than start again from scratch.

The rebuilding required yet another Act of Parliament and work was still not under way in July 1802.[40] The opportunity was taken to widen the bridge in rebuilding. The northern shops are around two feet larger than those by Adam and the overall width is now 58 ft. The different phases of building can be seen in the cellars of these northern shops, where a good deal of the earlier structure survives.

It was presumably John Pinch the elder who was responsible for the rebuilt north side; the smooth and self-effacing style is typical of his work (Fig. 15). Pinch was a Bath builder and architect, who designed terraces on the slopes of Lansdown, as well as New Sydney Place in Bathwick for Pulteney.[41] Pinch's design for the north side of Pulteney Bridge was a pale reflection of the mutilated Adam façade opposite. Adam's pavilions are echoed by token pediments perched meaninglessly on an otherwise almost featureless façade. The dentil work with which Adam had embellished cornices and pediments was dispensed with. One feels that the need to acknowledge Adam cramped Pinch's own vision without recreating Adam's. His one piece of self-indulgence was the bowed west end, similar to his later designs for the segmental end house of Cavendish Place and the bowed ends of Sion Hill Place on Lansdown. Seen in isolation (Fig. 16), this is the most attractive feature of his design, but it further eroded the homogeneity of the bridge. The architecture had in any case become subordinated to the shop-window displays.

Although Adam's concept had now been so weakened and had come within a whisker of being abandoned altogether, Pulteney Bridge could still arouse at least a half-hearted spirit of emulation. Thomas Bunn of Frome was a passionate admirer of Palladian Bath; his was the hand behind a late Georgian development of Frome consciously modelled on Bath. In 1821 Frome bridge was rebuilt with one side lined with two-storey houses over shops.[42] The architecture is unremarkable; this lone descendant of Pulteney Bridge is no child of Adam, which is scarcely surprising in the circumstances.

THE RIVER FAÇADES

Nineteenth-century prints and photographs of Pulteney Bridge from the river (Fig. 17) make a sad contrast with Malton's view. On the south, a timber-built vegetable market had blocked the view of the western pavilion by 1804.[43] Baldwin's alterations started the process of blocking and inserting windows that gave this façade a ragged look even before individual lessees took bolder initiatives. By 1819, chimney stacks were jutting out of the façade in three places.[44] Engravings then show little change until the middle years of the century, when shop-keepers on both sides of the bridge began building timber additions, cantilevered out over the river like those on the Ponte Vecchio. An engraving of 1864 shows the two on the south side,[45] while a watercolour of the north side (Fig. 18) was made to support a complaint by William Lewis to the Sanitary Committee in 1873:

About twenty years since the occupiers of the Bridge began to destroy the architectural features of the buildings by throwing out water-closets and additional rooms as shown in the drawing . . . each person seems to have disposed of the ordure of the closets in the cheapest manner

Fig. 17. *Nineteenth-century photograph of Pulteney Bridge (Bath Central Library)*

Fig. 18. *'View of the North side of Pulteney Bridge Bath showing the disposition of the Sewage and Rainfall. Note This shows the nuisance on one side of the Bridge only, the Circular apertures over the Abutments are used to throw out the Ashes and house refuse'. Watercolour donated to Bath Corporation by William Lewis's son Egbert (Victoria Art Gallery, Bath)*

possible by the simple expedient of cutting a hole through the floor and ejecting the contents . . . direct into the river below.[46]

As the leaseholder of property in Northgate Street with a garden down to the river, William Lewis was incensed. He felt that the practice was detrimental to the health of the persons living and working with him. Conveniences were commonly grafted onto the back of Georgian houses in the Victorian period. Since neither of the river façades was easily visible to passers-by, they were naturally treated as the backs of the shops. The north view having been more completely blocked, perhaps it is not surprising that few, apart from William Lewis, seemed to care what happened to it. Makeshift, box-like structures spread over Pinch's façade like a disease.[47] Floods and obscurity combined to take this side furthest away from Adam's original conception.

Once the vegetable market was removed in 1901, the southern façade was visible from the new Grand Parade, but the immediate result did not please every eye. The antiquary Leveson Scarth complained in April 1903, 'In passing across the North Parade Bridge today I observed with much concern two very conspicuous advertisements being newly painted at the back of two old houses built upon quaint and unique Pulteney Bridge.'[48]

Scarth was not the only Bathonian concerned to protect what remained of Adam's work. The west end of the south row of shops was now an awkward intrusion into Grand Parade and in 1902 the Corporation decided to cut it back. An anguished letter of protest in the *Bath Herald* pointed out that this would mean destroying the domed pavilion. 'Let our city councillors refuse to allow this vandalism.'[49] The Council sought a suitable comprise. In 1903 the end three shops were purchased for immediate demolition.[50] Then a new pavilion was built to the plans of Bath architects Gill and Morris, whose enthusiasm for interesting Adam features was perhaps a shade excessive. Recessed, arched windows were placed on three sides of the pavilion.[51] The whole western half of the façade was tidied up at this time, in the Council's first restoration.

The eastern half remained dilapidated. Bath City Council had purchased the eastern pavilion and the shop next to it in 1916, but it was not until 1937 that plans were made to remove the wooden extrusion on the back.[52] In January 1936, Pulteney Bridge had been scheduled as a national monument. The Council bought all the shops still in private hands and the following year the Surveyor of Corporate Property, John Francis Bevan Jones, carefully traced Adam's own plans and designed a restored façade.[53] War intervened; the Jones restoration was finally executed in time for the 1951 Festival of Britain. Jones initially fought shy of the Venetian window. His first designs in May 1937 offered the alternatives of a plain sash window in the centre or one with an arched head, both of the same size as the other windows. However, his final design in February 1938 was as close as was practicable to Adam's original, though he did not correct the error of Gill and Morris. Despite that oddity, the result is among the most celebrated of Bath's Palladian façades.

The north side remained neglected. Bryan Little wrote in 1961, 'An appalling penthouse clings to most of its length; the picture is of desecration and squalor, a noisome disgrace to the beauty of Bath.'[54]

SHOPS AND THEIR FRONTAGES

Unlike the Ponte Vecchio, which has been for centuries the preserve of goldsmiths and jewellers, Pulteney Bridge is not restricted to any particular type of trader. In the first half of the nineteenth century, cheese intermingled with chessmen, wedding cakes with chronometers. But in the middle of the century a business started which was to change the face of Pulteney Bridge. In 1848 William Duck leased no. 2 Pulteney Bridge as a music and instrument seller.[55] He then moved to no. 7 and by the summer of 1856 Duck's Musical Repository had expanded to no. 6 and had a new front unifying the two shops (Fig. 19). These gently arched plate-glass windows eventually became the integrating link for all the Pulteney Bridge shop fronts. The abolition of excise duty on sheet glass in 1845 approximately halved its price and in London small-paned shop windows were rapidly supplanted by sheet glass.[56] The fashion spread more slowly in provincial towns and Duck may have been the first shopkeeper on the bridge to adopt it. It was particularly suitable for the display of large instruments like pianos and harps, though the tiny shops on Pulteney Bridge were a perverse choice to house them. Expansion was really a necessity for such a business to operate comfortably. From 1859 to 1880, Duck gradually spread out westwards, extending his shop front as he went.[57]

Fig. 19. *This wood engraving first appeared in the* Bath Express *12 July 1856* (Bath Post Office Directory 1858)

Fig. 20. *Work in progress on the southern street façade* (Evening Chronicle 28 January 1976)

The 99-year building lease to William Henry Marks (and therefore all leases devolving from it) expired on Lady Day 1891. The then owner of the Bathwick estate, the 4th Duke of Cleveland, was evidently not happy with the state in which he was regaining his property. He issued a new lease to Duck, Son and Pinker requiring them to take down all the additions over the river and return to Pinch's bowed west end at the expiry of the lease.[58] However, in 1921 a large part of the Bathwick estate was auctioned and most of the shops on Pulteney Bridge were bought by Duck, Son and Pinker, who sold them to Bath City Council in 1936, subject to a 999-year lease on their own shop and long leases on several others.[59] Pulteney Bridge might be a national monument, but Bath City Council did not have freedom in its management.

At least the long front of Duck, Son and Pinker gave a unity to the northern street façade. By contrast, the southern street front was a patchwork of the preferences of individual lessees. This was the problem that the Georgian Group tackled in 1975 to mark European Architectural Heritage Year (Fig. 20). Their architect was John Vivian, who explained that 'the intention of the work has been to do enough to recall the architectural charm of the Adam design . . . without doing too much.' The end pavilions and central bay were restored 'as far as is reasonably possible' to the original design, while the shop fronts were rationalized and given windows with arched heads to echo the north side of the street.[60]

This was the second stage in returning the south range of the bridge closer to the original than it had been since 1792. Pulteney Bridge has become one of the architectural symbols of Bath; probably few realize how short a life Adam's bridge had, and how long it took to restore even half of it to a semblance of his original vision.

ACKNOWLEDGEMENTS

This study was commissioned by Bath City Council and I am most grateful for the help and encouragement of their conservation architect David McLaughlin. The Bath Preservation Trust had previously supported a study of the Bathwick Estate by Dr Michael Rowe, who was therefore able to suggest useful sources. My thanks also go to Francis Kelly for pointing out the bridges of Frome and York, to Trevor Fawcett for several references to Bath journals and to Phillip Hughes for measuring the present width of the bridge.

Illustrations are reproduced by courtesy of Avon Library Service, Bath Central Library (figs 1, 2, 16, 17, 19, 20); Bristol University Library (fig. 3); The Trustees of Sir John Soane's Museum (figs 4–9) Victoria Art Gallery, Bath (figs 10, 11, 18); Bath Preservation Trust and Unichrome (fig. 12); Bath Record Office (fig. 13); Lord Barnard and Bath Preservation Trust (fig. 14); Bath City Council (fig. 15).

ABBREVIATIONS

BL	British Library
BCL	Bath Central Library
BRO	Bath Record Office
HL	Huntington Library, California
PEP	Pulteney Estate Papers

NOTES

1 Walter Ison, *The Georgian Buildings of Bath* (London, 1948), p. 66.
2 BL K.37.29.a: A Survey of The Manor of Bathwick belonging to the Honble Willm Pulteney Esqur., 1727.
3 John Wood, *An Essay Towards a Description of Bath* (Facsimile 2nd edn 1765, Bath, 1969), p. 91.
4 BCL PEP MS 1809.
5 R. S. Neale, *Bath: A Social History 1680–1850* (London, 1981), pp. 227–28; BCL, PEP Bundle 3, MS 1750: Act of Parliament 1769 gives the date of death of General Pulteney.
6 BRO Council Minutes 6 February 1768.
7 BCL PEP AL 1788: Letter to Lord Darlington 23 August 1769.
8 BCL PEP MS 1750.
9 BCL Philip Hemmings, 'Bath: A Study in Planning and Development' 3 vols (Thesis 1982), 2, between pp. 5 and 6. The original is now missing from PEP.
10 BCL PEP AL 1788: Letter to Lord Darlington 23 August 1769.
11 HL PU 4: Letter from James Adam to William Johnstone Pulteney.
12 BCL PEP MS 1811; Sir John Soane Museum, Adam drawings vol. 38, discussed in Walter Ison, *The Georgian Buildings of Bath* 2nd edn (Bath 1980), p. 200.
13 Bruce Boucher, *Andrea Palladio: The Architect in his time* (New York, London, Paris 1994), p. 227.
14 Boucher, op. cit., pp. 215, 225.
15 Geoffrey Beard, *The Work of Robert Adam* (Edinburgh & London), p. 86.
16 York City Art Gallery, William Marlow, *Ouse Bridge, York c.* 1764; Leeds City Art Galleries, Thomas Hearne, *North-east view of Durham c.* 1781–84; British Museum, Thomas Hearne, *Elvet Bridge, Durham*, mid-1790s.
17 Eric de Maré, *The Bridges of Britain* (London, 1975), p. 123.

18 HL PU 4: Letter from James Adam to William Johnstone Pulteney 31 July 1770; BCL PEP AL 1787: Letter from William Pulteney to Lord Darlington 18 February 1771.
19 HL PU 57: Letter from the Mayor and Corporation of Bath to William Pulteney.
20 BCL PEP, bundle 6, MSS 1751, 1815.
21 BCL PEP, bundle 6, MSS 1811, 1818/1, 1820.
22 BCL PEP, bundle 6, MS 1753.
23 Bath Chronicle 13 Jan 1774, p. 3, col. 4.
24 Capt. Rowland Mainwaring, *Annals of Bath* (Bath, 1838), p. 337; Mowbray Green, *The Eighteenth Century Architecture of Bath* (Bath, 1904), p. 169; Ison, *Georgian Buildings*, p. 66.
25 Ison, *Georgian Buildings*, 2nd edn (Bath 1980), pp. 11, 201.
26 BCL PEP, bundle 6, MSS 1811.
27 Mainwaring, *Annuals*, p. 338.
28 Howard Colvin, *Biographical Dictionary of British Architects* (London, 1979).
29 *Gentleman's Magazine* June 1805, p. 587.
30 Ison, *Georgian Buildings*, 2nd edn, p. 13.
31 BL MSS, ADD MS 15546, nos 90, 123.
32 BRO Bathwick Estate Papers Deed Pk FF.
33 *Bath Herald* 28 Sept. 1799, p. 3.
34 *Bath Journal* Sept. 30 1799.
35 *Bath Chronicle* 19 December 1799, p. 3, col. 3.
36 *Bath Herald* 15 Nov. 1800, p. 3.
37 *Bath Chronicle* 20 Nov. 1800, p. 3., col. 2.
38 *Bath Chronicle* 19 March 1801, p. 3, col. 2; BRO Cotterell and Spackman map of Bath 1:500 1852–54; Wood-engraving of the Northgate Brewery in George Measom, *Official Illustrated Guide to the Great Western Railway* [1860].
39 Lionel Thomas Caswall Rolt, *Thomas Telford* (London, 1958), pp. 4, 11–12, 15–16, 24–25, 32–33.
40 HL PU 65 Nathaniel Bayley to Sir James Pulteney 30 July 1802.
41 Ison, *Georgian Buildings* 2nd edn, p. 44, pls 102, 103, 106.
42 Michael McGarvie, *The Book of Frome* (Basingstoke, 1980), p. 114.
43 Engraving dated November 1804 in *Bath Illustrated by a Series of Views from the Drawings of John Claud Nattes* (1806), pl. 3.
44 Engraving by H. S. Storer in Pierce Egan, *Walks Through Bath* (Bath, 1819).
45 *Illustrated London News* 24.9.1864, p. 308.
46 BCL Pulteney Bridge cuttings file, sheet stamped Lewis bequest 1934 (collection of William Lewis's son Egbert); BRO, Minutes Urban Sanitary Committee 1872–1885, pp. 3, 50.
47 OS map 1886; BRO Bathwick Papers 1889 plan of Duck, Son and Pinker's premises, Building Control Plans 1223; Avon Central Library drawing of Pulteney Bridge by Samuel Loxton *c.* 1900.
48 *Daily Chronicle and Argus* 6 April 1903; These advertisements are visible in Green, *Eighteenth Century Architecture*, pl. cv.
49 *Bath Herald* 11 April 1902.
50 BRO Deed Pk 2702B.
51 BRO Gill and Morris plans July 1904.
52 BRO Deed Pk 2750E; Plans M: 12 March 1937 J. F. Bevan Jones proposed alterations to nos 9 and 10 Pulteney Bridge.
53 BRO Plans M: J. F. Bevan Jones: tracings of Adam's plans in the Soane Museum, elevations of 1937/8, revision 3.10.50 and sections for mouldings 1951.
54 Bryan Little, *A Bath Portrait* 4th edn (Bristol, 1980), p. 71.
55 Duck, Son and Pinker, *One Hundred Years of Music* (Bath, 1948); Bath Directories.
56 Mark Girouard, *The English Town* (New Haven and London, 1990), pp. 227–28.
57 BRO Building Control Plans C31, E106–7; Bath Directories.
58 BRO Bathwick Estate Papers, bundles of leases and particulars of leases fallen in hand.
59 BRO Bathwick Estate Papers Particulars of Sale; Deed Pk 2702A: Draft conveyance of 1–8, 11–16 Pulteney Bridge to Bath Corporation 1936.
60 Bath City Council Building Regulation Application no. 750636, 13 August 1975; *Evening Chronicle* 18.3.76: letter to the editor from John Vivian.

15

Two masonry bridges
II. Telford's bridge at Over

J. Heyman and B.D. Threlfall

In contrast to the light footbridge of Clare College considered in Paper 7554, Telford's bridge at Over is a large-span masonry arch carrying heavy road traffic. The bridge has exhibited much movement and cracking since its construction 150 years ago, and is now a three-pin arch rather than a monolithic structure. However, the abutments have ceased to shift, and the bridge now illustrates the thesis that the cracked state is 'natural'; Over Bridge is satisfactory despite its gross deformations.

Introduction

Séjourné[1] listed in 1913 all masonry bridges in the world for which one span at least has a clear opening of more than 40 m. Eight bridges only are noted for the UK; the first four are:

1749	Pont-y-Pridd	William Edwards	42·67 m
1824	London Bridge	John Rennie	46·33 m
1826	Over	Thomas Telford	45·72 m
1833	Grosvenor (Chester)	Thomas Harrison	60·96 m

Of the remaining four, two are railway bridges (Victoria Bridge, Durham, 1836 and Ballochmyle, 1839) and two are over the Thames (Putney, 1882 and Kew, 1901).

2. Pont-y-Pridd is only 3·35 m wide; London Bridge has now been taken down. Thus Telford's Over Bridge at Gloucester is the oldest large-span road bridge remaining in England (at the moment it carries the A40 road over the western branch of the River Severn into Wales), and is second in span only to Grosvenor Bridge.

3. In describing the bridge, and in making calculations, a problem arises with regard to units. In general, descriptions are given in Telford's units of feet, but calculations are presented in SI units. This enables direct quotation without the necessity of 'translation'. The length of the eighteenth-century French foot was about 12·8 English inches; thus the clear spans of Perronet's Pont de Neuilly at 120 *pieds* were about 128 ft or 39 m.

Description and construction of the bridge

4. Telford, in his *Life*,[2] devotes ten pages to a description of the bridge and of its construction.

> 'For this bridge of one stone arch I introduced a form which, although a novelty in England, had, in 1768, been employed by an eminent French architect (Mr Perronet) in a bridge of five arches, of 128 ft span each, over the river Seine at Neuilly; and this appeared to be peculiarly applicable to

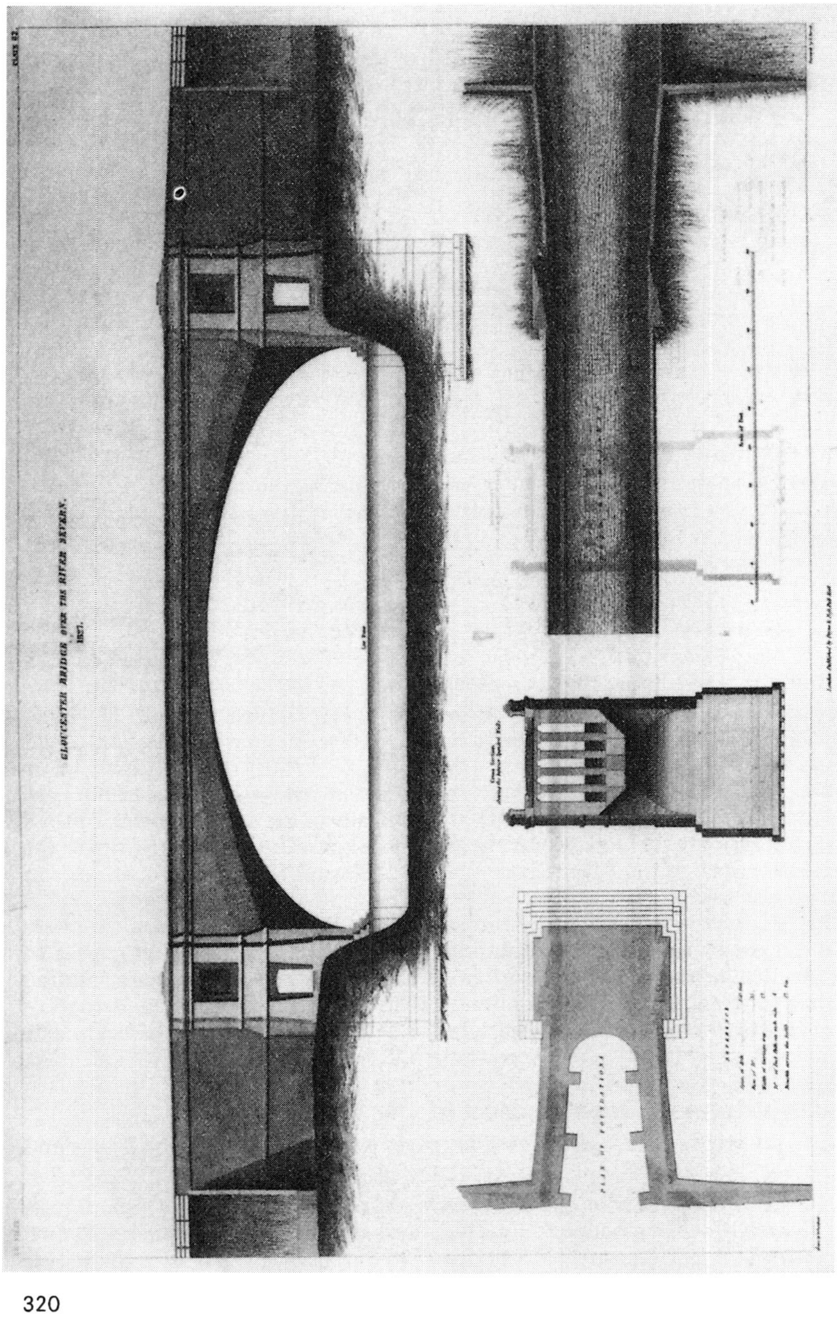

Fig. 1. Telford's engraving of the bridge at Over

the Severn, where the upland floods are considerable. By referring to the Plate [Fig. 1 in the Paper], it will be seen that this is accomplished by making the general body of the arch an ellipse, 150 feet on the chordline, and 35 feet rise, while the voussoirs, or external arch-stones, being in form of a segment, have the same chord, with only 13 feet rise. This complex form converts each side of the vault of the arch into the shape of the entrance of a pipe, to suit the contracted passage of a fluid, thus lessening the flat surface opposed to the current of the river whenever the tide or upland flood rises above the springing of the middle of the ellipse, that being at four feet above low-water, whereas the flood of 1770 rose 20 feet above low-water of an ordinary spring tide; which, when there is no upland flood, rises eight or nine feet.'

5. The cutting away of the soffit of the arch, as seen clearly in the photograph of Fig. 2, will greatly ease the passage of water in a flood. The difference of total head on either side of the bridge is expressed in terms of the losses incurred by the water in its passage, and these in turn are largely due to the generation of turbulence. A convergent–divergent 'pipe' will help to keep the flow smooth; in the limit, if the passage of the water is perfectly smooth, only surface friction will affect the flow.

6. Telford's handling of the soffit seems clumsy visually, certainly when compared with Perronet's Pont de Neuilly, Fig. 3. Perronet[3] allows the *cornes de vache* to cut the faces of the piers in smooth curves, and this in turn is related to his use of carefully specified eleven-centred profiles for the oval of each arch. Perronet states that he designed for a 23 ft flood; his arches had 30 ft rise, so that 7 ft clear was left beneath the keystone, and an elliptical profile would have left too narrow a width for the flood water. (All these dimensions are in French feet.)

7. Comparison between the engravings of Figs 1 and 4 shows the close correspondence between the general shape of Perronet's original and of Tel-

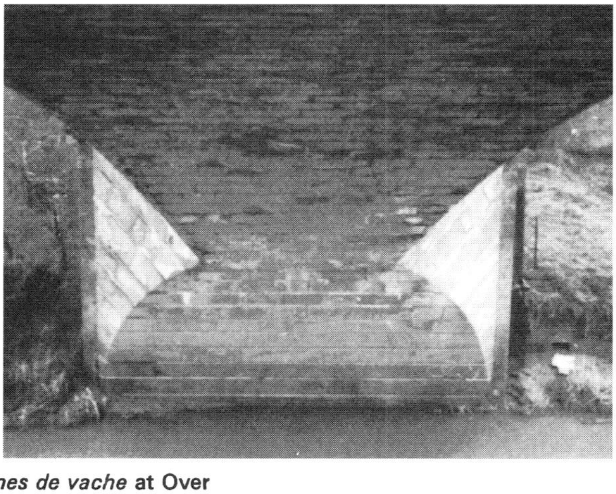

Fig. 2. *Cornes de vache* at Over

Fig. 3. Perronet's Pont de Neuilly, 1772

ford's copy. In profile, Telford's bridge appears to be a flat circular arch of 150 ft span and 13 ft rise, with voussoirs varying in depth from 4½ ft at the keystone to 5½ ft at the springings. In reality, as Telford stated, the arch is elliptical with a 35 ft rise. The central section of the span is solid, but the roadway is carried over the haunches on six parallel walls, as is seen in the cross-section in Fig. 1; the external spandrel walls are 3 ft 6 in. thick, and the internal walls each 2 ft thick.

8. The masonry is sandstone from the Forest of Dean. Warland[4] gives (for Grey Bed) a crushing strength of 569 ton/sq. ft; this agrees well with the value of 67 N/mm^2 given by *Specification* 1970.[5]

9. Telford started the building works on the eastern (Gloucester) side, and he gives the following stratification of the subsoil:

Feet	
0	Rich loam
11	A bed of soft blue silt
23	A layer of peat moss
28	Brown clay
33	Strong coarse indurated gravel
36	Finer gravel or coarse sand
44	End of borehole

Baker (Sir Benjamin Baker) worked on the bridge in 1880,[6] and an account of this work was given by Phillips[7] in 1907. Phillips believes that Telford was misled as to the stratification of the soil. Telford states that he founded the eastern abutment at a depth of 33 ft on the indurated gravel, but Phillips could find no sign of this gravel.

10. That Telford was misled is supported by a soil survey carried out in connexion with new road works. A borehole sited about 200 ft upstream from Over Bridge on the same (eastern) bank records:

Feet	
0	Topsoil and weathered soft clay
8	Soft brown and soft grey silty clay
17	Very soft peat and peaty clay
21	Soft blue and grey silty clay
24	Soft black clay
27	Soft red silty clay
32	Clean sand and pebbles
(33	Inrush of soil and water)
35	Soft blue clay
38	Hard blue clay (Lias)

11. At this station, therefore, a thin stratum of sand and pebbles overlies a further stratum of soft clay (of apparent cohesion in an undrained triaxial test of less than 600 lb/sq. ft; the stiff Lias has a cohesion of about 4000 lb/sq. ft). It is possible that Telford mistook the clean sand and pebbles (perhaps the old bed of the Severn, which has moved westward) for a thick bed of gravel.

12. On the opposite bank he found the ground in general to be much firmer, with gravel at 27 ft. This figure is confirmed exactly by a recent borehole opposite the one quoted in § 10.

13. Thus Telford excavated to 33 ft on the eastern bank, and to 27 ft on

Fig. 4. Perronet's engraving of the Pont de Neuilly

the western; otherwise the foundations to the two abutments are the same. Each consists of a course of large and flat-bedded rubblestone, on which a timber platform was carefully laid. This platform was made from Memel logs (that is, fir) at 3 ft centres, the square spaces between being filled with well-grouted rubble masonry, and the whole topped with close-jointed 4 in. planed beech planks; the finished platforms were 40 ft by 37 ft.

14. Telford's engraving, Fig. 1, shows no piles under the abutments and it seems fairly certain, from his own account, that none were actually provided. He blames himself for omitting piles and platforms for the 40 ft wing walls backing the abutments. These wing walls were to be 7 ft thick at their bases and to taper to 2 ft at their tops, and to be founded at 2 ft below the surface of the meadow (original specification of 22 February, 1826). However, a new document of August 1828 deepens the foundations to 8 or 10 ft below the surface of the meadow, and specifies 8 ft piles and a fir platform; it is these that Telford omitted with 'injudicious parsimony'.

15. Telford then describes the timber centering for the main arch:

> 'Instead of a complete trussed framing, depending on the two extremities, [the contractor] supported his centering by [seven] parallel rows of piles, fixed in the current of the river, each row connected by cross-braces and caps, and each supporting a rib, which formed the actual centering.'

This construction allowed 16 ft wide waterways to be kept open for barges.

16. The centering was very rigid compared with a timber truss of 150 ft span, and decentering in 1829 was followed by an immediate sinking of the crown of 2 in. Thrust was then imposed on the abutments for the first time, and the crown sank a further 8 in., making 10 in. altogether. All accounts, from Telford's own to the present, are agreed that the fault in the bridge lies in the eastern abutment, poorly founded on a stratum of soft clay, and poorly backed by wing walls built with an ill-judged parsimony. The western abutment appears to have held completely firm, but the eastern has moved, destroying the line of the bridge, as shown in Fig. 5.

17. Although Telford much regrets[2] the sinking of Over Bridge, he publicly pronounced himself satisfied, and quoted the fact that Perronet's bridge at Neuilly had sunk 13 in. on decentering, and had then moved another 10½ in., making 23½ in. total in a span of 128 English feet.

18. However, he was asked in 1830 to examine 'sundry openings which require to be stopped'. Having filled these in, he examined further cracks in 1831, but was again confident about the stability of the bridge. He was in

Fig. 5. Over Bridge, looking west

fact paid the balance of his fee in that year, but had to make further repairs in 1832 costing £500, of which he paid half.

19. Telford died in 1834, and the old bridge that Over Bridge replaced was pulled down in that year.

Later history

20. Telford's new bridge continued to cause concern. A major survey was made in 1864 in connexion with a projected new railway, and in 1880 alarm at the state of the bridge was sufficient for a report to be called for. Baker made a thorough examination; the pavement was taken up and the extrados of the arch inspected. It was found that the joints at the haunches were open by about 2½ in., and that these had been stuffed with iron wedges made from old cart tires, presumably in Telford's time. Correspondingly, the joints in the intrados at the crown were open, and a stick could be pushed up 'two feet from the soffit without touching anything in the open joint.'[7]

21. Baker summarized his findings some seven years later:[6]

> 'The abutments had gradually gone over, and had been continuing to go over for sixty years. The result was that certain barbarians were actually urging the magistrates to take down the bridge, because the soffit of the arch had got a reverse curvature. It certainly looked rather like tumbling down, and the cracks in the spandrels were big enough to walk through quite comfortably. He told the magistrates it would be a disgrace to the

country if they pulled down Telford's historical work and substituted a hideous iron lattice-bridge. It had failed because the wing-walls had been built (and many failures had since occurred from the same thing) on what Telford mistook for clay, whereas it was really indurated mud, and underneath that was peat. Mr Baker tried to get down and underpin in the wing-wall to solid ground, but came to water, and stopped, as pumping would certainly have brought down the bridge. What he then did was simply to build on the bad ground, underpinning as deep as he could and as wide as he could to reduce the load per square foot on the foundation. He suggested that the cracks should be left for eighteen months, and then cut out and rebuilt. He was glad to say that the measures adopted seven years ago had stopped the whole of the movement, and he saw no reason why that bridge should not remain perfectly satisfactory to all time.'

22. Baker's underpinning operations involved the placing of concrete under the wing walls of the eastern abutment, and it was then intended to support the concrete by means of brick 'flying buttresses' constructed in shafts. However, in sinking the first shaft, an inrush of water was encountered at a depth of 8 ft short of the intended total; this depth seems to correspond exactly with that of the modern borehole record on the eastern bank. The water could not be lowered easily, and Sir John Fowler, who was called in to advise, gave his opinion that the work should be abandoned. This advice was followed, and the shaft was filled with concrete.

23. The open joints in the intrados were wedged with Broseley tiles and cement, and the same stuffing was used for the extrados at the haunches. Baker believed that his measures had stopped the whole of the movement of the bridge; that this is substantially true is confirmed by more modern records.

24. Phillips worked with Baker in 1880, and he makes no mention of any movement between that date and 1908, when his monograph was published. The County Surveyor of Gloucester has kept records of levels from 1908. Since that time, the permanent downward depression of the arch appears to have increased by 0·05 ft (of which 0·02 ft may have been due to the passage of a 180 ton load in recent years). Some cracks have been attended to in this century, for example in 1928 and in 1934, but the main work in the last ten years has been the repair of the masonry parapets due to accidental damage.

25. A new type of cracking was first observed in 1958; shear cracks developed in the voussoirs at the crown at a depth about 1 ft below and parallel to the extrados, on the north face of the arch. It is not known how far these hair-line cracks penetrate into the arch ring. The cracks were observed to extend slowly for ten years, but there has been no further movement since 1969.

Calculations for the bridge

26. Even without the evidence of the shear cracks, it is clear from the historical records that the bridge is acting as a three-pin arch, with the line of thrust approaching the extrados at the crown, and the intrados at the haunches. For the purpose of analysis, the bridge was sliced into 2 m segments; half the arch is shown in Fig. 6. The weight of each slice was calculated from nominal measurements, allowing for the spandrel walls where appropriate; a unit weight of 24 kN/m^3 was used. A roadway, including surfacing (gravel, stone, clay etc.), of 0·64 m was allowed for.

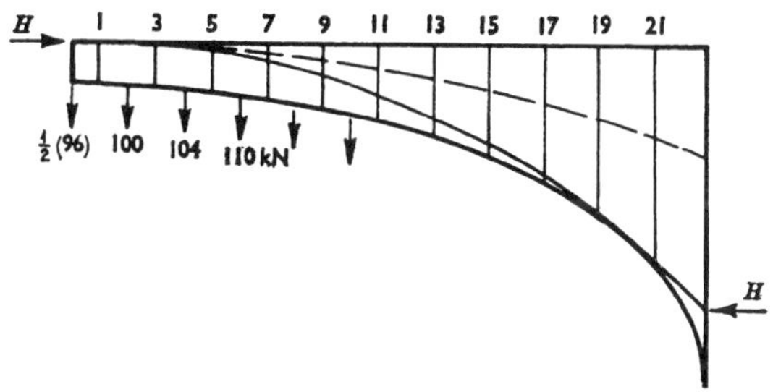

Fig. 6. Thrust line for half the arch

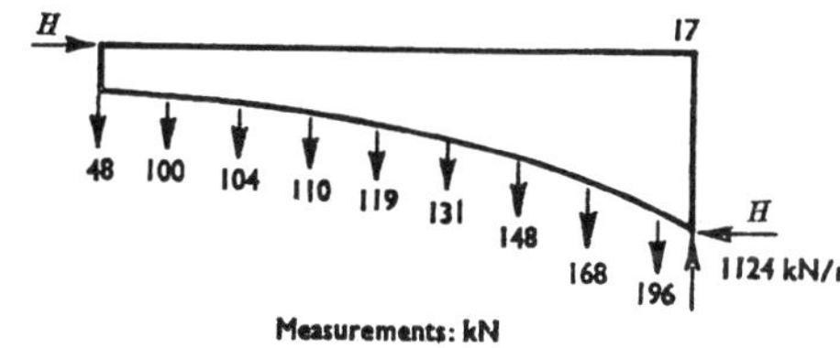

Fig. 7. Free-body diagram for 17 m of arch

Table 1

x (m)	0	1	3	5	7	9	11	13	15	17	19	21	22·86
y (intrados, m)	1·370	1·380	1·461	1·629	1·895	2·232	2·687	3·263	3·988	4·906	6·106	7·822	12·038
W (kN)	48	100	104	110	119	131	148	168	196	233	288	384	
ΣW (kN)	48	148	252	362	481	612	760	928	1 124	1 357	1 645	2 029	
Wx (kNm)	0	200	416	660	952	1310	1776	2352	3 136	4 194	5 760	8 428	
ΣWx (kNm)	0	200	616	1276	2228	3538	5314	7666	10 802	14 996	20 756	29 184	

27. In Fig. 6 the horizontal abutment thrust H is assumed to act at the extrados at the crown, and the first problem is to determine the section at which the thrust line touches the intrados. Table 1 was drawn up to aid the calculations; the origin is at the crown, and a unit (m) width of bridge is assumed. (Taking the total width of the bridge as 8·5 m, the last entry in the row for ΣW implies a total arch weight of $2 \times 2029 \times 8{\cdot}5 = 34\,500$ kN; this agrees well with the mass of 3500 ton usually quoted.)

28. The way in which Table 1 is used may be seen from Fig. 7. This free-body diagram has been drawn on the assumption that the haunch hinge lies 17 m from the crown, that is, that the line of thrust touches the intrados at that point. Using values directly from the table,

$$17: \quad 4{\cdot}906H = (1124 \times 17) - 10\,802; \quad H = 1693 \text{ kN/m}$$

29. Adjacent sections must now be investigated; the maximum value of H resulting from such a search will give the minimum value of the abutment thrust necessary for stability of the arch. (This is exactly the method used by Coulomb[8] two hundred years ago.) Repeating the calculations, therefore, for sections at 19 m and at 21 m:

$$19: \quad 6{\cdot}106H = (1357 \times 19) - 14\,996; \quad H = 1767 \text{ kN/m}$$
$$21: \quad 7{\cdot}822H = (1645 \times 21) - 20\,756; \quad H = 1763 \text{ kN/m}$$

30. Clearly the calculations are insensitive to the precise location assumed for the intrados hinges (this was noted by Coulomb). The thrust line is tangential to the intrados at about 20 m from the crown, as indicated in Fig. 6, and the present 'three-pin arch' thus has a span of about 40 m and not the full 46 m.

31. The total dead-load abutment thrust is estimated therefore as about 1770 kN/m (or about 15 000 kN for the bridge of width 8·5 m). However, this figure is based on an idealized and perfect geometry. Sinking of the crown may be allowed for merely by altering the coefficients on the left-hand sides of the above equations. As an example, if the total drop at the crown is taken as 0·5 m, then the equation for the thrust line to pass through the 19 m point on the intrados is

$$5{\cdot}606H = (1357 \times 19) - 14\,996; \quad H = 1924 \text{ kN/m}$$

This represents an increase of about 9% from the 'perfect' value.

32. The geometry of a three-pin arch is simple. Taking the span as 38 m, then a 1 cm spread in the distance between the abutments implies a drop at the crown of about $38/(4 \times 6{\cdot}106) = 1{\cdot}55$ cm; a similar calculation for a 42 m span gives $42/(4 \times 7{\cdot}822) = 1{\cdot}34$ cm. Thus the crown movement of about $1\frac{1}{2}$ cm recorded in the last 60 years indicates that the abutments have moved apart by 1 cm at most during this time.

33. The line of thrust cannot of course lie in the extreme surface of the masonry. Using the value of 1924 kN/m, and a crushing strength of 67 N/mm^2, the masonry will start to crush when a voussoir depth of only 2·87 cm remains in contact. A depth of 30 cm in contact would give a very satisfactory factor of safety of about 10; it is at this depth at the crown that the shear cracks are evident.

34. With the large mass of masonry (3500 ton) in the arch proper the effect of live load is correspondingly small, although not insignificant. If a load of 250 kN is placed at the crown of the 1 m width of bridge (i.e. a vehicle of 212

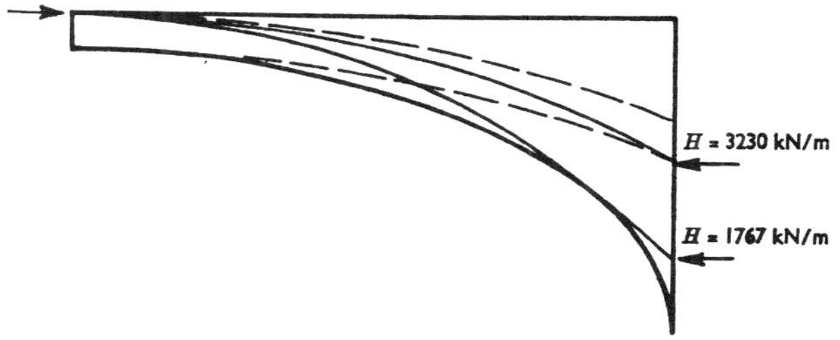

Fig. 8. Alternative thrust lines

ton at the centre of span of the actual bridge of width 8·5 m), the horizontal component of abutment thrust is increased by 423 kN/m (from 1924 to 2347 kN/m).

Summary and conclusions

35. As noted for Clare College Bridge in Paper 7554, the three-pin configuration is the 'natural' state for a masonry arch and, in principle, a three-pin arch is a perfectly satisfactory structure. Thus the early troubles with the eastern abutment of Over Bridge, which led to the marked drop at the crown, are not now significant; what is important is the present state of the deformed arch.

36. The numerical calculations of the previous section help to assess this state. Although the calculations have been made to several significant figures, some of the quantities are not known with corresponding accuracy. Weights have been estimated, for example, and nominal leading dimensions have been used, rather than those resulting from an accurate survey of the bridge.

37. Further, the cut away of the soffit has not been taken into account in calculating the position of the line of thrust. Fig. 2 indicates that the position shown in Fig. 6 might be valid for roughly the centre half of the carriageway, but that the two edge slices of the bridge, carrying the footpaths and parapets, are 'eaten into' by the *cornes de vache*.

38. Figure 8 shows the position of the thrust line if it is compelled to lie within the apparent depth of the voussoirs; the thrust on a 1 m width of bridge is nearly doubled. Thus for the bridge as a whole the presence of the *cornes de vache* may increase the total thrust on the abutments by about 20% compared with the value for a full elliptical arch.

39. The detailed analysis of the previous section indicates that mean stress levels are comfortably low, and safety factors correspondingly high. These general conclusions will remain valid even if the thrust is increased by say 20 or 25%; there is room for large variations in the quantities before the safety of the bridge is brought into question.

40. Similarly, there is no question of any geometrical instability due to an excessively thin arch ring (as there was for Clare College Bridge); on the contrary, the depth of masonry permits an enormous variation in possible lines of thrust. Thus further movement of either abutment could be accommodated easily by the arch, although such movement, if at all large scale, would obvi-

ously cause concern for the abutments themselves. However, there is no record of any significant movement in the abutments for the last 60 years.

41. If both the abutments and the structure of the arch are generally satisfactory, then the bridge will become finally unserviceable only by decay of the material. Phillips's view in 1908 was that the road surface should be well maintained to protect the extrados of the arch, and that the bridge should be inspected periodically for signs of excessive weathering.

42. Providing this is done, then it seems hard to quarrel with Baker's view that the bridge should remain perfectly satisfactory to all time. *Tempus destruendi et tempus aedificandi;* it may be hoped that the time to break down is not yet.

Acknowledgements

43. Mr Bernard M. Feilden, OBE, FRIBA, FSA, a member of the Ancient Monuments Board, arranged for a great deal of information to be made available to the Authors. Much of this information was freely provided by members of the County Surveyor's Department, Gloucestershire County Council; the County Archivist was also very helpful. The Authors are most grateful to all of these.

References

1. Séjourné P. *Grandes voûtes*, 6 vols. Bourges, 1913–16.
2. Telford T. (Rickman J. (Ed.)). *Life of Thomas Telford*, 2 vols (text and atlas), London, 1838.
3. Perronet J. R. *Description des projets et de la construction des ponts de Neuilly, de Mantes, d'Orléans*, etc., 2 vols (text and atlas). Paris, 1782, 1783.
4. Warland E. G. *Constructional masonry*, London, 1946.
5. Architectural Press. *Specification* 1970. The Architectural Press, London, 1970.
6. Baker B. Discussion on the Hooghly 'Jubilee' Bridge. *Min. Proc. Instn Civ. Engrs*, 1888, **92**, 116.
7. Phillips R. *Gloucester: Over Bridge: a monograph.* Gloucester, 1908.
8. Heyman J. *Coulomb's Memoir on Statics: an essay in the history of civil engineering*. Cambridge University Press, 1972.

Conversion factors

Imperial units	*SI units*
1 ton force	10 kN

16
Arch bridges

J.B.B. Owen

Isambard Brunel (1870) devotes chapter seven of his biography of Isambard Kingdom Brunel to his father's railway bridges and viaducts. He deals first with those bridges built of brick and masonry—the arch bridges which form the subject of this chapter. The first of these which he mentions is that over the River Brent at Hanwell (Fig. 21), later called the Wharncliffe Viaduct. He describes it as "the first of Mr Brunel's important railway works . . . a handsome brickwork structure" (as indeed it still is, particularly when viewed, over green fields, from the main road just west of the Viaduct Inn, Hanwell) "65 feet high, with eight semi-elliptical arches, each 70 feet span and 17 feet 6 inches rise". Brunel's general calculation book *(CB, 1837)*, which forms part of the manuscript collection of Bristol University, also contains under the heading "Brent Viaduct" several pages of drawings and calculations relating to these arches. These calculations are discussed later.

But to provide sustained news interest bridges must fail or be considered to be in a critical state. It is then not surprising to find that all Brunel's biographers deal in some detail with his Maidenhead bridge. Most gloss over the troubles which occurred with the eastern arch of this bridge, but Fowler (Mackay, 1900), when a young engineer, considered them most serious. Later he was responsible for retaining Brunel's arch shape when the bridge was widened. The calculation book *(CB, 1837)* also contains drawings and calculations for this bridge and, going westward along the main line, for Thorney Broad Bridge, which is over the River Colne just above Thorney Weir, and for the central quasi-Gothic arch over the Avon just below the weir above Temple Meads railway station at Bristol. The drawing

ARCH BRIDGES

T. L. J. Bentley

Fig. 21. Wharncliffe Viaduct at Hanwell

in the book, however, is of a segmental arch, which forms the central span of this bridge. Its elegant Gothic quoins, which are now almost hidden by latticed girders on both sides, complete the transformation of the central segmental arch into the Gothic arch of Fig. 22 (sketch-book, *SB, 1835, 5 29/5 35*; see also Bourne, 1846) which is one of the better of the many sketches of arches which abound in the Bristol manuscripts. The emery-paper and sandpaper stuck inside the covers of the bound volumes of the manuscripts provide some evidence of Brunel's workmanlike approach to his draughtsmanship. What a pity it is that the artistic flair so evident in the drawings of a two-span skew Gothic arch bridge (Brees, 1837, plate XLII) over the Floating Harbour on the approach to Temple Meads railway station, is now almost completely hidden by segmental arches which make the bridge wider on both sides.

I. Brunel (1870) indicates that the crown of a 120 foot span, 20 foot rise segmental arch bridge "over the New Cut at Bristol . . . sunk much more than expected",[32] but says nothing about the replacement of an arch bridge (the Somerset Bridge) over the Parrett which MacDermott (1927) describes. The drawing store of British Rail at Paddington has a drawing of this bridge, part of a priceless collection of Brunel's drawings which are at present housed there. A brief description of some of Brunel's arch bridges can also be found in Brees (1837), Simms (1838), Bourne (1846), Sekon (1895), MacDermott (1927) and Rolt (1957).

The problems with the Maidenhead bridge occurred in the spring of 1838. Considerable numbers of arch bridges were built in the mid nineteenth century but it was not until 1846 that W. H. Barlow, in a paper to the Institution of Civil Engineers, indicated the nature of the mechanisms which form when arches fail, and displayed models illustrating his arguments. These ideas were developed a little further by Snell (1846) and in the middle of the present century again by Pippard and Baker (1962) and others eventually to yield modern ideas of structural collapse. Studies by Ruddock (1974) suggest that failures of arches by rising at the crown are historically not uncommon, but Heyman (1969) considers that a design has indeed to be clumsy to fail in this way! In this chapter what can be gleaned of the nature of the Maidenhead and Somerset bridge troubles is outlined and the ideas underlying the calculations in the Brunel manuscripts are examined. It would appear that Brunel might have developed an almost direct and safe approach to arch design, which would nowadays be described as "statically admissible". Brunel's practice of using what Fowler describes as "very excellent and scientific centering" avoided disaster in their construction (Mackay, 1900).

According to MacDermott (1927), where the Maidenhead bridge was constructed the Thames was

> ". . . not quite 100 yards wide between low banks with a small shoal in the middle. Brunel designed a brick bridge of two of the flattest and largest arches that had ever been constructed in brickwork. Each arch had a span of 128 feet with a rise of only 24¼ feet, the centre pier being placed on the shoal. Four semi-circular arches at each end to provide for floods, one of 21 and three of 28 feet span, completed the structure. Work was begun early in 1837 by the contractor, one Chadwick, and before the end of the year the arches had been turned. In February 1838 the Directors reported that the bridge was in a very forward state and the centering would shortly be eased. It was eased in the course of the spring, and to the delight of the critics, who had all along declared that such flat arches could not possibly stand, the eastern arch soon showed signs of distortion. The western arch, however, annoyed them by standing perfect."

Mackay (1900) quoting from a letter of Fowler, dated 11th May, 1838, paints a gloomy picture of these "signs of distortion". Fowler writes:

> "The bridge I went to see is of a novel construction, being built of brickwork set in cement for two elliptical arches of 128 feet each, and is now, I am sorry to say, in a dangerous situation. The centering has been slackened and the arches have followed it for 5 inches at the crown, and now rest on them, but not as you would suppose at the crown of the arch, but at about 15 feet on each side, while the crown is 2½ inches clear of the centering. This settling has made the outline of the arch perfectly Gothic and totally destroyed its symmetry, but the worst of the matter is that the arch which has been turned in half-brick rings has separated at every other joint, and consequently the arch is further there weakened by the lower part of it leaving the upper part. The outside of the arch shows eight half-brick curves, but the inside has thirteen. The spandrel wall of one arch has cracked from top to bottom, so likewise have all the walls for backing up the inside of the arch, both occasioned of course, by the settling. The foundations are on rock, and therefore very unyielding. The centering is a very excellent and scientific one, the whole of the timbers being in a state of thrust. What will or can be done is too much for me to say."

There is no evidence, however, of panic action by Brunel. The defective arch was resting on its "very excellent and scientific centering". Earth was being taken over from the Buckinghamshire side by

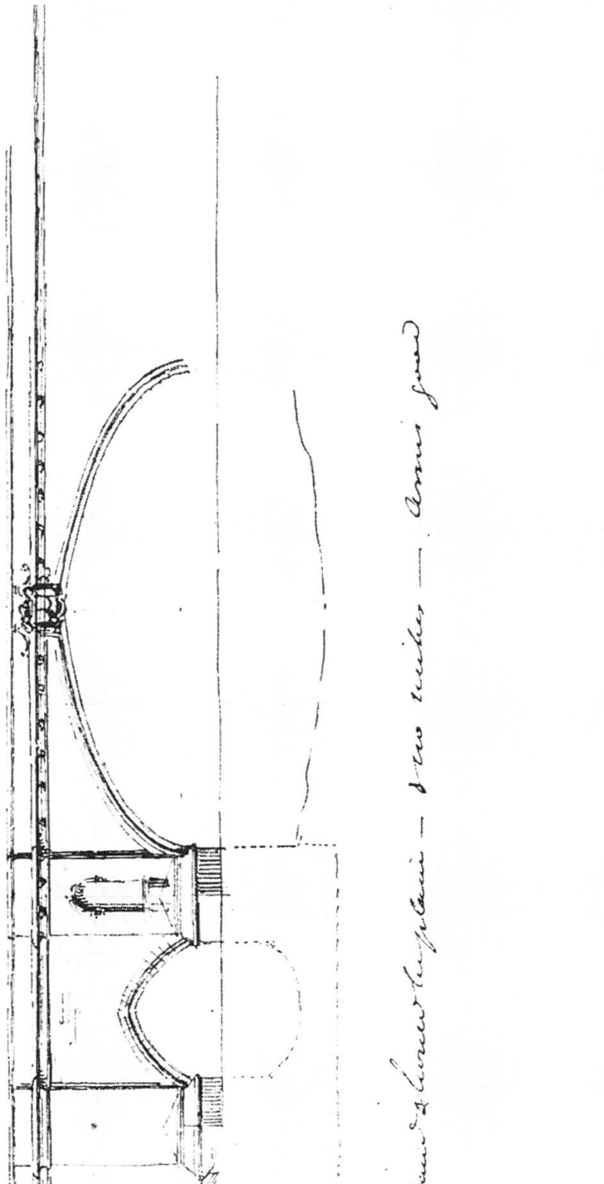

Fig. 22. Arch near Temple Meads station

temporary roads to form the embankment immediately west of the bridge, and in June Brunel directed that the repair of the bridge should be postponed until this was completed and the temporary roads removed so as not to "interfere with the works of the bridge". Rolt (1957) in his biography of Brunel makes light of the failure and so does MacDermott (1927) who writes:

> "The distortion of the eastern arch, about which so much fuss was made, consisted of a separation of about half an inch between the lowest three courses of bricks for about twelve feet on each side of the crown of the arch. It was caused by the centering having been eased before the cement had properly set, and was worse at the two faces than in the interior. In July the contractor admitted that he alone was to blame, and as soon as the earth tipping was finished, set about replacing the defective part. At this time the attack of the Liverpool opposition party on Brunel and the Directors was at its height, and the apparently insecure state of the bridge was of course used as a weapon. Hence it formed one of the matters on which Wood and Hawkshaw were asked to report. The latter considered that more weight was required on the crown of the arch and accordingly recommended that some thirty feet of brickwork should be replaced by stone. Wood, on his part, found that the trouble was due merely to the cement not having had time to set and concurred in the remedy proposed by Brunel. The necessary rebuilding was done and Gibbs records that the centres were eased on 8th October. They were, however, left in position, and the critics continued to declare that the bridge would fall down as soon as they were removed. Brunel had ordered that they should not be taken away till another winter had passed, but one night in the Autumn of 1839 a violent storm blew them down. To the confusion of the critics and the triumph of the Engineer, the arches stood for all to see, and needless to say, have been standing ever since. As a matter of fact, they had been quite clear of all support since the spring of 1839."

It is significant, however, that Brunel kept the centering in place. In June 1838 Brunel reported (MacDermott, 1927) that:

> "The Contractor is now anxious to remove the centerings, as he feels confident that the arches have reached their maximum settlement, and proposes to remove and alter the external surface only of the distorted part, and thus prevent its being apparent to the eye. To this I have objected. I cannot say that I think the work unsafe, or that the Bridge would not stand very well in its present state, but the defect has been caused entirely by the fault of the Contractor, and while the centerings are in place he has still the means of removing this defective portion, and rendering the eastern arch as perfect as the western now is, and such as he contracted to make it. I propose therefore to direct Mr

Chadwick to tighten up the centres of both arches and to remove and replace so much of the eastern arch as upon close examination during the progress of this work I may consider defective . . ."

T. M. Smith (1846) in the spring of 1838 (the time when the centering on the Maidenhead bridge was eased) presented a paper to the Institution of Civil Engineers in which he describes the failure, some 85 years earlier, of a William Edward's 140 foot single-span masonry arch bridge, over the Taff at Pontypridd, which probably had a rise of 35 feet:

> "From the great rise of the arch, it was necessary to raise the roadway at each end, in order to facilitate the access. The quantity of material in the centre of the arch was so small, in proportion to that laid over the spandrils, that the pressure upon the haunches forced up the crown, and the whole was again reduced to ruins."

Perhaps this upward movement of the crown may have prompted Hawkshaw's recommendation that "more weight was required on the crown" of the Maidenhead bridge (MacDermott, 1927). It seems that Edward added such weight in his successful single-span arch over the Taff at Pontypridd and also introduced three cylindrical holes to lighten the spandrels, and in his bridge there is some evidence (Owen, 1939) that this may well have been necessary. Although not externally apparent in the Maidenhead bridge, Brunel's design also took the weight off the haunches. Drawings available in the British Rail drawing store at Paddington indicate that longitudinal walls were used "as ribs" in the spandrels which are not solid. These ribs are presumably the "walls for backing up the inside of the arch" which Fowler indicates were "likewise" cracked from "top to bottom" (Mackay, 1900). It is significant that although the use of these longitudinal ribs should have increased the effective depth of the arch ring considerably, they did not prevent failure and they thus provide an exception to Ruddock's (1974) conclusion that "the parallel wall method of construction had solved the problem of instability of arches".

By 1846 the Maidenhead bridge had been accepted, and Brunel receives honourable mention by Sir John Rennie in his presidential address to the Institution of Civil Engineers:

> "Brick has been much used for bridges . . . and, latterly, it has been carried to a far greater extent by Brunel in his bridge across the Thames at Maidenhead, for the Great Western Railway . . . they are built wholly of brick and Roman cement."

Sir John also remarked in his address that:

> "Additional strength has been given to brick structures, by the intro-

> duction of bands of thin hoop iron between the courses; this improvement was first generally introduced by Sir M. I. Brunel."

The descriptions of the Maidenhead failure by Fowler and others suggest, however, that such bands were not used in the Maidenhead bridge. Examination of the under side of the Brunel arches at Maidenhead indicates alternate header and stretcher rings, and would suggest a construction of sensibly successive layers of rings of the arch of the kind which would permit the arch ring to separate as described by Fowler (MacDermott, 1927). It may be, however, that the mystery surrounding the Maidenhead bridge failure in 1838 eventually set in motion the investigations of W. H. Barlow and Snell which were published in 1846 and which really dealt with what would nowadays be looked on as the collapse of arches by their becoming mechanisms.

Barlow early on in his paper states (correctly) that the essential element of equilibrium, when the only force acting is gravity, is "that the horizontal forces in any part of the curve (or the arch) are equal to each other". He states that this had not been pointed out previously although it was probably known. Indeed it is a fundamental concept used by Brunel in his manuscripts of 1837. Commenting on Barlow's paper, Brunel remarks that:

> "In a very large arch, with a small rise, the line of pressure must be confined within very narrow limits."

He should have emphasised, as Barlow pointed out, that the thickness of the ring rather than the rise governs the accuracy required, but in emphasising the need for accuracy he made an important point which explains in part why, in his manuscripts of 1837, he eventually uses numerical methods rather than the less accurate methods of graphic statics of his contemporaries.

The calculations in his general calculation book (*CB, 1837*) give only an occasional comment on the procedure used, but from what is done the underlying principles and assumptions can be largely inferred. In strength calculations for arch bridges the arch ring proper is usually assumed to be the only part providing its strength. This is a safe assumption not too far from the truth if the centering is struck before the mortar has set. Some assumptions have also to be made on the nature of the loading on the arch. Brunel initially assumes for the Brent[33] and Maidenhead bridges that the vertical loading on the arch is the weight of elements of the bridge obtained by dividing the arch "at right angles to the line of pressure" or at "the angle of" an arch "joint" (which might not be precisely the same thing) and then vertically through the spandrel, as in Fig. 23, which although for

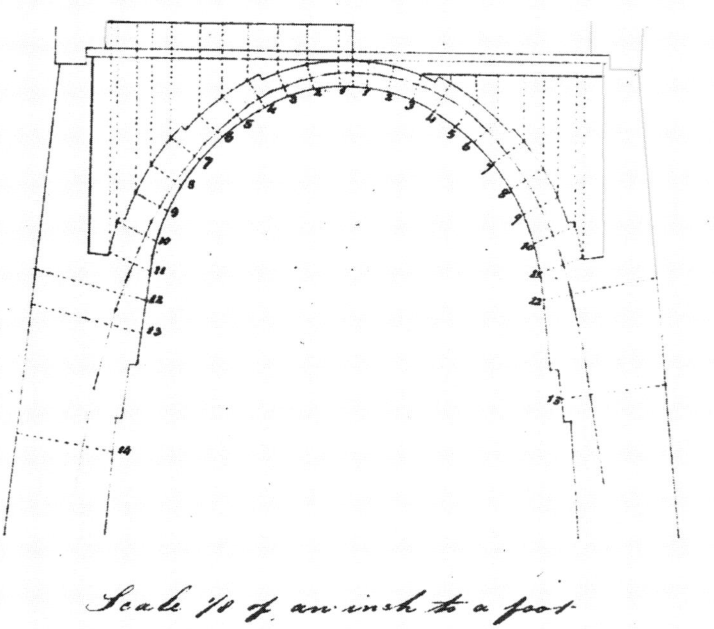

Fig. 23. Arch analysis for Hanwell

only one span may well be a drawing on which were based the six arches in the approach to the Wharncliffe Viaduct. An assumed "angle" at a particular point of the arch ring enables Brunel to calculate by statics the horizontal thrust in the arch. The "thrust line" in, i.e. the funicular polygon for, the arch can now be drawn starting with a horizontal at the crown and diverting this downwards, where it meets the line of the vertical load from one of the elements into which the bridge has been divided. This is clearly seen in the left hand element between 6 and 7. For static equilibrium to be obtained, the point where the polygon changes direction in this element should be the centre of gravity of this element and this would hardly seem to be the case here. The point where the polygon changes direction is critical in determining its shape. The polygon on the left hand side of Fig. 23 comes very close to the intrados at points 6, 7 and 8 and Brunel may not have been satisfied with the result, for he proceeds to calculate what the load distribution should be to make the thrust line

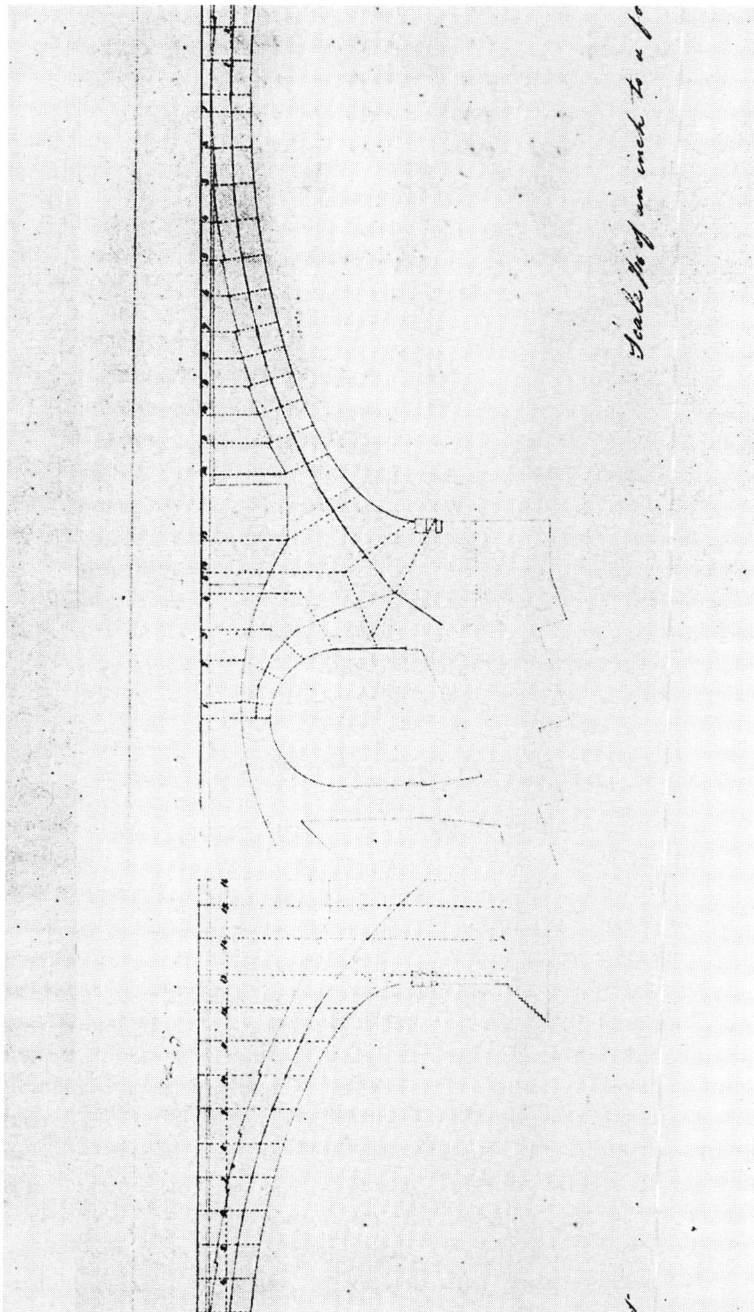

Fig. 24. Arch analysis for Maidenhead

coincide with the centre line of the arch. Still not satisfied perhaps, he starts again and divides the arch into elements by vertical lines running through the whole depth of the bridge (as on the right hand side of Fig. 24 for the Maidenhead bridge). The weights of these elements enable him, without making any further assumption, to calculate "the weights of the half arches", i.e. the shear on vertical sections of the bridge.

At this stage a most novel feature is introduced which avoids the necessity of finding the centre of gravity of each of the elements into which the bridge is divided. The shear is represented by a polynomial of the form

$$W = Ay + By^2 + Cy^3 + Dy^4$$

where y is measured horizontally from the centre line of the arch and A, B, C and D are constants chosen so that the polynomial is correct at four points on the span. Brunel checks that the error at other points is negligible. For the Brent viaduct (short span arches near Hanwell station) he quotes the result

$$W = 42y - y^2/7 + y^3/6$$

This weight has to be resisted by the vertical component of the arch thrust, the horizontal component of which is H. Simple statics then gives the result that

$$H\,\mathrm{d}x/\mathrm{d}y = 42y - y^2/7 + y^3/6$$

and integration gives

$$Hx = 21y^2 - y^3/21 + y^4/24 + C$$

where C is constant. Brunel now chooses to place the thrust at the mid depth of the arch at the centre of the span, where x and y are zero. This makes C zero. He also decides that the thrust will pass through the mid arch ring depth at about quarter span, where the earlier calculations indicated that it approached the intrados. Substitution of the values of x and y for this point then give $H = 441$ and so the equation

$$441x = 21y^2 - y^3/21 + y^4/24$$

from which the shape of the thrust line can be immediately drawn. In important parts of the bridge he thus establishes that it is possible to have a thrust line which is very close to the centre of the arch ring. A modern arch designer might well take such a calculated value for the horizontal thrust in the arch and adjust flat jacks in the abutment until this value was obtained. Brunel effectively demonstrates that if this thrust could be obtained then the arch would be safe. The horizontal component of the thrust in the arch represented by the number 441 he notes is equivalent to a "pillar 24½ feet high".

At this stage he turns to the Maidenhead bridge (Fig. 25) (and the calculations again seem to have been "copied for Brodie, 29 April 1851"). The first calculations with the thrust line made normal to the "voussoirs" (left hand side of Fig. 24) give the horizontal thrust as equivalent to a pillar height of 213 feet and a thrust line which is very close to the centre line of the arch ring. This is followed by the statically more acceptable polynomial approach, already described, in which the thrust line is made to pass through mid arch ring depth at the centre line and a point close to the end of the span (right hand side of Fig. 24). The horizontal thrust he obtains is the equivalent of a pillar of 205 feet (10.2 tons per square foot)—a very low stress compared with the probable crushing strength of the bricks which could be expected to be well over 60 tons per square foot. The thrust line (Fig. 24) is again everywhere near the centre line of the arch ring. There is nothing here except the intensity of the pressure (which is almost ten times that obtained for the Brent viaduct but still low in relation to the strength of the bricks) which would suggest that trouble would arise.

These results indicate why when trouble did arise Brunel did not attempt to load the crown of the Maidenhead arch as suggested by Hawkshaw and why, with good centering, he chose to wait for the weight of the bridge to consolidate the abutments (in case of any movement there) and for the mortar to set. There would be no trouble if he could but obtain his calculated horizontal thrust from the consolidated arch and abutments.

Brunel's manuscript calculations on the Maidenhead bridge stop here. He proceeds next, using only the more acceptable polynomial approach, to check Thorney Broad (88 foot equivalent pillar), the large spans of the Brent, i.e. Wharncliffe Viaduct proper (84 foot equivalent pillar) and the Avon bridge (156 foot equivalent pillar); the thrust lines for the latter again run close to the centre line of the voussoirs.

There are no calculations in the manuscripts relating to the Somerset bridge over the Parrett near Bridgwater; this bridge also gave Brunel some trouble. It was:

> "a masonry arch of 100 ft. span with a rise of only 12 ft., nearly twice as flat as the much criticised brick arches at Maidenhead. This was begun in 1838 and finished in three years" (MacDermott, 1927).

6

Maidenhead Bridge

The arch stones divided at rt Ls to the line of Pressure & the load divided vertically

Left side of Fig: Page 8 — H = 160837 = 165½ lbs on sq in: = 10 tons 13 cwt on sq ft

Equivalent Pillar 213 feet high

Copied for [illegible] 29 April 1851

No	Wt of each Piece	Wt of half Arch	Angle of the line of Pressure
1	2220	2220	0°·47′
2	4749	6969	2°·29
3	4877	11846	4·13
4	5648	17494	6·13
5	6003	23497	8·18
6	6603	30100	10·36
7	6951	37051	12·58
8	5564	42615	14·50
9	5998	48613	16·49
10	7196	55809	19·8
11	7406	63215	21·27
12	10190	73405	24·32
13	28208	101613	32·17
14	22499	124112	37·39

Unit of wt = 1 [illegible]

Small Arch H = 20520

Copied

1	3000	3000	8°·19′
2	8374	11374	29·0
3	8704	20078	44·23
4	16410	36488	60·39
5	20068	56556	70°·3

Wt of Large Arch = 124112
" " Small = 56556 — H in Large Arch = 160837
Part of Pier = 12741 — " " Small = 20520
Sum = 193409 — Diff = 140317

Thence 54°·2 = vertical angle of the resultant
Resultant = 83 tons 1 cwt for 1 ft in breadth

Fig. 25. Arch analysis for Maidenhead

ARCH BRIDGES

A drawing of it is still available at Paddington. This had an intrados which was an arc of a circle of 110 feet radius but otherwise, in general concept, was superficially similar to Brunel's other arch bridges. Longitudinal voids occurred in the spandrels which had longitudinal ribs. The arch ring was 3 feet deep at the centre line. From these drawings the distribution of weight over the span can be estimated and from this the simply supported bending moment obtained. (This is sensibly what Brunel did by deriving and integrating his polynomials.) A value of the horizontal thrust H can then be chosen so that the relieving moment, namely H times the vertical height of the arch, reduces the net moment to any desired value. The author found that the arch shape is such that the thrust line cannot follow the arch centre line as closely as is the case with Brunel's calculations for the Maidenhead, Thorney Broad and Avon bridges.

With the Somerset bridge, Brunel again kept the centering in place as long as he dared and it may be that continued movement of the bridge relative to the centering and calculations like those just described, but of which there is no record in the manuscripts, resulted in the following being MacDermott's account of Brunel's replacement of this bridge:

"This part of the railway had not been open long when a public outcry arose at the obstruction to the River caused by the centering of the Somerset Bridge, which, owing to a slight movement of the foundations, Brunel had not ventured to remove. At last the Directors were obliged to insist, and in August 1843 Brunel reported: With regard to Somerset Bridge, although the Arch itself is still perfect, the movement of the foundations has continued, although almost imperceptibly, except by measurements taken at long intervals of time; and the centres have, in consequence, been kept in place. Under existing circumstances, it is sufficient that I should state in compliance with a Resolution of the Directors, measures are being adopted to enable us to remove these centres immediately, at the sacrifice of the present Arch.

"Six months later the Directors stated that 'a most substantial bridge has been built over the River Parrett without the slightest interruption to the traffic.' Between the same abutments Brunel had substituted the timber arch which did duty till 1904, when the existing steel girder bridge was erected."

The closeness of Brunel's thrust lines and the centre lines of the arches for the Maidenhead, Avon and Thorney Broad bridges suggests that a deliberate design approach to achieve this might have been developed, along the following lines (at least this might be expected from the advice given to one of his assistants in a letter dated 30th

December, 1854, in which Brunel writes, " . . . all forces should pass exactly through the centre of any surfaces of resistance . . . especially . . . in anything resembling a column or strut" (I. Brunel, 1870)).

In the initial stages of a design the line of rail and the position of the springings would be available. From these the approximate span and the rise of the arch would be found. Then from the many rules available or Brunel's experience the thickness of the arch ring at the crown would be chosen. With an allowance for filling over the crown, the intrados of a possible arch could now be sketched and starting from the extrados at the crown, the extrados of the arch also sketched. This would probably give an arch ring, increasing in thickness towards the roots. The positions where it became practicable to start the longitudinal walls to lighten the spandrels would now become apparent and a possible bridge would thus be available in outline. Division of this by vertical lines into, say, twelve sections would give the weight distribution over the span, whence from an appropriate polynomial the horizontal thrust and a possible thrust line for the arch could be obtained, starting from the crown and choosing the thrust line to be at mid arch ring depth there and also, say, at quarter span. The ring shape could now be adjusted in depth so the thrust line would be close to half ring depth over the whole span. In the Maidenhead bridge (Fig. 24) this might have been accomplished by just changing the starting point for an extra ring of bricks in the arch. The shape of the intrados would remain unaltered and there would be no difficulty (on paper) in altering the ring depth in this way. It would probably result in a small change in the weight distribution over the span and it may well be that Brunel's calculations in the manuscripts were his final check to ensure that the thrust and arch centre lines did not deviate appreciably.

The rapidity with which such a process of design converged would depend on the skill of the designer, particularly in choosing the initial shape for the arch. In the case of the Somerset bridge it may well be that this process was not used. If it had been, the form of the intrados near the roots would have been altered and the circular arch shape would have been modified.

This design approach is fundamentally sound in that it is an endeavour to produce an arch structure which is entirely in compression and so potentially a Maxwell (1869) optimum structure. The arch of minimum weight of material (cf. Owen, 1965) would have a considerably greater rise, but in the main Maidenhead arches the choice of rise seems to have been limited to the small value Brunel was forced to use by the terrain and line of rail. The design approach outlined is simple and avoids the complications and design tables which have not been a very sound or desirable feature of many of the texts on arches which have appeared from time to time.

W. H. Barlow's paper of 1846 and the discussion following it contain comments which may be of relevance in relation to the trouble which arose when the centering of the Maidenhead bridge was eased in 1838. "Lines of thrust", writes Barlow, "must not be brought too near the extremities of the voussoirs and in brick arches, particularly those turned in separate rings, a much greater latitude must be allowed." Barlow also emphasised that it was desirable "to form brick arches as much as possible in one bonded mass, using the best cement", and Bidder reinforced Barlow's remarks by commenting that Barlow's paper "showed the impropriety of constructing brick arches in separate super posed rings: the line (of thrust) would in almost every instance, travel out of the ring in which it commenced and in case of fracture the rings would fail consecutively but if . . . well bonded . . . would be more satisfactory. All the best arches were now built with full bond." Brunel, who was present at this meeting, made no reference to bonding, although his father had done work on this as indicated by Rennie (1846). Brunel chose to emphasise the need for arches to be considered as elastic homogeneous bodies (a study which was pursued later both in this country and abroad and found to be appropriate and relevant provided the nature of the support at abutments is known).

Fowler some years later widened Brunel's Maidenhead bridge by adding on each side arches which follow the profile of Brunel's bridge, but the arch ring depth is only that of the edge of Brunel's original arch. The aesthetic effect so obtained is to suggest that the arch ring is more slender than it actually is. Fowler used "superior bricks" (a needless compensation for the reduction in mean arch ring depth), "Portland cement and clean Thames sand possessing a degree of cohesive strength and regularity in quality . . . unknown in the days of lime and Roman cement" (Mackay, 1900). It is significant that Fowler did not change the shape of the intrados so he must have been satisfied with whatever he considered to be the position of the thrust line, but in spite of thinking that the piers were on "rock foundation", when he wrote his letter of 11th May, 1838, his foundations were extensively piled. He too would have known that it was essential to ensure that the abutments provide the necessary horizontal thrust. Isambard Brunel (1870) tells us that it was "originally intended that the foundation of the bridge should be on the chalk, which was at a short distance below the surface; but it was found to be very soft, and Mr Brunel therefore decided to place the foundations of the bridge on a hard gravel conglomerate overlying the chalk." The centre of the arches on the landward side of the main arches, he continues, were loaded with concrete and:

> ". . . were struck, and an active thrust opposed to the main arches before their centerings were eased. The line of pressure of each main arch was diverted downwards by the thrust of the flat arch adjoining it without the necessity of employing a great mass of brickwork in the abutment."

It thus appears that the following might well be an explanation of what happened to the Maidenhead bridge when the centering was eased. The western span stood firm because the mortar in the joints had set sufficiently firmly and the western abutment did not effectively move laterally. In the eastern span the "Roman cement" may not have had adequate time to set; the contractor, according to MacDermott (1927) and Rolt (1957), seems to have accepted this, although Brunel states (according to MacDermott) that the contractor considered that the abutments had moved. Some spanwise outward movement of the eastern abutment and the squeezing-out of mortar, which had not set between bricks, could both have had the effect of reducing the horizontal thrust until it reached such a value that the thrust line at the crown approached the outer ring in which the arch was turned. If this was not bonded adequately, it could have failed as a bent strut not bonded to the ring next to it, i.e. by the crown of the arch moving upwards. The next ring down could then have failed in the way described by Bidder (W. H. Barlow, 1846) as the mode of failure of an unbonded arch. (Was he describing what he thought to be the nature of the Maidenhead bridge failure?) Thus arch rings may have buckled upwards at the crown in turn and this process resulted in the arch settling on the centering "but not at the crown" just as Fowler describes (Mackay, 1900).

It could have resulted in the bottom three courses of bricks separating to "the extent of half an inch" for 12–15 feet on either side of the crown (Rolt, 1957). In the meantime rotation could have taken place about a fourth "hinge" near the intrados near the springing at the abutment where Fowler reported cracking "from top to bottom". The mortar in the longitudinal ribs would not have set to provide the added strength, bond and stiffness which Ruddock (1974) correctly indicates to be the advantage of this form of spandrel lightening.

From the satisfactory performance of the western arch Brunel knew that the basic design of the similar eastern half would be satisfactory if it could only develop the necessary horizontal thrust. Maintaining the centering in place was an insurance against disaster and allowing the bridge to continue in use to convey spoil to the western embankment would have retained some load on the arch ring and the abutments and helped consolidation, and on 8th October (MacDermott, 1927), when the centres were again eased, there was no further trouble. But in view of the earlier trouble a wise engineer would have retained the eased centering in place and used measurements taken between it and the arch to check whether further settlement was taking place. This would not be, as Rolt (1957) suggests, "excessive caution" or indicate "an impish sense of humour". The advice Brunel passed on to his assistant was "that you cannot keep centres or shores up too long" (I. Brunel, 1870).

In the case of the Somerset bridge (MacDermott, 1927), Brunel indicates that careful measurements of its movement were taken, possibly from the centerings which he again kept in place. This time slight movement persisted and, although the masonry arch appeared satisfactory, since he was being compelled to remove the centering he prudently replaced the bridge, presumably lest continued outward movement of the supports would eventually result in the horizontal thrust not being large enough to ensure the safety of the arch.

Notes

32. Simms (1838) and Bourne (1846) indicate that the brickwork at the Wharncliffe Viaduct "upon striking the centering" followed it by "from 1¾ to 2¾ inches, averaging 2 inches".
33. In his calculation book (*CB*, 1837), under the title "Brent Viaduct" Brunel deals with both the approach (Hanwell embankment) and the main Wharncliffe arches of the crossing over the River Brent. MacDermott (1927) also refers to a Brent viaduct in Devon, only the pier of which still remain and which is not far from a lovely Brunel viaduct at Rattery, which he also mentions.

References

Barlow, W. H. (1846). On the Existence (Practically) of the Line of Equal Horizontal Thrust in Arches, and the Mode of Determining it by Geometrical Construction. *Minut. Proc. Instn Civ. Engrs*, **5**, 162–182.

Bourne, J.C. (1846). *The History and Description of the Great Western Railway*. Bogue, London.

Brees, S.C. (1837). *Railway Practice*. Williams, London.

Brunel, I. (1870). *The Life of Isambard Kingdom Brunel, Civil Engineer*. Longmans, London.

Heyman, J. (1969). The Safety of Masonry Arches. *Int. J. Mech. Sci.*, **11**, 363–385.

MacDermott, E.T. (1927). *History of the Great Western Railway*. G.W.R. Co., London.

Mackay, T. (1900). *The Life of Sir John Fowler*. Murray, London.

Maxwell, J.C. (1869). On Reciprocal Figures, Frames and Diagrams of Forces. *Trans. R. Soc. Edinb.*, **26**, 1–40.

Owen, J.B.B. (1939). Edward's Single Span Arched Bridge at Pontypridd, 1755. *Proc. S. Wales Inst. Engrs*, **55**, 20–59.

Pippard, A.J.S. and Baker, Sir John (1962). *The Analysis of Engineering Structures*, 4th edition. Arnold, London.

Rennie, Sir John (1846). Address of Sir John Rennie, President to the Institution of Civil Engineers. *Minut. Proc. Instn Civ. Engrs*, **5**, 19–122.

Rolt, L.T.C. (1957). *Isambard Kingdom Brunel*. Longmans, London.

Ruddock, E.C. (1974). Hollow Spandrels in Arch Bridges: A Historical Study. *J. Instn Struct. Engrs*, **52**, 281–293.

Sekon, G. A. (1895). *A History of the Great Western Railway*. Digby, Long, London.

Simms, F.W. (1838). *Public Works of Great Britain*. Weale, London.

Smith, T.M. (1846). Account of the Pont-y-tu-prydd, over the River Tâfe, near Newbridge, in the County of Glamorgan. *Minut. Proc. Instn Civ. Engrs*, **5**, 474–477.

Snell, G. (1846). On the Stability of Arches, with Practical Methods for Determining, According to the Pressures to which they will be Subjected, the best Form of Section, or Variable Depth of Voussoir, for any Given Intrados or Extrados. *Minut. Proc. Instn Civ. Engrs*, **5**, 439–474.

17

Valley crossings on the Old Croton Aqueduct

L.D. Lankton

On June 2, 1835, the New York City Board of Water Commissioners hired Major David Bates Douglass as Chief Engineer of the Croton Aqueduct. (See Map, Figure 1.) On October 11, 1836, the same Commissioners fired Douglass, after realizing that they had mistakenly put a "mere theorist" in charge of planning and building the most extensive water-works project ever undertaken in the United States. After working for a year and a half, Douglass, an excellent surveyor, had routed most of the aqueduct's forty-one-mile long line, but he had completed no final plans for its structures. The Commissioners decided that Douglass lacked "that practical knowledge" necessary to complete the aqueduct successfully, so they ousted him in favor of John B. Jervis.[1]

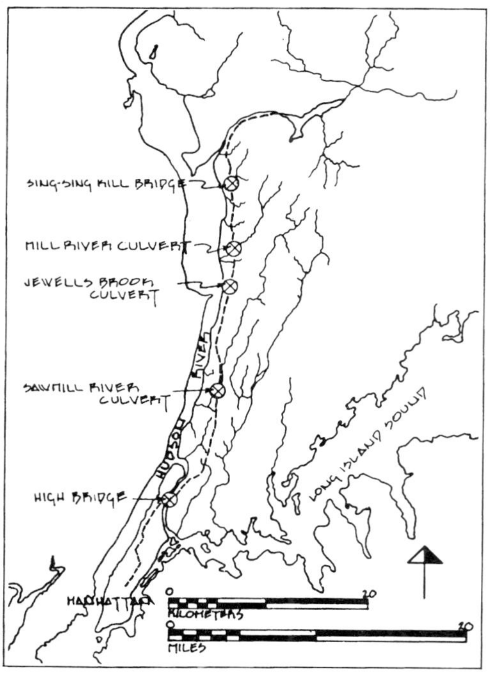

Figure 1. *Map. Dashed line represents original route of the aqueduct. Drawn by Robert Levy. Courtesy of HAER.*

Unlike Douglass, who was a Yale graduate and a former West Point professor, John Jervis had no university training and no formal schooling at all after he was fifteen. But Jervis had something that his predecessor lacked–a wealth of experience in designing and constructing large works.[2] Jervis had learned his engineering on the Erie Canal between 1817 and 1825, and he had subsequently served as Chief Engineer of several canals and of the Mohawk and Hudson Railway.

When he accepted the chief engineership of the Croton Aqueduct, Jervis accepted the most difficult charge of his career. The aqueduct, running from the Croton River in upper Westchester County to central Manhattan, called for innovative structures of sufficient variety to tax the skill of any civil engineer. Jervis built a dam on the Croton, an enclosed masonry conduit, and two reservoirs on Manhattan. He designed gatehouses, waste weirs, ventilators,

[1] Stephen Allen, Chairman of the Water Commissioners, was the man most responsible for Douglass' dismissal. For Allen's side of the story, see "New York Water Works," MS in the Stephen Allen Papers, New-York Historical Society. Also John Travis, ed., "Memoirs of Stephen Allen," typescript, NY Historical Society; and Allen, letter, New York Morning *Courier and Enquirer,* November 12, 1840. For Douglass' contrary view, see his letter in the New York *Times & Star,* October 30, 1840.

[2] The best published work on Jervis' career is Neal FitzSimons, ed., *The Reminiscences of John B. Jervis* (Syracuse, 1971). In manuscript form, Jervis' professional papers are maintained as a special collection by the Jervis (Public) Library in Rome, NY.

Figure 2. *Arch of 88 feet over Sing-Sing Kill in Ossining. Jervis originally planned for an 80-foot arch. He increased the span to found the abutments on rock. Taken from Fayette B. Tower,* Illustrations of the Croton Aqueduct *(New York, 1843).*

Figure 3. *Mill River Culvert near North Tarrytown. Taken from Tower.*

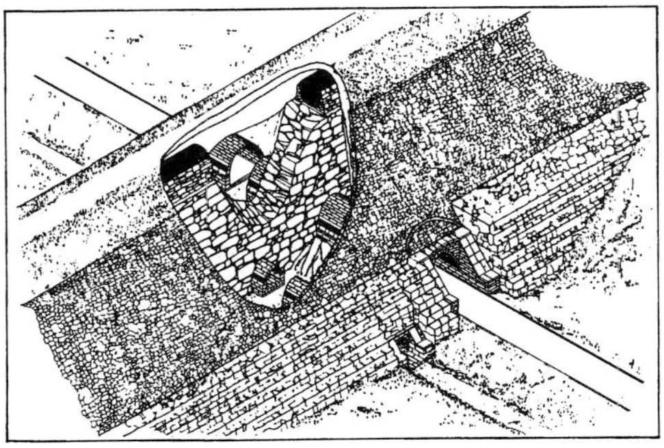

Figure 4. *Jewells Brook Culvert at Irvington. HAER drawing by Peter Basmajian; edited by David Schaaf.*

and culverts.[3] Where the grade line of the aqueduct fell below natural ground level, he chose to tunnel or to excavate an open cut. Where the grade line passed a valley, he decided whether to carry the channel on a bridge, an embankment, or an inverted syphon of cast iron pipes laid along the valley floor.

Jervis opened the Croton Aqueduct in 1842, and, with some interruptions, it delivered water to New York City until circa 1955. During its operating life, the aqueduct underwent numerous changes, particularly on Manhattan. As the city's water needs changed and as supplementary water resources were introduced to the city, many structures were demolished. Today, virtually none of Jervis' work survives on the Island. Yet from the Harlem River northward, the survival rate is very high. Long stretches of the line are intact and contain a variety of structures. Of the surviving structures, the valley crossings command the most attention because they were some of the largest, most difficult and most expensive works that Jervis engineered.

The aqueduct's major valley crossings included Sing-Sing Kill Aqueduct Bridge in Ossining (Figure 2); Mill River

[3]Illustrations of all these various works are found in Theoph Schramke, *Description of the New-York Croton Aqueduct* (New York, 1846), and in Fayette B. Tower, *Illustrations of the Croton Aqueduct* (New York, 1843). Both of these authors worked in the aqueduct's engineering department–Schramke as a draftsman and Tower as a first Assistant Engineer. Also, original engineering drawings survive in the Jervis Library and in the Engineering Societies Library in NYC. Finally, HAER executed drawings of extant examples of these types of structures in 1975.

Culvert near North Tarrytown (Figure 3); Jewells Brook Culvert near Irvington (Figure 4); Sawmill River Culvert in Yonkers (Figure 5); High Bridge over the Harlem River (Figure 6); an inverted syphon on the island at Manhattan Valley (Figure 7); and the crossing at Clendenning Valley, between 96th and 101st Streets (Figure 8). All of these structures survive, except for the last two—although a few have been altered.[4] Jervis used diverse means to cross depressed ground and by examining his work on three structures—Sing-Sing Kill Aqueduct Bridge, Mill River Culvert, and High Bridge—we can discover the reasons for this diversity.

Before examining the three structures in detail, some general observations must be made. While working on the Croton Aqueduct, Jervis knew he was not building just another canal or railroad. These things could occasionally break down, but the aqueduct was another matter entirely. It was to serve as a lifeline to New York, and the lifeline had to be durable and reliable. Jervis was very cognizant of the Roman aqueducts that had functioned for centuries. He felt that the Croton Aqueduct, too, had to be built not just for now, but for ages to come.[5]

[4]In 1937 (according to FitzSimons, *Reminiscences*, between pp. 94-95), a steel span replaced five of the masonry arches in High Bridge. The original road arch seen in the Sawmill River Culvert illustration was enlarged in c. 1875, and at some unknown time the base of the structure received stepped buttresses of stone laid in cement.

[5]Jervis emphasized durability and longevity in numerous letters to the Water Commissioners and to his assistant engineers. For example, see his reports on Sing-Sing Kill Aqueduct Bridge and on the crossing of the Harlem River. (These reports are cited below.)

Figure 5. *Sawmill River Culvert in Yonkers. A 20-foot road arch and two 25-foot culverts over the river. Taken from Tower.*

Figure 6. *High Bridge over the Harlem River. Not completed until 1848. Constructed of gneiss ("bastard granite"). Taken from Tower.*

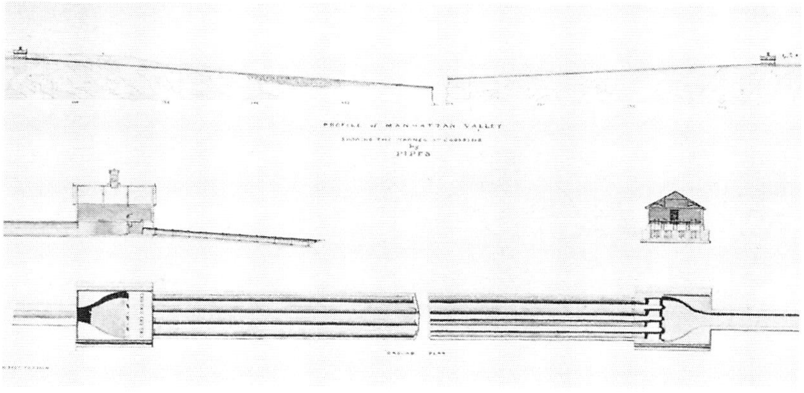

Figure 7. *Inverted syphon at Manhattan Valley. Cast-iron pipes across a depression 4180 feet wide with a maximum depth of 105 feet. Original drawing found in Jervis Papers.*

Industrial Archeology

Figure. 8. *Crossing at Clendenning Valley, a depression 2000 feet wide with a maximum depth of 50 feet. The road and pedestrian arches accommodated 98th, 99th and 100th Streets. Taken from Tower.*

Durability was a major goal, but Jervis wanted to achieve it as economically as possible. Frugal by nature, the Chief Engineer scrutinized the engineering department's minor expenditures for drafting paper and other office supplies. While watching over the nickels and dimes, he also sought significant economies when planning large structures. Because he insisted on balancing economy and durability, Jervis preferred embankments to aqueduct bridges, and his emphasis on economy sometimes led him to adopt an inverted syphon in place of an embankment or a bridge.

Jervis shunned aqueduct bridges wherever possible. Because they demanded materials and workmanship of the highest quality, they were extremely expensive. An even more serious problem was their susceptibility to failure, if any of the water under transport leaked into the masonry and then froze. Jervis feared that in New York's harsh winters, "leakage amounting to only a sweating of the . . . bridge masonry would tend to disintegrate even the most durable stone."[6]

Research into the history of aqueduct bridges showed that they had exhibited a strong tendency to decay. The failures included some of the ancient Roman bridges, which had been built in a much milder climate:

> It may be observed that even in Rome, that portions of their aqueducts, which are elevated on bridges of masonry, have often required extensive repairs . . . , and those portions of the Roman aqueducts, which have stood, undisturbed, the test of time, are placed underground, and therefore, not exposed to material atmospheric changes.

English aqueduct bridges, too, had suffered:

> In the stone aqueducts for the English canals they formerly adopted the plan of lining the inside with well puddled earth [to guard against leakage]. This earth was found to heave by frost, and this produced the same derangement in the masonry as had been experienced when the masonry only was depended upon. It was not infrequently the case that a portion of the masonry in a few years required to be supported by strong bars of iron, or taken down and rebuilt.

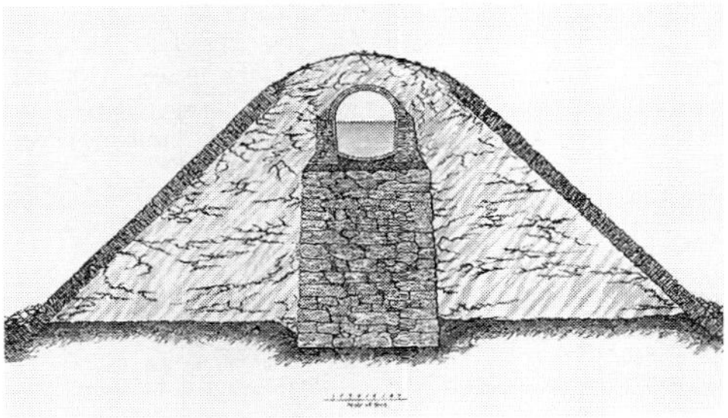

Figure 9. *General plan for embankments, showing the conduit atop a foundation wall laid dry. Taken from Tower.*

As for aqueduct bridges in the United States, Jervis found them "quite too leaky, to promise the durability required in the Croton Aqueduct." The Little Falls Aqueduct on the Erie Canal had been in service for only twelve years before exhibiting "decided marks of injury from frost."[7]

Instead of relying heavily on aqueduct bridges, Jervis adopted a general plan for crossing all but the widest and deepest valleys on embankments (Figure 9). Eventually, he

---

[6]FitzSimons, *Reminscences of JBJ*, p. 128.

[7]Jervis, "Report on Sing-Sing Kill Aqueduct Bridge," February 8, 1837, Jervis Papers, Jervis Library.

Figure 10. *Typical masonry culvert, across Rider's Brook, measures 6 feet by 6 feet. HAER field photograph, 1975.*

employed this construction along a total of six miles of the aqueduct's run. The Chief Engineer seated the enclosed masonry conduit atop a heavy foundation wall with sides that carried a slight batter. The wall was composed of "large stones laid in a rough but compact manner, the interstices between the stone[s], and to level up the courses, to be filled with fine broken stone, so as to give firmness and stability to the work."[8]

A foundation wall was more expensive than rubble stone simply dumped into low areas as fill. Jervis believed the additional expense was necessary to prevent slides or uneven settlement that would crack the conduit's masonry. At the same time, the foundation wall, laid without mortar, was much less expensive than another plan which Jervis had considered. Over the aqueduct's run, he calculated that the dry wall would cost some half-million dollars less than a wall "built of solid hydraulic masonry."[9] Since it contained no mortar, the wall required a heavy earthen embankment on both sides to assure that it "kept in place." Where the height of the embankment demanded it, the earth was kept in place by a stone protection wall.

Because many valleys contained streams, culverts were an essential part of Jervis' embankment plan. The Chief Engineer proposed a system of standardized culverts, ranging from two to twelve feet wide. These culverts generally had inverted arches on the bottom, vertical side walls and arched tops (Figure 10). Once his engineers had completed plans for the standard culverts, Jervis chose one of a proper size for any given valley. If a stream required a culvert larger than twelve feet, then Jervis prepared a special plan for it.

Jervis preferred embankments because they were safer and generally less expensive than bridges, but he recognized that their usefulness was limited. Some valleys between the Croton and the Distributing Reservoir on 42nd Street were too deep or wide to cross with an embankment. A broad-based, trapezoidal embankment would require too much space and would impinge on surrounding structures. Or it would require quantities of earth and stone so immense as to be uneconomical. Where an embankment would not do, Jervis could fall back on an aqueduct bridge, protected in some way from leakage and frost. Or he could resort to an inverted syphon formed of cast-iron pipe. Unlike the masonry conduit, an inverted syphon required no tall, supportive structure. Instead of maintaining the aqueduct's grade line, it ran down one side of a valley and up the other. The water filling the descending pipes created a head, or pressure, that forced the water to rise on the opposite side of the valley.[10]

[8] Jervis, "Report to the Board of Water Commissioners," December 27, 1836, Jervis Papers.

[9] FitzSimons, *Reminiscences of JBJ*, p. 162.

[10] The "inverted syphon" derived its name from the fact that it resembled the bent tube of a syphon, turned over. It did not, however, function like a true syphon.

The greatest advantage of an inverted syphon was its relatively low initial cost. Because it more-or-less followed natural terrain and required no tall foundation wall or high arches, it reduced labor and materials costs.

Jervis, however, had to weigh these savings against certain liabilities. An inverted syphon's cast-iron pipes were subject to corrosion. They promised a shorter working life than the masonry conduit, and higher maintenance costs.[11] A greater liability was that wherever Jervis used an inverted syphon, he sacrificed considerable elevation—a commodity extremely valuable in its own right. Every foot of elevation sacrificed over the aqueduct's forty-one miles resulted in an equal loss of elevation for its Distributing Reservoir on 42nd Street. Jervis had to keep the reservoir as high as possible, so that its head was sufficient to distribute water to all parts of lower Manhattan without the aid of pumps.[12]

The masonry conduit was free-flowing and gravity-fed. Water in the conduit did not run under pressure; it ran downhill. The slope or declivity was gradual. The conduit sacrificed only thirteen-and-a-quarter inches of elevation per mile. Still, it was capable of discharging, according to hydraulic formulae, sixty million gallons daily.[13] For an inverted syphon to match that discharge, hydraulic formulae told Jervis that he would have to sacrifice significantly more elevation per mile. In the hypothetical case of crossing a depression seventy-five feet deep and a half-mile wide, the masonry conduit, supported on an embankment or bridge, would descend just over six-and-a-half inches. An inverted syphon, over the same run, would descend perhaps three feet.[14] Jervis would have to terminate the pipes that far below their starting height to achieve the desired flow rate. Because he could not afford such a loss in elevation at all valleys, Jervis employed inverted syphons judiciously.

With these general points in mind, we can now examine the particular factors that determined the forms of three extant valley crossings on the Croton Aqueduct.

Sing-Sing Kill Aqueduct Bridge

The kill (or brook) that ran through Sing-Sing (now Ossining) was small, but it had carved a substantial valley. When Major Douglass routed the aqueduct through Sing-Sing, he saw this valley as a major obstacle on the way to Manhattan. He sought the easiest passage across it, but he still left Jervis with the task of carrying the aqueduct over a depression 536 feet wide that fell to some 70 feet below grade. And Jervis faced more than natural obstacles as Sing-Sing; he faced man-made ones as well.

In crossing the valley, the aqueduct's line intersected two village roads that had to be spanned, because Jervis had no authority to move them. The first road posed no major problem. The line met it almost at a right angle, shortly after entering the valley. The second road was situated in a "peculiar manner," as Jervis called it.[15] The road crossed Sing-Sing Kill on a wooden bridge as it ran to a nearby water-powered mill. The aqueduct's line intersected this road at a sharp angle directly over the wooden bridge. In addition to the two roads, there was a house that stood between them. The line passed immediately behind the residence and cut it off from the owner's garden. The crossing could not impinge upon the house, and at the same time it had to allow the owner a "convenient pass" to all his property.[16]

Had it not been for the man-made obstacles, Jervis undoubtedly would have crossed this valley with a massive

[11] The durability of cast-iron pipe was still very much in question when the aqueduct was built. For expertise in the use of such pipe, Jervis turned to Frederick Graff, Superintendent of Philadelphia's Fairmount Water Works.

[12] The total fall from the surface of the fountain reservoir on the Croton River to the surface of the Distributing Reservoir was 51 feet.

[13] Jervis used formulae developed by Bossut, Dubuat, Prony, Eytelwein, Langsdorf and Robison. These were available to him in John Robison, *A System of Mechanical Philosophy* (Edinburgh, 1822); Charles S. Storrow, *Treatise on Water-Works for Conveying and Distributing Supplies of Water* (Boston, 1835); Olinthus Gregory, *Mathematics for Practical Men* (London, 1825); Thomas Tredgold, *Tracts on Hydraulics* (London, 1826); and in the 1832 edition of the *Edinburgh Encyclopaedia*. All these volumes were in Jervis' personal library.

The thirteen-and-a-quarter inch declivity of the masonry conduit predominated, but in a few locations the slope was even more gradual.

[14] This case is indeed hypothetical. The exact loss of elevation would of course depend on the number and diameter of the pipes used in crossing the valley.

[15] Jervis, "Report on Sing-Sing."

[16] See *Acts of the Legislature of the State . . . in Relation to the Subject of the Introduction, Supply and Use of Croton Water* (New York, 1861), p. 14. According to an act passed May 25, 1836, the Water Commissioners had to "erect and sustain convenient passes across or under the aqueduct whenever said aqueduct shall intersect the land . . . belonging to an individual . . . for the farming and other purposes of the land thus intersected."

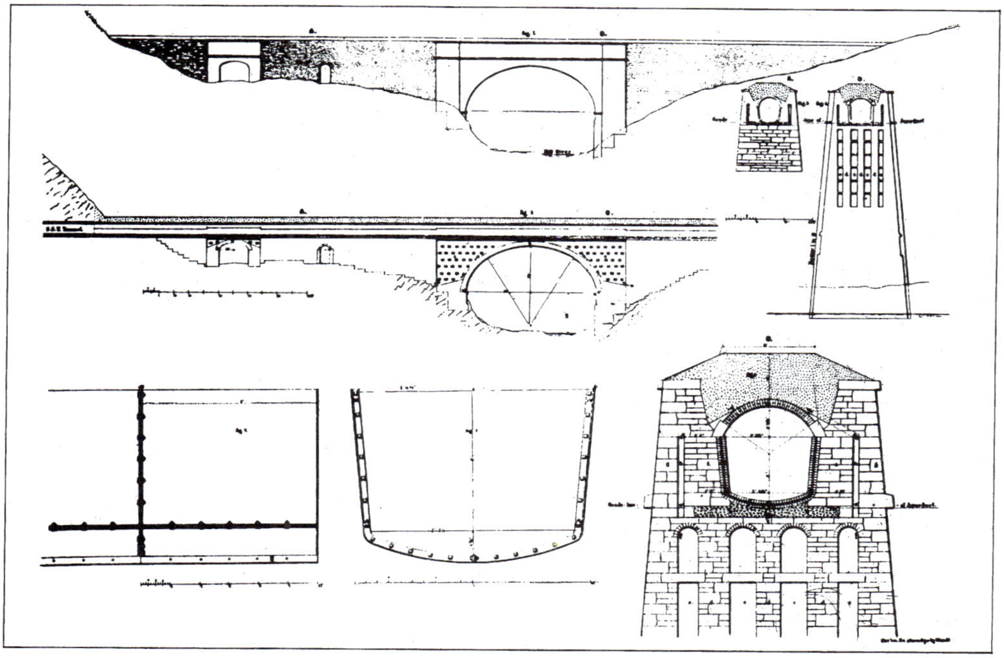

Figure 11. *Sing-Sing Kill Aqueduct Bridge. Taken from T. Schramke,* Description of the New-York Croton Aqueduct *(New York, 1846).*

embankment having one or two culverts to straddle Sing-Sing Kill. But the proximity of the house dictated a narrow structure, and the second road could not be accommodated by culverts. A large arch was needed to span the wooden bridge and the kill. Jervis abandoned his general plan for embankments and designed a structure adapted to this particular site (Figures 2, 11-13).

The Chief Engineer did not want to use an inverted syphon here, or anywhere in Westchester County. Because inverted syphons consumed too much precious elevation, he reserved them for use only from the Harlem River southward into the City. Also, due to his prejudice against aqueduct bridges, Jervis did not want to cross the valley of Sing-Sing Kill with a long arcade. Instead of supporting the masonry conduit on a long line of arches, he supported it across most of the valley on a foundation wall with face stone laid in cement. He "broke" this wall with a low twenty-foot arch that passed the first road, and with a nine-foot arch that allowed the home-owner easy passage across his

Figure 12. *Road arch at Sing-Sing. HAER field photograph, 1975.*

Figure 13. *88-foot arch across Sing-Sing Kill. HAER field photograph, 1975.*

property. He also "broke" the wall with an aqueduct bridge, a single eighty-eight-foot arch that spanned the second road.

Forewarned of the danger of leakage and frost, the Chief Engineer took great care to protect the masonry in this large arch. The first step was to try to make the conduit water-tight. Jervis searched the literature and discovered that Thomas Telford, "an eminent English engineer," had placed a cast-iron floor in an aqueduct bridge on the Ellesmere Canal. Thirty years later, "the aqueduct was tight, and in all respects in good condition." Following Telford's lead, engineers on the Glasgow and Union Canal had lined three aqueduct bridges with cast iron on the bottom and sides, and these structures, too, had escaped injury from frost. On the basis of these precedents, Jervis decided that "the aqueduct over heavy arches, after being made of the best hydraulic masonry, should be lined with cast iron, made impervious to water."[17]

The lining (made of bolted plates 5/8" thick) went between layers of brick on the sides and bottom of the conduit.[18] These plates offered the best known protection against leakage, but what if they failed? Jervis built for such a contingency. If the conduit leaked, the water that ran down to the arch barrel would be carried outside the structure by small copper drain pipes. On top of this measure, the Chief Engineer took another significant step to preserve the large arch. He reduced its superincumbent mass, believing that the less load on the arch, the longer it would last.

The deck of the Sing-Sing Kill Bridge—the masonry conduit, lined with cast iron, filled with water, and topped with earth for frost protection—placed on the arch a heavy load that Jervis could not reduce. He could, however, reduce the dead-load imposed by that part of the bridge that supported the deck and carried its load down to the arch.[19] In most masonry bridges of the period, builders used an earthen or rubble fill to support the deck. Jervis did not follow this practice. Instead, he supported the deck on interior spandrel walls, tied together with cross-walls. These walls left large, open spaces in the bridge and thereby reduced the dead-load on its arch. The hollow spaces (also found in the parapet walls) had other functions, besides weight reduction.[20] Their dead air insulated the structure against frost, and the openings hastened the drainage of any water from the masonry.

Mill River Culvert

At Mill River (now Pocantico River), running in Sleepy Hollow near North Tarrytown, Jervis avoided an aqueduct bridge and realized considerable savings. Here, in an isolated, wooded area no man-made obstacles interfered with the crossing. Jervis had only to carry the conduit across a hollow that extended for over 600 feet and fell to 72 feet below grade.

[17] Jervis, "Report on Sing-Sing."

[18] Jervis later stated that large pipes, with faucet and spigot joints, would have been better than the bolted plates, because the plates failed to allow for expansion and contraction with temperature changes.

[19] Carl W. Condit, *American Building* (Chicago, 1968), p. 73. Also see FitzSimons' "Introduction" to *Reminiscences of JBJ,* p. 11.

[20] The hollow spaces in the parapet walls were suggested to Jervis by R.F. Lord, an engineer on the Delaware & Hudson Canal.

Major Douglass had tentatively proposed crossing the deepest part of Mill River valley with a bridge having five arches, each with a seventy-foot span. When Jervis first examined the site, he also thought a bridge was required:

> The great elevation of the grade line above the bottom of the valley, and the fact that a stone bridge had been proposed [by Douglass] . . . , had given me an impression in favor of a bridge. Accordingly, I had a plan made for a stone bridge, with 5 arches, each of 60 feet span, and an estimate made of the probable cost of the work. It appeared probable that a bridge with 60 feet arches . . . would be the most economical.[21]

The Chief Engineer estimated that a five-arch bridge would cost $142,700. To test its economy, he next calculated the cost of a bridge having six arches with reduced spans of fifty feet. As it turned out, he thought the second bridge would cost some $140,000, or slightly less than the first. But because he was not fond of aqueduct bridges, Jervis estimated the cost of crossing the hollow with an embankment having a double culvert (two arches of sixteen feet) to accommodate Mill River. The Chief Engineer arrived at a figure of only $97,000 for the embankment, so he dispensed with all plans for a bridge at Mill River.

The Chief Engineer chose an embankment because it was less expensive, easier to construct, and less susceptible to settlement, frost, and other "contingencies that ultimately may derange, or impair the uniform efficiency of the aqueduct." Yet in one respect Jervis regretted his decision; a bridge would have been "a much superior work, in point of architectural beauty." One part of Jervis, the pragmatic engineer, wanted an embankment for the sake of economy and stability. Another part of him, the proud, status-conscious engineer-architect, wanted the aesthetic structure. But as almost always happened on the Croton project, the pragmatic engineer won out, due in part to the fact that the line crossed Mill River in an isolated spot. As Jervis noted, "The location does not appear to me, one that would justify the extra cost of a bridge merely to improve the architectural appearance of the work."[22]

Jervis, no doubt for the sake of economy, ultimately chose to straddle Mill River with a single, twenty-five-foot culvert under the embankment, instead of the double culvert he had initially envisioned. It took about two years to build Mill River Culvert (Figures 3, 14-15). Laborers first cleared

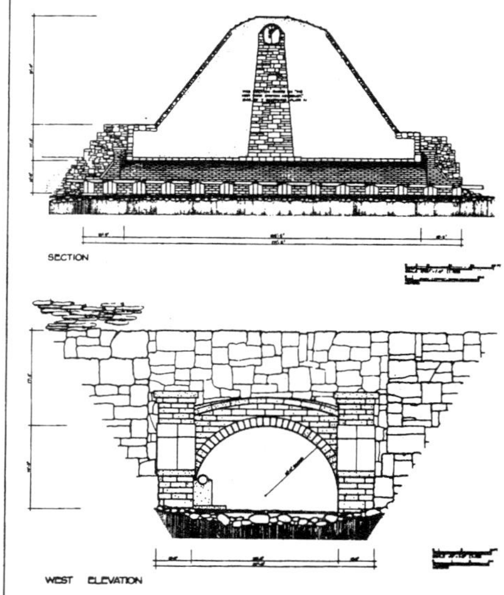

Figure 14. *Section and partial elevation of Mill River Culvert. HAER drawings by Roz Li and Paul Aferiat, 1975. (The pipeline seen in the section is a modern instrusion.)*

all timber, vegetable matter, and loose, spongy earth from the valley floor.[23] Then they prepared the foundation of the culvert. After completing the culvert arch, they began laying the dry foundation wall for the conduit.

While raising the wall, workers simultaneously carried up the contiguous earthen embankment, flanked by inclined stone protection walls. At all times they kept the embankment at least two feet below the foundation wall, so that inspectors could examine how it was being laid. When the wall reached the requisite height, it was capped with a layer of concrete. Skilled masons then laid the conduit and when they were finished, laborers carried earth up and over the conduit's top arch.

[21] Jervis, "Report on Crossing Mill River Valley," June 5, 1837, Jervis Papers.

[22] *Ibid.*

[23] See Jervis, "SPECIFICATIONS of the manner of constructing the general work for the Croton Aqueduct," April, 1837, Jervis Papers.

Figure 15. *Mill River Culvert. HAER field photograph, 1975.*

The stepped stone buttresses seen at the base of the embankment were not in the original plans. After some particularly tall embankments had been completed, such as this one and the one at Jewells Brook, Jervis and his resident engineers recognized that buttresses were needed to resist horizontal thrusts and to prevent the steep embankment from sliding.[24]

High Bridge over the Harlem River

At the Harlem River, the aqueduct encountered a depression 1450 feet wide, measured along the aqueduct's grade line. Down from grade, the valley fell to a maximum depth of about 150 feet. Along the valley floor (composed of bedrock, boulders, sand and mud) the Harlem River ran in a channel from 560 to 620 feet wide, depending on the tides. At flood tide, the river's surface was 112 feet below grade.

Major Douglass had intended to maintain the aqueduct's grade across the Harlem by means of a high masonry bridge. John Martineau, who served the Water Commissioners as a consulting engineer in 1834-1835, had suggested crossing the Harlem with an inverted syphon.[25] Ever since they had received Martineau's suggestion, the Commissioners had favored a low syphon bridge, because it appeared much less

[24] The stepped buttresses do not appear in any original drawings for the aqueduct, but since they appear in Tower's 1843 illustration, they were definitely added while Jervis was Chief Engineer.

[25] See reports of Douglass, Martineau and the Water Commissioners in *Board of Aldermen Document No. 44* (New York, 1835).

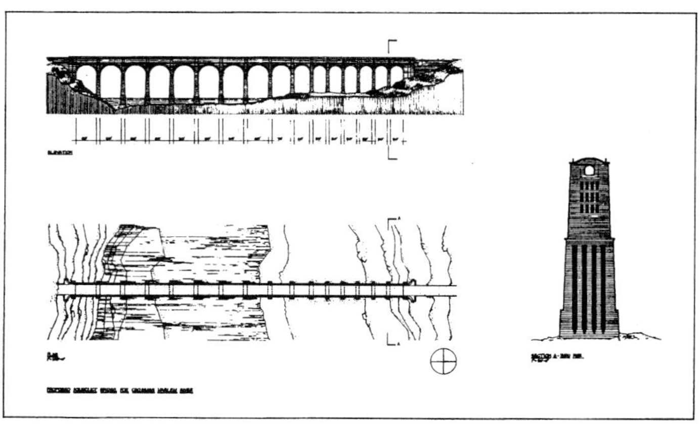

Figure 16. *Jervis' original plan for high bridge over the Harlem River. Structure not built. Drawn by Paul Aferiat after an original drawing in the Engineering Societies Library.*

expensive. To test its economy, they instructed Jervis to provide them with two plans for the crossing: a high bridge and a syphon bridge.

Jervis' high bridge plan is shown in Figure 16.[26] It was to have had sixteen semi-circular arches supported on piers, seven standing in the river and the others on table land. The foundations of the river piers rested eighteen to thirty-two feet beneath the Harlem's surface at flood tide. The height of the structure, from the lowest foundation to the tops of the parapet walls, measured 163 feet.

In this design, the Chief Engineer attempted to "effectually combine stability, permanence, symmetry and economy." For Jervis, combining these qualities proved most difficult when selecting the arch spans. He preferred small arches on aqueduct bridges, yet in this bridge his preference for small arches had to yield to his desire to cross the Harlem on as few piers as possible. Because of the bridge's height, the piers would be very costly. Of even greater concern, a contractor would have a difficult time preparing their foundations.

On the table land, about half the piers would stand on rock, but the others would have to stand on wooden piles driven deeply into sand and capped with concrete. Founding the river piers would prove even more difficult. The exact nature of the river bed was unknown, and to reach it a contractor would have to erect a $14,000 coffer dam for each pier, evacuate up to thirty-two feet of water from within the dam, and then remove a heavy layer of mud or sand from the river bottom. Unless bedrock was discovered at a reasonable depth, some river piers, too, would have to stand on piles.

After weighing the merits of small arches against the merits of fewer piers, Jervis called for smallish fifty-foot arches over land, and for eighty-foot arches over the river. Between the arcades of eighty- and fifty-foot arches, Jervis planned for one seventy- and one sixty-foot arch. These were apparently intended to improve the structure's symmetry or balance, by muting the contrast in size between the river and table land arcades.

Jervis' high bridge shared several internal features with the Sing-Sing Kill Aqueduct Bridge.[27] He lined the conduit with cast iron plates; he left insulating spaces in the parapet walls; he reduced the dead-load on each arch by using interior spandrel walls, instead of solid fill. In addition, although he planned to construct the piers under the fifty-foot arches of solid hydraulic masonry, he intended to leave hollow spaces in the larger piers for the following reasons:

> The interior masonry not being dressed as well as the exterior, is liable to settle more, and eventually force the face stone to bulge outward, and injure, if it does not destroy the work. A second reason is, the tendency that a large mass of masonry has to prevent the uniform and early hardening of the cement.[28]

All things considered, the high bridge represented a great engineering challenge to John Jervis, and the solution would not come cheaply. He estimated that the high bridge would cost $935,000. Next, he turned to the possibility of crossing the Harlem with iron pipes supported on a low masonry bridge.

Before designing the bridge itself, Jervis had to determine how many pipes of what size were needed to assure a daily delivery of sixty million gallons. He began with the idea of a single large pipe but finally decided upon multiple pipes, each with a diameter of thirty-six inches:

> A single pipe, sufficiently large to carry the whole quantity of water, would be accommodated on the most narrow bridge. There are, however, objections to this: a single pipe would place the successful action of the aqueduct on its good condition; consequently, interruption would be involved in any necessary repairs . . . ; and very large pipes would be more liable to imperfections than smaller ones. Water pipes of cast iron have not, that I am aware of, been larger than three feet diameter. The principal iron mains, in the water works of London, are of this size; and the same are used in part of the water works of the City of Glasgow. . . . I can see no reason why this particular limit has been adopted, unless experience has decided it to be the most economical. There certainly can be nothing impracticable in going to four feet, so far as . . . making the casting[s] is concerned, for experience in casting cylinders for steam engines has demonstrated this; and were there any particular necessity for this dimen-

[26] The original drawing of this version of High Bridge is in the Engineering Societies Library in NYC. The written report on this plan is found in Jervis, "Report in relation to the Plan for Crossing Harlem River," December 12, 1837, Jervis Papers. It was published under the same title in "Semi-Annual Report of the Board of Water Commissioners," *Board of Aldermen Doc. No. 55* (New York, 1838), pp. 389-406.

[27] The Sing-Sing Kill Bridge was the first bridge on the line which Jervis engineered, and it indeed served as a model for others. At Clendenning Valley, for instance, Jervis again used a foundation wall to carry the aqueduct across most of the depression, and where he used road bridges at Clendenning Valley, they had the same internal structure as the Sing-Sing Bridge.

[28] *Doc. No. 55*, p. 392.

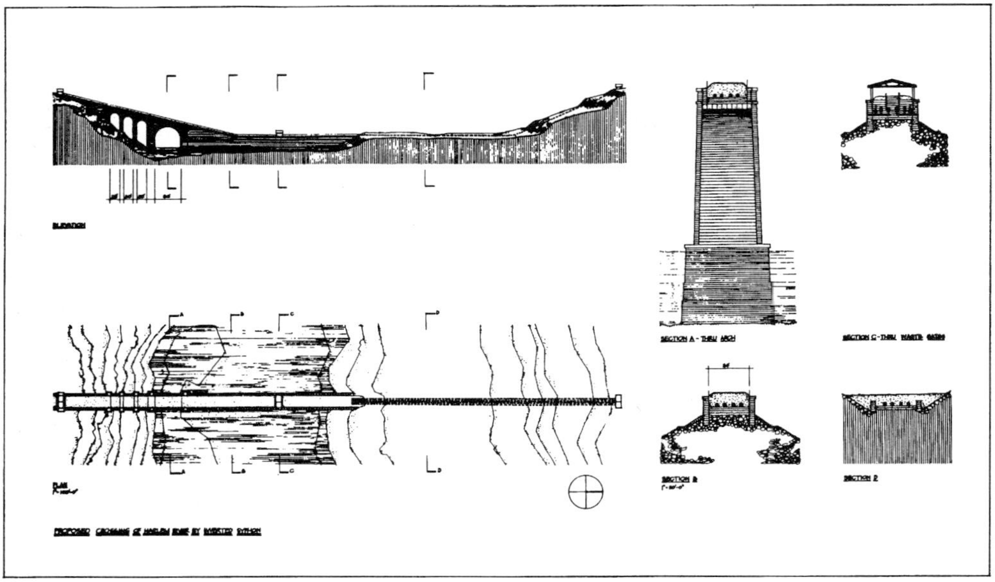

Figure 17. *Jervis' plan for a syphon bridge across the Harlem River. Structure not built. Drawn by Paul Aferiat after a plate found in* Board of Aldermen Doc. No. 55, *1838.*

> sion, I should have no fear that it might be successfully accomplished. But in view of all the circumstances of the case under consideration, I have arrived at the conclusion, that three feet pipe will be most appropriate.[29]

Given the length and diameter of the pipelines, the depth to which they fell in crossing the valley, and the fact that Jervis terminated them some 28 inches below their start, hydraulic formulae predicted that each 36-inch pipe would discharge about 15 million gallons per day. Consequently, Jervis made the bridge wide enough to carry four pipes (Figure 17). But as a cost-cutting measure, he recommended that the Water Commissioners initially lay only two pipelines across the bridge, because their discharge would "probably be sufficient for the next fifty years." The city could add the third and fourth pipes when needed.

Jervis planned to commence the pipes in an influent chamber where the entrance to each pipe was protected by a cast iron gate. The nine-foot long pipe sections (generally having a wall thickness of one inch), descended into the valley on a squat foundation wall that closely followed natural terrain. In the center of the valley the foundation wall leveled off at four feet above the Harlem's flood tide. Here Jervis located a stopcock and waster on each pipe that could be used to wash sediment out of the line. After passing this level stretch, the water rose towards Manhattan, passing over a semi-circular arch spanning eighty feet and rising fifty feet above the river. The water then passed over an arcade of three arches of thirty-five, thirty, and twenty-five feet. From the abutment of this arcade, the pipes again ran on a foundation wall until reaching an effluent chamber, where the water discharged back into the masonry conduit.

Jervis estimated that the syphon bridge would cost $426,000—or a full half-million dollars less than a high bridge. Given this savings, he strongly recommended that the Water Commissioners approve the construction of an

[29] *Ibid.*, p. 399. If Jervis had opted for a single, large pipe, the plan for the syphon bridge would have sacrificed less elevation in crossing the valley. The wetted surface of the large pipe would have been smaller than that of four 36-inch pipes. The pipe would have presented less friction to the flowing water. Hence, the friction head loss would have been reduced.

inverted syphon.[30] The high bridge would be "greatly superior, in point of architectural magnificence," and it would maintain "two feet greater elevation." But those two points of superiority did not justify its much higher cost.

The Water Commissioners concurred with Jervis' choice. In reporting on the inverted syphon, they stressed the structure's economy. They emphasized that it could be constructed more readily, because its contractor would not have to sink numerous piers, a type of work "attended with many unforeseen difficulties and casualties."[31] And they expressed the belief that the syphon bridge would be safer from the dangers of settlement, water leakage and frost. The Commissioners, like Jervis, presented a strong case for the inverted syphon and they let a contract on the work. But it was never built. It was stopped after a long debate over the Harlem River crossing had been carried on in the press, in New York's Common Council, and in the State Legislature.

An anonymous author laid out the debated issues in a letter published in the New York *American* on March 9, 1838. This letter faulted the syphon bridge on several counts. Stylistically, it was unimpressive; it would "deprive the work of all that would render it an ornament to the city and the age in which we live." Technically, the inverted syphon was an experiment. It might fail to deliver the desired amount of water, and its pipes might burst from the pressure of the water under transport. But the syphon bridge's most serious fault, according to this letter, was that it would block off most of the Harlem and close it to all but those vessels small enough to pass through its eighty-foot arch with its fifty-foot rise.

Stung by this criticism, Jervis and his Principal Assistant, Horatio Allen, fired off their own letters to the press. They countered all of the charges made against the syphon bridge—particularly the charge that it would ruin navigation on the Harlem River. As Jervis noted, the river was *already* impassable:

> There is at present no navigation west of McComb's Dam, nor has there been any for near[ly] forty years. It is now obstructed by the dam and bridge near the Spuyten Duyval Creek . . . , and by the Fordham Bridge on one side of the aqueduct line, and by McComb's Dam and Harlem Bridge on the other side.[32]

The opponents of the syphon bridge admitted that the Harlem had long been useless as a shipping lane, but they hoped to improve the river. They wanted to dredge its channel and remove existing obstacles, or circumvent them with canals. They hoped that the Harlem would become an important connector between the Hudson and East Rivers, that it would spawn and support businesses and industries on the northern end of Manhattan. A percentage of these visionaries backed their hopes with investments. They had purchased land on both sides of the Harlem and were waiting for a boom in the region's development.[33]

John Jervis did not share the speculator's dreams, but he recognized that they formed an influential group. So in his own letter to the ***American***, he attempted to appease their interests. He offered to increase the span of the river arch to 120 feet and its rise to 65 feet.[34] This effort failed. On May 3, 1839, the Chief Engineer's opponents won their way, when the State Legislature passed "An Act Prescribing the Manner in which the Croton Aqueduct shall pass the Harlem River." The act stipulated that:

> The Water Commissioners shall construct an aqueduct over the Harlem River, with arches and piers; the arches in the channel of said river shall be at least 80 feet span, and not less than one hundred feet from the usual high water mark of the river, to the underside of the arches at the crown; or they may carry said aqueduct by a tunnel under the channel of the river, the top of which shall not be above the present bed of the said river.[35]

For the Chief Engineer, it was literally time to go back to the drawing board, and in less than a month he had a new plan for crossing the Harlem. His plan met, but did not exceed, the State Legislature's requirements.

Jervis did not adopt the option of a tunnel under the river. He thought the construction of a masonry-lined tunnel large enough to accept an inverted syphon with four thirty-six-inch pipes would be a very uncertain venture.[36] To document the problems that might plague a Harlem River tunnel, he cited the history of Marc Brunel's tunnel under the Thames. After twelve years' work, that tunnel

---

[30] *Ibid.*, p. 406.

[31] *Ibid.*, pp. 371-372.

[32] Clipping, Jervis Memoranda Book, entry for March 17, 1838. The manuscript text of this letter is found as "Navigation of Harlem River," March, 1838, Jervis Papers.

[33] For further discussion, see Nelson Blake, *Water for the Cities* (Syracuse, 1956), pp. 152-156; and FitzSimons, *Reminiscences of JBJ*, pp. 125-127.

[34] A drawing of the proposed 120-foot arch is in the Engineering Societies Library.

[35] *Acts of the Legislature . . . Croton Water*, pp. 20-21.

[36] "Report of the Chief Engineer on Plans for Crossing Harlem River," *Board of Aldermen Doc. No. 10* (New York, 1839), p. 152.

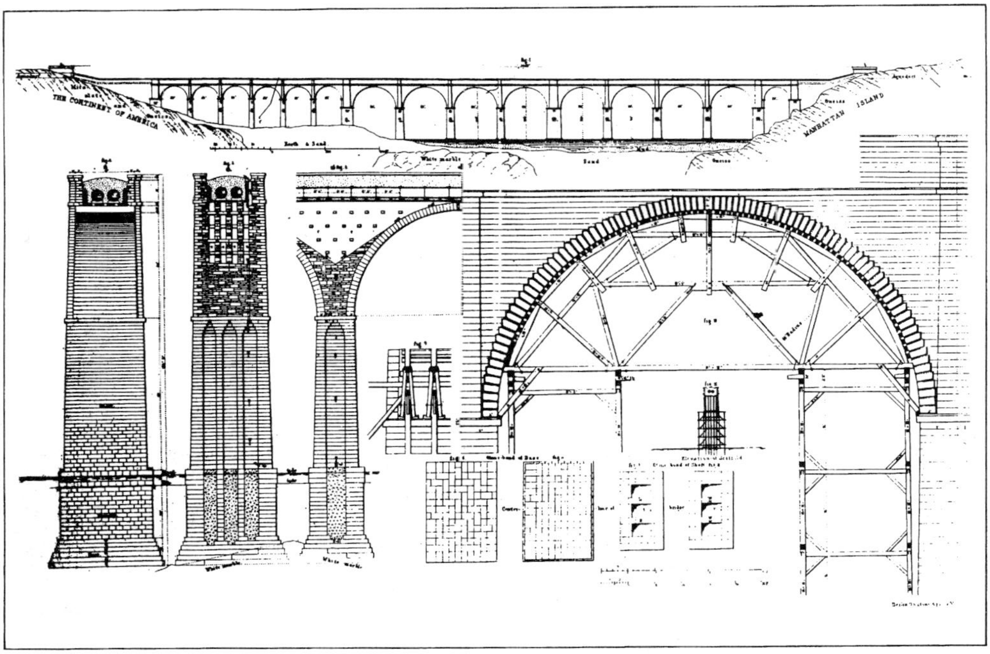

Figure 18. *Final High Bridge Plan. Taken from Schramke.*

was not completed, and its final cost was foreseen as four times the original estimate.

Jervis felt that under the best circumstances, a contractor might drive a tunnel under the Harlem in four years at a cost of $424,000. But because he feared that serious problems would be encountered, he added fifty per cent for contingencies. This raised the estimated cost of a tunnel to $636,000, and there was no guarantee that its real cost would not go even higher. In addition, Jervis believed that a tunnel would incur high maintenance costs, because corrosive salt water would inevitably percolate through to the iron pipes. So Jervis decided against a tunnel, and this decision left him with no alternative to a high bridge.

Interestingly, although Jervis had publicly argued against a high bridge, he privately relished the challenges it presented and the credit it would bring him when successfully completed. He wrote, "I cannot say by any means that I regret this—as you know Engineers are prone to gratify a taste for the magnificent when there is good reason for the execution of prominent works."[37]

Jervis already had a high bridge plan on hand, but that design exceeded the height requirement set by the State Legislature, so he chose not to use it, at least not *in toto*. He designed a second, somewhat lower bridge (Figures 6, 18-19) that retained some important internal features seen in the first high bridge plan, such as hollow piers and the use of interior spandrel walls.[38] The second bridge, with its arches, piers, pilasters, parapets, and water table also exhibited a style consistent with that of its predecessor. Nevertheless, the Chief Engineer's second bridge differed significantly from his first.

[37] Jervis to J.J. Abert, December 29, 1838, Jervis Letter Book.

[38] The modified High Bridge is discussed in *Doc. No. 10.*

Figure 19. *High bridge across the Harlem River, with original masonry arches in the foreground. HAER photograph, 1975, by Stephen Zane.*

Of primary importance, it was twelve feet lower. Jervis dropped the arches to the minimum allowable height of one-hundred feet above high water. Because the bridge now fell short of maintaining the aqueduct's grade line, Jervis dispensed with the masonry conduit between its parapet walls and substituted a shallow inverted syphon that could carry water under pressure. He calculated that two forty-eight-inch pipes could handle the masonry conduit's discharge of 60 million gallons daily, so he designed the bridge's deck to accept pipes of that number and size. But he chose to economize by initially laying two less-expensive thirty-six-inch pipes, deeming them sufficient to meet the city's water needs for years to come.

In December, 1837, Jervis had estimated that a high bridge over the Harlem would cost $936,000. In June, 1839, he estimated that his modified high bridge would cost $837,000. So by reducing the structure's height, the Chief Engineer anticipated a savings of about $100,000. Still, his modified high bridge could hardly be called an economical means of carrying two pipes over the river. Jervis could have greatly reduced the bridge's cost by substituting timber construction for part of the masonry, but he was loath to do it. After all, his main objective all along had been to achieve durability, so he could hardly carry the aqueduct onto Manhattan by means of a timber bridge subject to rapid decay and vulnerable to catastrophic fire.

Conclusion

John Jervis carefully evaluated alternative means for crossing all the major valleys encountered by the Old Croton Aqueduct. He judged a design's suitability on the basis of numerous criteria. These criteria can be identified, even if they cannot be ranked exactly in order of their importance. A design certainly had to conform to the geological and topographical peculiarities of a given site, and it had to accommodate any immovable man-made obstacles. Also, it had to comply with any applicable laws passed by the State Legislature. Jervis wanted designs that promised durability and constant performance, but these qualities were to be purchased at the least possible expense, in terms of both original and maintenance costs. Factor endowments were considered, such as the availability of materials and labor. Wherever possible, Jervis used local stone (gneiss or granite) in designs that could be expedited. From the standpoint of hydraulics, he wanted to achieve a given flow rate and yet sacrifice as little elevation as possible. In populated or highly visible areas, the Chief Engineer also considered aesthetics, or the architectural appearance of the work.

The aqueduct that John Jervis built from the Croton River to Manhattan bore evidence of his experience, his design and research skills, and of the balance that he had struck between durability and economy. Its extant valley crossings, in particular, are more than just means to carry water across depressed ground. They are tangible evidence of an early American engineer's thinking and values.

18
The Canton Viaduct

E.D. Galvin

Much has been written about the Boston and Providence Railroad (B&P), and accounts dealing with its origin usually mention the stone viaduct built in Canton, Massachusetts. While "the viaduct" has never been ignored, it has never been the subject of more than local attention when the early days of railroading are discussed. Yet together with B&O's much more noted Thomas Viaduct, the Canton structure now stands as one of the earliest surviving multiple-arch stone railroad bridges in the U. S. Furthermore, its design and history make it unique in its own right. What I shall describe are the reasons for the viaduct's being built, its construction, alterations and its emergence as a symbol for an entire community.

Before the B&P's incorporation on June 22, 1831, several preliminary surveys were conducted in order that the best possible route could be chosen between the two cities. It was not lost upon the people of 1830 that the presence of a rail line passing through their town or village couldn't help but aid the unending problem of transporting people and goods. By the time the company was formally incorporated, the surveyors were well along with their task, and in a matter of weeks a decision as to the planned route was expected.[1] Although the Corporation had until January of 1834 to file the location of the road and until January 1, 1837 to complete it, it was eager to commence construction.

The company's directors were confronted with the fact that both Boston and Providence lay in geologic depressions, separated by a ridge which had it highest point in Sharon, roughly midway between the two terminals. The Neponset River with its extensive flood plain and valley lay to the north of the high ground between the two cities. With an uncommon foresight the directors agreed that the route of the road should be as short and as straight as possible, even if this meant bypassing such established towns such as Dedham and cutting through a cemetery in Attleboro.[2]

As a result, the company decided that the assault upon the Sharon hill be made from the eastern edge of the Neponset Valley, passing close to Canton. While perhaps shorter, it had some drawbacks: A westerly approach to Sharon hill would have taken the road through Dedham, with a more gentle ascent of the hill. The Sharon route required several cuts and many embankments. It also meant crossing the east branch of the Neponset River (then called the Canton River), which lay in a wide and relatively deep valley—a site so difficult that the company directors originally thought it would be necessary to use inclined planes on each side.

As an aside, however, it should be noted that although the Canton route was unquestionably more difficult than one through Dedham, there were undoubtedly "other considerations." For example, one of the six members of the railroad's Board of Directors was Joseph Warren Revere, son of Paul Revere. In the later part of his life, Paul Revere had moved to Canton and established a copper works and brass foundry on the site of the Revolutionary War powder mill. The Revere Copper Works—now operated by Joseph W. Revere—lay within a quarter of a mile of the new railroad's route through Canton.[3]

Mercifully, the original idea of using inclined planes to get across the Neponset (Canton River) Valley did not last long. After a fatal accident involving several well-known businessmen on the Granite Railway at Quincy, Mass.—which used inclined planes—all thought of inclines was discarded and it was decided to build a stone viaduct over the valley instead. The firm of Dodd and Baldwin contracted to build the structure, according to these specifications set down by the Boston & Providence's engineer:

> It is understood that the said Viaduct shall commence at a point 10.5 feet northeast of station marked 670, of the center line of the Boston and Providence Railroad and extend 613.5 feet on said center line to a point three feet southwest of station marked 676, crossing the mill pond of the Stone Factory, so called: that the said Viaduct shall terminate at each end by an abutment and circular wing walls; that the basement wall shall, in every part be laid on a solid foundation —shall extend entirely across the base of the structure and project from one to two feet beyond the exterior face of the superstructure, as may be directed.
>
> The said basement wall shall be constructed of the best dry masonry and the stone of such dimensions as shall be approved by the Agent or his assistant; the said basement wall shall in every part commence at least three feet below the surface of the earth except where solid rock shall be encountered, and shall be raised to such elevation as the Agent or his assistant may direct.
>
> The superstructure shall consist of two walls extending the entire length of the viaduct, connected at intervals of 27½ feet by buttress 5½ feet thick extending transversely across the walls, and projecting 4 feet beyond their faces. The main walls to be 4 feet thick, 4½ feet below the grade of the road, at all points, and to have a battre on their exterior faces

of 1 foot in 48 feet, or 1 inch to 4 feet—the interior faces to be perpendicular, and to have a clear of 4 feet. . . .
(Engineer's Report, June 1, 1834. Baker Library, Harvard Un.)

Captain William Gibbs McNeill was hired and charged with the construction of the railroad. The foundation stone of the viaduct was laid on April 20, 1834, commencing an eventful two years of building. Construction of the railroad itself was started at both the Boston and Providence ends, working toward Canton.[4] The viaduct proved to be the railroad's largest and most difficult project, and was the final link in connecting the two cities.

Initially, it was thought that a local quarry in Canton could provide all the necessary stone for the structure. As it turned out, however, the quarry—located near Dunbar Street in Canton—contained stone which was not suitable as finish stone, thus relegating it for use only in the foundation and for backing. This stone was drawn by oxen over more than two miles to the viaduct. It was necessary to obtain face stone from a quarry in Sharon, on the westerly slope of Rattlesnake Hill. The stone was hauled over the road some three miles by oxen or horses to the railroad near what used to be Sharon Heights station, then loaded on a flat car where it rolled downhill by gravity another four miles down to the viaduct site. An elderly white horse named Charlie hauled the empty car back to Sharon once its load of granite had been removed.[5]

There were two major groups of men who built the viaduct: The stone cutters and masons were all Scotch, while those engaged in the rough stone laying and construction of the gravel roadbed were Irish. The Irishmen built their shacks at Dunbar's Ledge, about a mile and a half below Canton Junction, and the Scotchmen ate their porridge and drank their "barley beer" at Dunbar's quarry, previously mentioned, near the Canton-Sharon town line. The Irish, who have a tradition of discriminating tastes, drank a distilled product of rye and the local lore surrounding the events that accompanied the construction of the viaduct are substantiated by the fact that the militia had to be called on at least two occasions to settle riots between the two groups. Legend has it that the quarrels were climaxed when one of the Irish leaders disappeared after one of the traditional Saturday evening brawls. It was never determined whether he had simply left town in disgust at the fighting or had been "done in" and was making his way to sea via the Neponset River.[6]

The Scotch artisans were Free Masons and every faced stone on the exterior bears the mark of the man who cut it. The taste and ability of the workman is evidenced by the size, shape and style.

CONSTRUCTION OF THE VIADUCT

The foundations of the viaduct are some eight feet below the surface of the ground. Most of the interior of the viaduct is hollow with the side walls giving the appearance of not being joined together except for intervals of masonry. The exterior has pilasters connected at the top by segmental arches with the top of the viaduct being finished the full width over the pilasters.

The bridge originally was planned for a single track. The top of the viaduct was rounded by ballast and the rails were placed on the crown.

The "Stone Bridge," as it was also referred to, is 615 feet in length, 22 feet in width and the rails are about 70 feet above the surface of the river. Originally, there was a single highway passway, a semi-circular arch 22 feet wide. Six rounded arches of 8 feet 4 inches each allow the river to pass through the viaduct. As the structure neared completion, a stone bearing the names of the railroad officials was set in the western end of the parapet. The inscription reads:

This viaduct erected by the
B&P R.R.
T. B. Wales, President
Directors: J. W. Revere,
C. H. Russell, J. P. Loring,
C. Potter, J. G. King

Years later, during the course of renovations, this stone was tossed off the viaduct and was left for a long while in a field at the bottom of the embankment. Though broken, it was subsequently raised back to the top of the viaduct, but placed on the Canton Junction end where it rests today.[8]

The spanning of the vale was accomplished with the use of an earth embankment from just west of the present location of Canton Junction station to a point where the viaduct began. The embankment rose more than sixty feet above the natural terrain. Together, the viaduct and embankment significantly changed the landscape in the vicinity of Canton and Canton Junction. The homes of the town's people of that day originally looked down a gently rolling slope to the Fowl Meadow, which separated the towns of Canton and Norwood. The newly-erected barrier shut out forever the magnificent view across the Neponset River Valley from some of the higher points in Canton.[9]

As sections of the Boston and Providence were completed they were opened for traffic. The line started operations from Boston to a point in Dedham in August of 1834. On September 12, 1834, trains began running to Canton, where the practice began of using a transfer around the viaduct building site to horse-drawn coaches for Providence.

The locomotive *Whistler* hauled these early trains between Boston and Canton, and until the completion of the viaduct, this was the only steam-powered section of the Boston and Providence.[10]

Late in the spring of 1835, when the viaduct was nearing completion, the following article appeared in the June 6 *Providence Journal:*

> Last Tuesday by invitation of the Directors we made, in company with a party of gentlemen, the first passage that has ever been made over the whole length of the Boston and Providence Railroad. It was in contemplation to have taken the new engine that had arrived from Philadelphia only the day before, but some of her pipes were not in order, and we finally set off from the depot on India Point at a quarter before one o'clock in the afternoon with two cars, each propelled by two horses. The application of the horses afforded us a most fortunate opportunity for inspecting the grand structure over which we passed. The road has been graduated in such a manner that the inclination is no where more than 37 feet to the mile, and it is for a short distance only, that it is any where sufficient to perceptibly retard the speed of the cars and it is finished with a neatness very gratifying to the eye.
>
> The Viaduct at Canton though yet unfinished, is a stupendous work. A view of it, many times repays the trouble of passing round. The excavation and embankments in Canton are also worthy of minute attention; they testify in strong language to man's dominion over nature, and his ability to overcome any obstacle to any undertaking that is not morally or physically absurd. The project of cutting through these rocky heights and crossing the valley of the river by the Viaduct was a very bold one. A hesitating mind would have surmounted this by a stationary engine or some less formidable way. But any other would have detracted very much from the facilities which give value to such a road. The road has been constructed under the direction of Major McNeill, and it will stand for ages an enduring monument of the high talents and high attainments of its accomplished engineer. Among curiosities on the way is the bog in Mansfield where the road sunk during its formation to a depth of 40 feet—and it is also a curious fact that 16½ miles of this road are on a perfectly straight line. After examining the work in Canton, we took the engine at twenty minutes past five and were landed in Boston about six o'clock. The party accompanied the directors to the Tremont House and enjoyed their overflowing hospitality with keene appetites and grateful hearts. In the evening we were indebted to Thomas B. Wales, Esq. President of the Railroad Company for

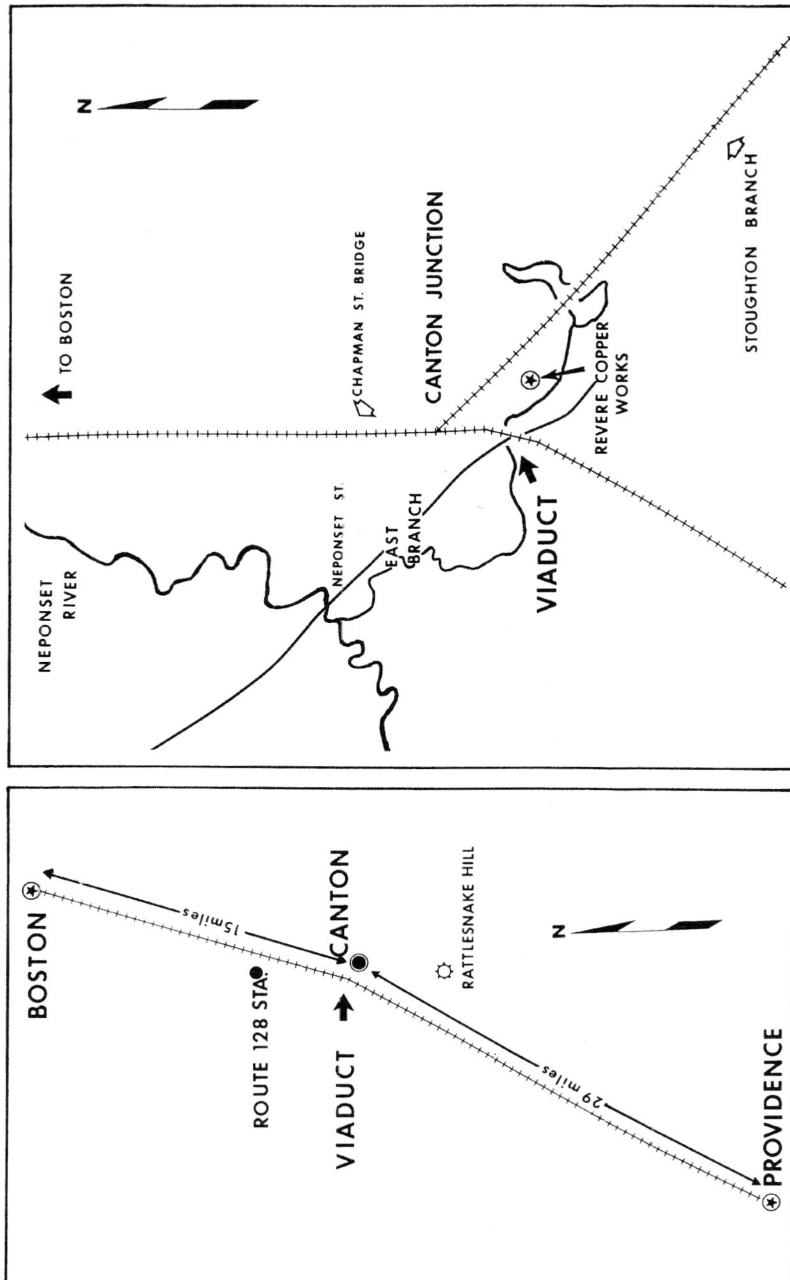

Vicinity Map.

The Canton Viaduct was built in 1834-1835 by the Boston and Providence R.R. This scene dates about 1880. (Canton Historical Society)

View of original and new vehicular arch (1952). (Photo by Author)

BULLETIN 129

New Haven train No. 170, the Colonial, passing over the viaduct Feb. 1968. (Photo by Author)

Aerial view of the Colonial on the viaduct taken by the author in September 1968.

> an elegant and hospitable entertainment at his house (that was 24 Winter Street). At fourteen minutes past six, Wednesday morning, the cars began to move from the depot at West Boston, and at fourteen minutes before seven were exactly at the fifteen mile stone, having come fifteen miles at the rate of 30 miles per hour, two of the miles of which we passed over in three minutes forty seconds, and one mile in one minute and forty-seven seconds, or at the rate of 34 miles per hour.

Five weeks later the viaduct was finally finished. During the last few days prior to completion, a great debate occurred (as you might expect) as to who should have the privilege of being the first to cross the viaduct by rail. The need to call the militia out again was averted when all agreed that the unsung hero of the viaduct's construction would have the privilege. So with great ceremony and laughter Charlie, the old white horse who had hauled the flat car between the viaduct and Sharon, was placed upon the flat car and hauled across the viaduct, thus becoming the first passenger to cross the structure.[11]

Tuesday, July 28, 1835, saw the first regular passage over the viaduct. The *Boston Advertiser* of July 30, 1835, reported the event:

> We understand that the magnificent Viaduct at Canton is so far completed that the locomotives with their trains of cars pass from Boston to Providence without interruption. The train which left this city at four o'clock on Tuesday afternoon arrived in Providence in an hour and forty-seven minutes. It returned the same evening with about a hundred passengers in two hours and three minutes.

It was only twenty-five years later that the wisdom and foresight of McNeill began to evidence itself as the traffic on the Boston and Providence rose to a point where double tracking of the B&P's main line became necessary. The viaduct, with its single set of rails set in the crown of the structure, underwent its first major renovation. Robert Rogers of Canton, a great-great grandson of Paul Revere and a local historian, wrote early in this century:

> In 1860 when the necessity of a double track became evident, the parapets were removed, large wooden timbers, somewhat longer than the width of the deck, were placed across the structure, on the ends of which a heavy wooden fence was built and the tracks laid, the outer of which rails are close to if not actually on the side walls. A hard pine timber was securely bolted on the top of the walls as guard rails. These timbers ran the entire length of the bridge on both sides. This stood for some time when in 1880 the wooden timbers were removed

> and iron girders replaced them with a strong iron fence or railing. This whole structure being fastened together and securely anchored to the stone work.[12]
>
> In 1910 the Railroad Company, at this time known as the New York, New Haven, and Hartford Railroad, reinforced the arches on both sides of the structure with concrete.
>
> Late in October in the year 1912, Mr. George T. Sampson, one of the railroad engineers, had one of the large stones near the ground removed and invited me to accompany him on an examination of the interior of the viaduct. Previous to our arrival three platforms had been erected inside the structure which were connected by ladders. We crawled through the opening and supplied with electric flashlights and a camera, we had no difficulties as we proceeded with the examination which proved satisfactory. There were no broken or cracked stones and the joints were still full of good hard mortar, all of which I think speaks well for these oldtime engineers and masons when you consider that this bridge had been built almost eighty years, and for one of the first railroads that was operated in this country, before they had a chance to learn by experience what would be the effect of the rapidly moving trains or the effect of the constantly increasing loads it has been called upon to carry. For you must remember that when this structure was built it was built to carry a single track and a train with an engine weighing perhaps thirteen tons. That would appear today to be a toy locomotive and be but little larger than the old fashioned stage coach, while today the engine will weigh a hundred tons or more and many of the loaded freight cars will weigh from seventy to one hundred tons.[13]

The thorough inspection of the viaduct stemmed from a series of minor problems which began around 1909. In the December 24, 1909 issue of the *Canton Journal* is an article titled "More Trouble with the Viaduct at Canton Junction:"

> Last Saturday morning, Mr. Wilbert H. Jerauld, section foreman for the New Haven road discovered a bad break in the top of one of the arches of the old stone Viaduct. It was the fourth arch to the east of Neponset Street and is shown in the picture on the heading of the Journal. A large block of granite had dropped out of place and the arch was badly cracked and ready to fall. A flagman was stationed on the tracks to slow down all the trains. A work train arrived during the afternoon with heavy timbers that were lowered over the broken arch. Stone masons have been at work since Sunday morning repairing the damage. This is the second arch that has given way

under the constantly increasing load and it is said that the whole bridge was condemned some time ago. It is probable that the old Viaduct that has been one of Canton's treasured landmarks since 1842 will have to be taken away and replaced with a bridge of modern steel construction in the near future.

The obituary for the viaduct was of course a bit premature. No evidence has ever come forth as to its being condemned and the article's author manages to combine this use of rumor with the 1842 date error in the viaduct's construction. Less than five years later, in March of 1914, however, foreman Wilbert H. Jerauld, who spotted the break in the arch in 1909, was added to the long list of people who were killed or maimed on the viaduct. The *Canton Journal* of March 6, 1914 reports:

> Wilbert H. Jerauld a section foreman was struck by an express train and instantly killed. He was working on the Viaduct and stepped from the tracks on which an outbound train was approaching and got in the way of the express. The engine threw the foreman high in the air and his body was hurled sixty feet to Neponset Street below. The accident happened within 100 yards of his home just opposite the Viaduct.

The structure's layout, in fact, made it quite dangerous to anyone using it as a short cut. The viaduct is on a slight curve that is banked as well, and anyone on the bridge looking towards the opposite end has difficulty determining on which track a train is approaching. Section Foreman Jerauld's death was a rarity in that two trains converged on the viaduct while he was on it. More typical of viaduct deaths was that of Paul John:

> After serving through the entire war with the German forces, Paul John of Chestnut Street, Sharon, one of the most popular employees of the Plymouth Rubber Company (formerly the site of the Revere Copper and Brass Foundry, ed.) was killed instantly when struck by a Boston to New York express while crossing the Canton Viaduct last Sunday.
>
> The dead man, one of the most industrious workers in the Canton factory, had worked late on Saturday and when through, began his walk home to Sharon along the tracks of the New Haven. He was nearly across the long bridge over the Neponset River when the New York bound train swung around the curve at the Canton Junction end and thundered down upon him.
>
> Instead of leaping to safety on the other track, Johns, probably terrified, attempted to outrun the train. Slamming on the locomotive brakes and blowing the whistle the engineer attempted to bring the train to a stop, the cars began to

buckle and sway from side to side on the bridge due to the tremendous and sudden change in speed. The attempt was in vain, and the engine struck the man about 30 feet from the end of the bridge, hurled him against the rail, fracturing his skull, and probably killing him instantly. At the moment of impact the train was traveling at about ten miles an hour.

(*Canton Journal*, October 24, 1930)

With all the various accidents on the viaduct involving humans or animals, there is no evidence or indication that any car or locomotive has ever toppled from the viaduct to the street or river below. This is either the result of engineering or fate, as several accidents—some of them severe—have taken place just north of the viaduct, usually between Canton Junction Station and the Chapman Street Bridge, which is only 1300 feet away. Most recent was the massive freight derailment of December 15, 1968, just north of the Chapman Street bridge where 40 cars wedged themselves in the cut at right angles to the track. Thirty of the cars were contained within an area of 75 yards, which indicates the stacking effect of the accident. The main line of the New Haven was closed to traffic through this point for six days, eleven hours and twenty-eight minutes.

Again, there was a freight derailment in June of 1965, under the Chapman Street Bridge. In February of 1963, the "Federal" from Washington roared across the viaduct with the last passenger car's trailing wheels on the ground, kicking up stone ballast and smashing wooden platforms and windows at Canton Junction Station. In February, 1927, another freight derailment had cars climbing the embankment and blocking the tracks at the Chapman Street Bridge. Crewmen of the train knew that the "Knickerbocker Limited" was due to go by Canton at 1:23 p.m. By the time the people at Canton Junction called the Readville tower, the "Limited" had passed the tower. It was left to the wrecked train's fireman, who wisely ran up the tracks and managed to flag the "Limited" down a few hundred yards from the blocked right of way, thus saving many lives. All of these trains had just crossed the viaduct, headed downhill at good speed toward Boston and disaster.

With the emergence of the automobile and truck, it quickly became evident that the one original arch in the viaduct would be unable to handle the highway traffic of Neponset Street which was the main road connecting Canton with Norwood. As early as 1899 newspaper articles appeared terming the single, small existing arch a death trap and calling for a second arch to be cut in the viaduct.[14] We can only hope that those who supported such a move in 1922 were patient, as it was to be thirty years before the second arch was actually built. This second vehicular opening altered the symmetry of the structure for the first time, as the 1952 arch was lower than the original arch and was lined with concrete. "Progress" had exacted its first toll from the venerable stone and mortar.

All things are subject to change, but few as slowly as this bridge of a hundred and thirty-eight years. Four owners later, it still etches itself in the minds of all those who pass by it. Its legend grows with every drop of water. When Robert Rogers made his journey into the viaduct interior, he reported that stalactites were forming. When the viaduct was opened for its second arch, the formation was rediscovered, validating what an old man had seen many years ago, and couldn't quite get people to believe.

Today the viaduct continues as a marvel of endurance which will unquestionably outlast its present owner, the Penn Central. As it approaches its 140th birthday it remains in main line service, though the main line is considerably reduced in importance. Few other such "living" tributes to the skill of the railroad designers and builders of the 1830's can be found today.

CANTON VIADUCT–NOTES

1. Alvin Fay Harlow *Steelways of New England* (2nd ed.) New York: American Book-Stratford Press, 1946 p. 105.

2. *Ibid* p. 107

3. "Historic Stone Bridge" *Canton Journal* August 11, 1933, p.7 (Reprinted from *Norwood Messenger* August 8, 1933).

4. *Canton Journal*–August 11, 1933 p. 7.

5. *Ibid*

6. *Ibid*

7. Robert Rogers *The Canton Viaduct* (unpublished–Canton Historical Society).

8. *Canton Journal* August 11, 1933 p. 7.

9. *Ibid*

10. McNeill christened the first engine in honor of his brother-in-law. Harlow *Steelways of New England* p. 108-109.

11. *Canton Journal* August 11, 1933 p. 7.

12. The wrought-iron parapet was designed and constructed by the Edgemoor Iron Company in 1877.

13. Robert Rogers *The Canton Viaduct* (unpublished–Canton Historical Society).

14. *Canton Journal* February 24, 1899.

19

The fall of Barentin Viaduct, 10 January 1846

D. Brooke

'About six o'clock on Saturday morning,' reported a Paris newspaper, 'a fearful noise - aroused the inhabitants of Barentin. The twenty-eight pillars [i.e. piers] which supported the arches of the viaduct fell one after another, and almost instantaneously.' No-one died but if 'those enormous pillars had not providentially fallen as they did (in a straight line), had they fallen but a few yards to either side, a great number of persons must have been entombed, whilst in their sleep.'[1] The destruction of railway viaducts, usually as a result of inferior workmanship and materials, was not a rare event in the history of railway building. The 1840s also saw the collapse of these structures at, amongst other places, Knaresborough (1848) and Congleton (1848) and on the Stalybridge Branch of the Manchester, Sheffield & Lincolnshire Railway (1845). Nevertheless, for the partnership of contractors William Mackenzie and Thomas Brassey, the fate of the first Barentin Viaduct was the most dramatic single event to occur during 10 years' work in France.

In May 1843 Mackenzie and Brassey completed their first major assignment in France with the opening of the Paris & Rouen Railway between the Paris & St Germain line in Colombes, Paris, and the south bank of the Seine opposite the City of Rouen. Their task had been accomplished in a little over two years and was a highly successful demonstration of the application of British railway contracting methods. It was also the first French component in the creation of rail communication between Paris and Channel ports and thus the improvement of connections with London. Attention now turned to a line between Rouen and Havre. Although the Paris & Rouen had entailed building four bridges over the Seine and several tunnels (the longest being Rolleboise, c. 1 mile, 1,300 yards), natural obstacles on the Havre route posed even greater problems and the first contract on the line, the Rouen and Barentin, was especially demanding. It began with Eauplet Bridge over the Seine and in the following 14 miles involved a total of two and a half miles of tunnel under Rouen, a substantial tunnel at Pissy-Pôville (1 mile 800 yards) and viaducts at Malaunay and Barentin. This final 'work of art' had 27 arches, a length of 1,640 feat and a maximum height of 112 feet; its dimensions thus slightly exceeded those of a contemporary viaduct, Digswell, at Welwyn on the Great Northern Railway. The plans, specifications and descriptions of the first Barentin Viaduct indicate that its piers were set on concrete foundations followed, for twenty-one piers, by masonry bases; above that level everything was in brick. The unusual and perhaps questionable feature of its design was the rejection of entirely solid piers in favour of those with interiors containing four 'chimneys' or flues measuring 101 x 90 cm throughout their height in order to reduce their weight. This design, which was also used for Malaunay, obviously had the approval of Joseph Locke, the railway company's engineer, and, under the French system, his opinion must have been endorsed by the State civil engineering administration, the Ponts et Chaussées. The estimated cost of the Rouen and Barentin Contract (£525,118)[2] was over half that of the entire line to Havre.

Mackenzie and Brassey obtained this contract in October 1843 by negotiation with Locke, and work from Maromme to Barentin proceeded steadily through 1844 and 1845 under two experienced agents, John Jones and Miles Day. James Guthrie obtained the subcontract for the masonry bases of the piers. In contrast to his supervisory practice during the building of the Rouen line, Mackenzie gave these men a very free hand and visited the Havre contracts on only twelve days in 1845.[3] His attention in that year had turned to railway promotion and, in particular, was focussed on obtaining the concession for laying the track of the Orleans to Tours line.

The collapse was followed by much speculation and several enquiries. Frissard, a leading inspector of the Ponts et Chaussées, immediately began an investigation and the Minister of Public Works set up a Commission of Enquiry which appears to have found no explanation. The most useful action taken by the agencies of the French State was the inauguration later in 1846 of a programme of experiments to test the stability of not only a strengthened Malaunay Viaduct and the new Barentin but also the third and longest viaduct on the line, Mirville, near Bolbec.

According to his diary, Mackenzie's immediate reaction on receiving news of the fall on Monday, 12 January, was:

> Mr Illidge informed me that on Saturday last Barentin Viaduct fell to the ground a heap of Ruins - fault Bad Mortar, we told Mr Locke Mortar was bad and proposed to use Hydraulic for Mortar and bear half the expense of the extra, he said he would allow nothing but we were quite at liberty to use it if we pleased the result is now to be seen.[4]

The following day with Brassey he met Locke, who 'looked sad and was low in spirits', and a visit to the site on 18 January confirmed Mackenzie's opinion concerning the mortar: 'Bricks good mortar infamous,' he commented.[5] The specifications for the contract stipulated that throughout the work the acceptable constituents of mortar were common lime and sharp river sand; hydraulic lime, which as a result of containing more clay than did common, was much harder and resistant to water had to be used only for concrete work. The railway press speculated that the crucial damage to the mortar had been done by a sharp frost followed by a sudden thaw at the time of the collapse. The mortar may have been weak but, as Mackenzie believed, the bricks were not and in the following Spring gangs of men had the task of cleaning bricks recovered from the ruins for use in the new viaduct.

Locke's verdict on the matter came in his report to the Rouen & Havre Company General Meeting of October 1846[6] which took place shortly after the first locomotive passed over the new viaduct. The significant weakness in the structure, he said, came from the state of the masonry base of Pier No. 5. Cracks had appeared in the bases of several piers but the fissures in two sides of No. 5 were especially conspicuous and, although they did not appear to be getting worse, they had been regularly inspected as a precaution. The validity of this interpretation, he claimed, was shown by the fact that 'All the arches fell in the two directions towards the pier No. 5. And this is a sure indication that the two arches adjoining it were the first to yield.' In relation to conjecture concerning another explanation, he believed that insufficient ballast had been loaded on the track to cause the catastrophe. Thus:

> According to these circumstances, the fall of the viaduct must be attributed to the mason-work in stone, forming the base of the pier, for, as far as we can form a judgment, there has never been any other apparent cause.

Nevertheless, the suspicion remains that a design which may have given only a small margin of strength and the use of inferior mortar played a significant part in the loss.[7]

Mackenzie and Brassey immediately accepted responsibility for what had occurred because, as they admitted, in order to accelerate construction they had persuaded Locke to build the bases of the piers in stone when he planned to use brick.[8] This confession won them widespread praise and, whether it was their intention or not, deflected attention from Locke's role in the matter. Within a week of 10 January, the assistant engineer George Neumann and the architect William Tite were drawing up plans for a new viaduct - this time with solid piers. Further, the new Barentin and the strengthening of the Malaunay Viaduct were made special assignments of agents John Milroy and George Goodfellow, respectively. Malaunay was reinforced by bolting plates onto the surface of the piers[9] and, amongst several precautions, the bases and heads of the forty-eight piers of Mirville were enlarged. Mackenzie did not blame any of this employees for what had occurred, although in general, as his diary shows, he was never reluctant to comment on their deficiencies. Altogether, the matter appears to have caused him less anxiety than did the discovery in the same period of financial losses at the partnership's iron works at Pontaudemer.

While the new Barentin was in the course of construction, engineers of Ponts et Chaussées applied their programme of stringent tests to every bridge and viaduct on the line, including that over the Seine at Rouen. This process initially consisted of loading them with sand and gravel, and Malaunay Viaduct finally withstood a weight of approximately 2,000 tons for 6 weeks without weakening. Next, dynamic forces were applied through rollers and very heavy locomotives travelling at c.20 mph.[10] All came through this multiple trial unscathed but, during a visit to Mirville in February 1847, Mackenzie had some disturbing thoughts after seeing:

> the Rolling weights going on with 4 Engines abreast 2 upon each road with 25 behind each two Engines - running backward and forward about 20 Miles P<u>r</u> hour up to tonight they have run 84 times across, and have to cross yet 116 times more in all <u>200</u> - The Viaduct is very weak and vibrates a great deal thus far all is safe - I wish it had been much stronger.[11]

It is evidence of the apprehension felt by the company and investors at this trial that daily reports on the state of the viaduct were sent to the Paris Stock Exchange.[12]

In addition to its rebuilding work, the partnership applied to Locke for compensation. A letter to the Directors of the Rouen & Havre Company in October 1846 claimed that reconstruction had cost 1,342,000 francs (c.£53,150) and added the specious argument that, if their claim was accepted, this additional expense to the company would be partly offset by a commitment in the original arrangements to pay the contractors an additional 255,000 francs for an early completion of the line.[13] With the assistance of Locke, compensation of approximately £20,000 was finally awarded.[14]

The destruction of Barentin did not delay the opening of the Havre line by more than about six months[15] nor did it prevent Mackenzie and Brassey from obtaining further work in France, some of which was under French chief engineers. Ahead lay their role in the completion of the lines from Amiens to Boulogne and Orléans to Bordeaux and, after Mackenzie's death in 1851, Brassey's contribution to the Lyons and Avignon. Nevertheless, it did damage their reputation in a minor way and encouraged a growing body of opinion in France, which expressed itself most violently at the time of the 1848 revolutions, that British influence in French railway affairs, whether in construction or finance, was not essential and should be curtailed. Criticism of their workmanship might have been more strident if the lines built by the Ponts et Chaussées had not contained defects. In the Autumn of 1846 the Orléans to Tours and Orléans to Vierzon lines proved most vulnerable to flooding by the Loire.

Although two arches at its eastern end were destroyed by bombing in the Second World War, the second Barentin Viaduct has withstood the test of time and today carries frequent, heavy and fast-moving traffic of the S.N.C.F.

REFERENCES

1. Quoted in the *Illustrated London News*, 17 January 1846.

2. Based on an exchange rate of 25.25 francs/£1; this rate is used throughout this article.

3. There is no evidence concerning the time spent by Brassey on this line.
4. *Diary of William Mackenzie,* 12 January 1846, the Mackenzie Collection, Institution of Civil Engineers.
5. *Ibid,* 18 January 1846.
6. *Herapath,* 14 November 1846, pp. 1435-7.
7. The most comprehensive retrospective attempt to explain the fall appears in G. R. Burnell Ed., *Supplement to the Theory, Practice, and Architecture of Bridges,* John Weale, London 1850. This claims that the type of stone used in the bases of the piers (chalk) and the mortar were inherently too weak to resist the weights imposed on them. In addition, the flues in the piers should at intervals have had their sides tied or bonded, in the same way as Nature strengthens a reed or cane, in order to prevent buckling.
8. *Railway Record,* 14 February 1846, p. 178.
9. *Diary of William Mackenzie,* 23 January 1846; an entry on 20 May 1846 refers to the piers being reinforced by cast iron cramps.
10. *Railway Record,* 12 December 1846, p. 1319, and *Herapath,* 14 November 1846, pp. 1435-7, and 23 January 1847, pp. 90-1.
11. *Diary of William Mackenzie,* 23 February 1847.
12. *Railway Times,* 2 January 1847, p. 14.
13. Letter Mackenzie and Brassey to the Directors of the Rouen & Havre Railway, 29 October 1846, *Mackenzie and Brassey: Letter Book, Rouen & Havre and Paris & Rouen, 17 July 1846-49,* pp. 66-8. This letter is in Brassey's literary style.
14. Favrin to the Directors of the R&HR, 3 March 1847, *ibid.,* p. 99. Favrin was Mackenzie and Brassey's secretarial agent in Paris.
15. The rebuilding of Barentin from a heap of ruins in 10 months is evidence of the exceptionally high standard of the contractors' staff.

The writer wishes to thank Mike Chrimes, Chief Librarian, Mary Murphy, Archivist, and Carol Arrowsmith of the Institution of Civil Engineers for their assistance with the preparation of this piece.

Also, the writer is indebted to the Social & Economic Research Council defraying the costs of research.

20
The safety of masonry arches

J. Heyman

Summary—The paper describes the application of limit principles, derived from the plastic analysis of steel structures, to the masonry arch. Most voussoir arches have low stresses, and the lines of thrust lie well within the masonry, so that conventional factors of safety on stress have little relevance. However, a "geometrical" factor of safety may be defined by consideration of the arch of minimum thickness, of the same shape as the real arch, which would just contain a proper line of thrust. If the real profile of the arch is then discarded, and the calculations referred to an arch having the profile of the ideal line of thrust, it is shown that a simple analysis may be made of the effect of travelling point loads. The ideas are applied to the analysis of two actual bridges: (a) the stone arch between the western towers of Lincoln Cathedral, and (b) the Ponte Mosca in Turin. The paper is concerned only with the action of the masonry, and not with the design of the abutments to resist the thrust of the arch. However, the effect of geometry changes due to yielding of the abutments is discussed.

INTRODUCTION

THE VOUSSOIR arch bridge at Pontypridd, completed by William Edwards in 1751, collapsed during the course of construction after the removal of the centering.[1, 2] (The bridge was rebuilt satisfactorily in 1755, and is sometimes considered[3] to be a masterpiece.) There was no question of failure of the abutments, or of any overstressing of the masonry; the bridge was of the wrong *shape*.

Boistard made a series of twenty-two tests on model voussoir arch bridges in 1800; the results are incorporated in a volume edited ten years later by Lesage.[4] The voussoirs were made from two bricks cemented together at a slight angle (about 3°) and assembled dry to give arches of various shapes over a span of 8 ft (Boistard, working in Paris, recorded his measurements in the antiquated system of feet and inches). In the first test the arch was semi-circular (of radius $R = 48$ in.); the radial thickness of the voussoirs was $t = 4$ in., giving $t/R = 1/12$. This arch collapsed immediately on decentering. In this model test stresses were extremely low and the abutments were firm; collapse must again be viewed as a geometrical problem rather than as a problem in strength of materials.

The present paper is concerned with an analysis of the safety of masonry arches based upon their geometry. That is, it will be assumed that failure of an arch is not initiated either by failure (crushing) of the material or by failure of the abutments. The analysis of some actual bridges indicates that the stress levels likely to be encountered in this form of masonry construction are low, and it may be concluded that for usual spans and shapes, there is virtually no danger of collapse due to overstressing of the material.

The question of strength of abutments is not touched upon here; no general conclusions could be drawn without knowledge of the nature of the material forming the foundations, and the analysis would then lie in the field of soil mechanics or of rock mechanics.

SOME HISTORICAL NOTES

In fact, the classic problem of arches in the eighteenth century was precisely that of their thrust against the abutments. Detailed investigation of the voussoir arch itself was only of interest insofar as it led to some method for estimating the value of that thrust. Thus Boistard's tests[4] at the end of the century were directed only partly towards the observation of actual arch behaviour; he also wished to determine experimentally minimum abutment requirements and to obtain some information on the forces on centering during construction.

Much earlier tests had been made by Danyzy[5] reported in 1737 by Frézier,[6] and the theoretical work of La Hire,[7, 8] Parent,[9] Bélidor[10] and Couplet,[11] between 1695 and 1730, is all directed towards the problem of determining arch thrust. The work of all these writers shows a progressive development and is based upon the notion that information about abutment thrusts can be obtained by considering the limiting collapse state of the arch. In the course of his analysis Couplet gave a remarkably complete solution to the problem and he had clear notions of lines of thrust on the one hand and of mechanisms of collapse on the other; as will be seen, these two ideas give the key to the solution of the problem as a whole.

However, Couplet's contribution was ignored, and it fell to Coulomb[12] in 1773 to take up the problem almost afresh. Coulomb's work, embedded as it was in a paper which also discussed the bending of beams and laid the foundations of the science of soil mechanics, was in turn neglected, although Bossut[13, 14] made some additional contributions, and the whole was assembled and digested in Gauthey's massive treatise[15] on bridge construction.

Lamé and Clapeyron[16] seem to have rediscovered independently, however, the idea of the collapse mechanism; in their work of 1823 on the stability of a dome they do not mention Coulomb. Nor indeed do they refer to the much earlier Italian work of the *tre mattematici*[17] or of Poleni,[18] who had made assessments of the stability of the dome of St. Peter's using similar techniques. Poleni, in particular, had "sliced" the dome into segments for the purpose of statical analysis, a technique identical to that proposed by Lamé and Clapeyron for the dome of St. Isaac in St. Petersburg.

The division by Poleni of a dome into individual lunes shows an extraordinary structural insight matched only in this sort of work by the early (1697) paper on arches by Gregory.[19] Gregory found that the catenary was the proper shape of a uniform arch carrying only its own weight, and stated that an arch of any other shape could stand only by virtue of a catenary being contained within its thickness. Interpreting this statement in terms of lines of thrust, it is, as will be seen, a precise preformulation of the lower bound theorem of plastic theory.

An intuitive use of this theorem is implicit in many of the early design methods. Yvon Villarceau's inverse method,[20] for example, of which more will be said below, arranges for the centre line of the voussoirs to coincide with the equilibrium line of thrust. Fuller accounts of this and of some of the other early work on arches may be found in the books by Straub[21] and by Timoshenko.[22]

More recently, Fleeming Jenkin[23] made an ingenious model in which lines of thrust became "visible" under various loading patterns on a voussoir arch; this work is also reported in one of the first modern texts, that by Ewing[24] of 1899. Just before World War II Pippard *et al.*[25, 26] made tests on model arches, and observed the "hinging" of one voussoir upon another; the results were interpreted in the light of elastic theory, rather than in terms of the plastic theory that was just developing.

LIMIT PRINCIPLES FOR MASONRY

The first application of the modern theorems of limit design to the problem of masonry construction seems to have been made by Kooharian,[27] acting on a suggestion by Drucker (see also Prager[28]). As has been noted elsewhere,[29] the ideas of simple plastic theory can be applied to masonry if it is assumed that the material has the following properties:

(i) *Stone has no tensile strength.* Although stone itself may have some tensile strength, the joints between voussoirs may be dry or made with weak mortar. Thus the assumption is equivalent to the statement that no tensile forces can be transmitted within a mass of masonry. In accordance with the principles of the limit theorems (see below) this assumption is "safe"; it may be too safe, that is, unrealistic, if, for example, spandrel masonry interlocks with the voussoirs in an arch so that tensile forces can be transmitted locally by such keyed stones.

(ii) *Stone has an infinite compressive strength.* This is equivalent to the assumption that stresses are so low in masonry that there is no danger of crushing of the material. The assumption is obviously unsafe, but the errors introduced are very small in the analysis of any real structure of the type considered here.

(iii) *Sliding failure cannot occur.* It will be assumed that friction is high enough, or that the stones are effectively interlocked, so that they cannot slide one on another. An example of possible sliding failure has been given elsewhere,[30] but it is not usually of importance.

With these assumptions of material behaviour, and with the usual structural assumptions of small deflexions, the ideas of plastic theory developed for steel frames (see, for example, Baker *et al.*[31]) may be "translated" to apply to masonry arches. The *equilibrium condition* will be satisfied by constructing a line of thrust in equilibrium with the loads acting on the arch (including self-weight). The *yield condition* will be satisfied if such a line of thrust lies everywhere within the masonry.

There is no question here of a "middle-third" rule, which is essentially an elastic concept. The fact that the thrust line lies within the masonry at all ensures that forces can be transmitted by purely compressive action from one cross-section to the next. At the limit, when the thrust line lies in the surface of the masonry, the force will be transmitted just at the surface, and the whole of the remainder of the cross-section will be at zero stress. The thrust line cannot pass outside the cross-section, since this would require tensile forces to be developed.

The *safe theorem* of plastic theory states that, to demonstrate that an arch is stable, a thrust line must be constructed in equilibrium with all the loads acting on the arch, and lying wholly within the masonry. The power of this theorem lies in the fact that the thrust line considered need not be the *actual* thrust line; the existence of any one line gives complete assurance that the arch cannot collapse. The safe theorem has always, in fact, been appreciated by the practical designer, who is accustomed to making calculations to derive "reasonable" systems of forces for the purposes of design.

As an immediate corollary of the safe theorem, any small settlement of the arch, which may lead to a considerable shift of the actual thrust line, cannot bring about collapse, since a thrust line satisfying the theorem before settlement will also be satisfactory after settlement. This corollary relies, of course, on the usual structural assumptions that deflexions are small; if the settlements are so large as to alter sensibly the geometry of the whole structure then the statement is no longer necessarily correct. This point is examined numerically in one of the examples given below.

At the limiting state when the thrust line lies in the surface of the masonry there is a possibility of *hinging* action of one voussoir on the next. If sufficient hinges form to turn the arch into a *mechanism* (in the usual sense, for example a four-bar chain for the arch), then collapse is possible. Thus to demonstrate that an arch can become unstable it is necessary to construct a pattern of hinges which corresponds to a mechanism of collapse. There are many masonry structural elements (for example, a plain flying buttress of the usual type with a straight extrados) for which no such pattern of hinges is possible; for such structures, collapse cannot occur except by gross overloading (leading to crushing of the material) or by decay of the stonework. Similarly, many other structures can collapse only by an arrangement of hinges which corresponds to an extremely unlikely pattern of loading, so that here too a mechanism of collapse is, for practical purposes, impossible.

SOME SIMPLE EXAMPLES

Inglis,[32] as an illustrative example on "mainly graphical methods", considered the problem of a voussoir arch supporting a level roadway, Fig. 1; the fill is supposed to be of uniform density and to exercise a purely vertical load upon the arch ring. If the origin of co-ordinates is taken in the road surface on the centre-line of the arch, and the equation

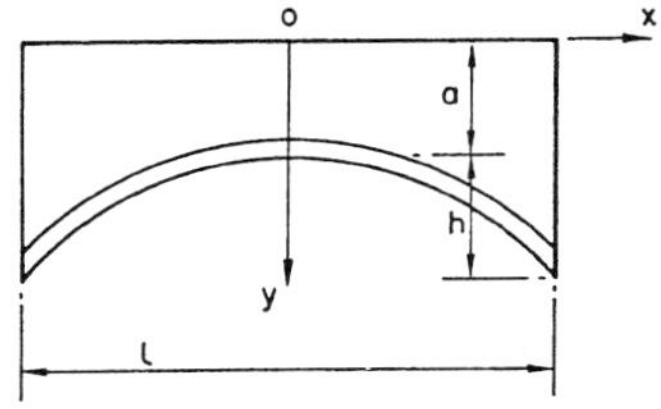

FIG. 1. Voussoir arch carrying level roadway.

of the arch centre-line is $y = y(x)$, then the intensity of load per unit horizontal length is proportional to y. The problem posed by Inglis is to determine the arch centre-line so that it coincides with the line of thrust resulting from the loading; the equation is

$$y = a \cosh\left[\frac{2x}{l} \cosh^{-1}\left(\frac{a+h}{a}\right)\right]. \tag{1}$$

Equation (1) represents a family of curves with parameter h; that is, a practical design might specify the span l and the total height $a+h$, but the value of h itself could be chosen by the designer. Shallow arches have, of course, high thrusts; the horizontal component of the abutment thrust of the arch conforming to equation (1) is inversely proportional to the square of $\cosh^{-1}(a+h)/h$.

Yvon Villarceau[20] developed an elaborate method of arch design which is essentially similar to that of the simple example just given. However, Yvon Villarceau allows for much more complex loading systems and gives tables for the numerical step-by-step solution for any given problem. The centre-line of the voussoirs is made to coincide with the calculated thrust line; Yvon Villarceau does not pretend that he is calculating the actual line of thrust, but only a possible one. Moreover, by putting the line of thrust centrally through the voussoirs, he is in some sense picking the "safest" of all the safe designs. Both Inglis and Yvon Villarceau were using intuitively the safe theorem of plastic theory, basing their designs on the satisfaction of the equilibrium and yield conditions, and not appealing to any notion of a mechanism of collapse.

Yvon Villarceau's calculations seem now to be unnecessarily precise, but his tables are certainly still valid. He himself made theoretical redesigns of several existing bridges, including that of the Pont d'Iéna (Lamandé, 1809). This five-span bridge is of about 3 m rise in each 25 m span, and the ratio h/a is about 2·0. The exact profile calculated by Yvon Villarceau differs from the profile given by the "Inglis formula", equation (1), by a maximum of about 20 mm; it is, of course, impossible to build a masonry arch whose profile can be guaranteed to this limit after decentering and subsequent settlement.

As has been noted, the best shape for a voussoir arch carrying its own weight only is that of the inverted catenary. The actual thickness of the arch, providing the centre-line conforms to the line of thrust, is not theoretically relevant. Similarly, the expression for the "Inglis arch", equation (1), does not involve the thickness of the arch ring. In practice, however, it is clear that a very thin arch, while theoretically stable, will be at the mercy of accidental superimposed loadings, which will distort the line of thrust so that it no longer follows the centre-line of the arch. Alternatively, small geometrical imperfections may distort the centre-line so that it no longer follows the line of thrust. In either case some limit will be reached at which the line of thrust can only just be contained within the masonry. At this limit sufficient hinges will be formed (at the sections where the line of thrust reaches the surface of the masonry) to transform the arch into a mechanism.

These ideas can perhaps be best illustrated by an analysis of the type of circular arch tested by Boistard[4] and which collapsed on decentering. Fig. 2(a) shows a "thick" circular arch for which a possible line of thrust is indicated (it is assumed that the arch is carrying its own weight only). A portion of the arch such as that sketched in Fig. 2(b) will be in equilibrium under the forces shown; by symmetry, the direction of thrust at the crown is horizontal. The line of thrust in Fig. 2 lies well within the masonry, and the arch can be reduced in thickness without violating this condition.

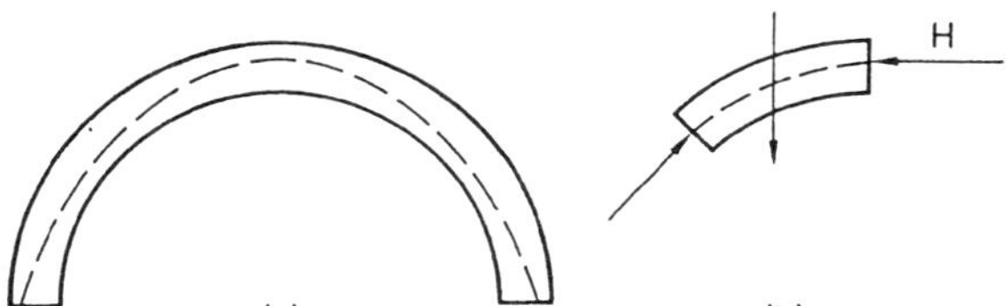

FIG. 2. Semicircular arch.

In Fig. 3 is sketched, approximately to scale, the limiting case beyond which further reduction in thickness is impossible. The line of thrust shown in Fig. 3(a) touches the extrados at the crown and passes through the extrados at the springing; it touches the intrados at points on the haunches. The corresponding mechanism of collapse (with one more hinge than is theoretically needed, due to the symmetry of the arch) is shown in Fig. 3(b).

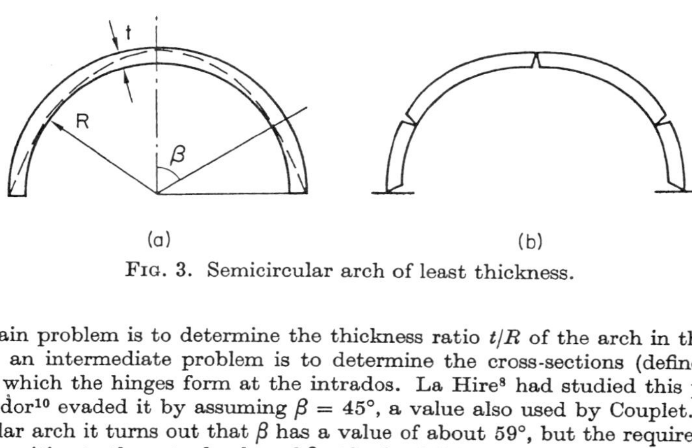

FIG. 3. Semicircular arch of least thickness.

The main problem is to determine the thickness ratio t/R of the arch in this limiting condition; an intermediate problem is to determine the cross-sections (defined by β in Fig. 3) at which the hinges form at the intrados. La Hire[8] had studied this problem in 1712; Bélidor[10] evaded it by assuming $\beta = 45°$, a value also used by Couplet.[11] For the semi-circular arch it turns out that β has a value of about 59°, but the required ratio t/R is very insensitive to the actual value of β. (Coulomb[12] had noted this insensitivity of the calculations to the precise position of the hinge at the intrados.)

The problem represented by Fig. 3 is essentially geometrical. A solution may be obtained by considering the statical equilibrium of portions of the arch in Fig. 3(a), inserting forces similar to those shown in Fig. 2(b). It may be confirmed that the correct value of β results from the solution of the equation

$$\beta \cot \beta \left[\frac{2\beta \cos \beta + \sin \beta \cos^2 \beta + \sin \beta}{2\beta \cos \beta + \sin \beta \cos^2 \beta - \sin \beta \cos \beta}\right] = \frac{\pi}{2}. \qquad (2)$$

Equation (2) may be solved numerically to give $\beta = 58° 49'$; the required thickness ratio t/R is then calculable from

$$\frac{t}{R} = 2\frac{(\beta - \sin\beta)(1-\cos\beta)}{\beta(1+\cos\beta)}. \quad (3)$$

The horizontal component of the abutment thrust [i.e. H in Fig. 2(b)] may be shown to be proportional to $\beta \cot \beta$.

Alternatively, the analysis may be made, although with much more complex algebraic manipulation, by writing the work equation for a small displacement of the mechanism of Fig. 3(b). No work is dissipated at a masonry hinge of the type considered here, so that the total work done by the loads acting on the arch (self-weight) must be zero. Thus the work equation will give a single relationship between t/R and β. The value of β must be chosen to maximize the value of t/R, and this condition leads eventually to the correct solution.

The whole analysis may be repeated for an incomplete circular arch of angle of embrace 2α, Fig. 4. Equation (3) is unchanged, but the value of β is determined from a new equation (2), in which the right-hand side, $\pi/2$, is replaced by the expression $\alpha \cot \frac{1}{2}\alpha$. For α small, the solution of the revised equation (2) is approximately $\beta = \alpha/\sqrt{2}$ and indeed the ratio β/α lies between 0·71 and 0·65 over the whole range $0 \leqslant \alpha \leqslant \pi/2$. Fig. 5 shows the required minimum thicknesses of arch for various values of α.

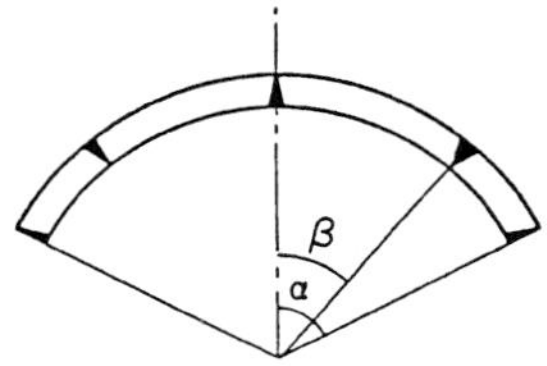

FIG. 4. Incomplete circular arch.

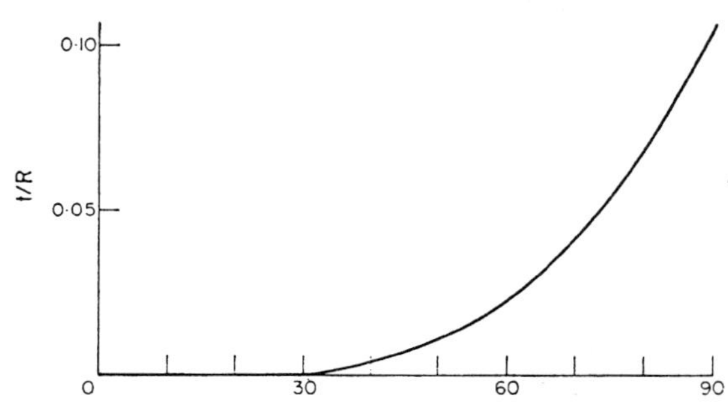

FIG. 5. Least thicknesses of incomplete circular arches.

For $\alpha = 90°$, corresponding to the first test of Boistard,[4] the required value of t/R is 0·106. (Couplet[11] had calculated $t/R = 0{\cdot}101$, using $\beta = 45°$ instead of the correct value 58° 49′.) Boistard's arch had $t/R = 1/12$, so that the observed collapse on decentering was to be expected.

THE SAFETY OF ARCHES

For the ideal problem of a circular arch just discussed, the limiting values of Fig. 5 represent arches which will just stand under their own weights. If the arches are made slightly thinner, they will immediately collapse. Conversely, arches made slightly thicker will never collapse. However, as noted above, an arch which is only marginally thicker than the theoretical minimum may collapse if it is geometrically slightly imperfect, and it will be sensitive to slight variations in superimposed load.

One measure of safety in an arch, therefore, might be taken as the ratio of the actual depth of the arch to that of the theoretical minimum necessary to contain the line of thrust. For example, Boistard's model circular arch had $t/R = 1/12 = 0{\cdot}0833$. For an arch embracing 120° ($\alpha = 60°$, Fig. 4), the theoretical minimum value of t/R is 0·0226 (Fig. 5). Thus Boistard's model in this case might be said to have a "geometrical factor of safety" of $833/226 = 3{\cdot}58$.

The establishment of a reasonable value for the geometrical factor of safety, however, is not in general sufficient to ensure an adequate margin for the stability of a given arch. As a trivial example, the centre-line of the voussoirs might have been designed to coincide exactly with the line of thrust (e.g. Inglis[32] or Yvon Villarceau,[20] or the inverted catenary of Gregory[19]). In this case, the geometrical factor of safety for given loading is infinite, whatever the thickness of arch, although a thin rib, even of the correct shape, will be in a practical sense potentially unstable. Thus a second measure of safety might be some measure of the resistance of an arch to a disturbing load (say a superimposed point load).

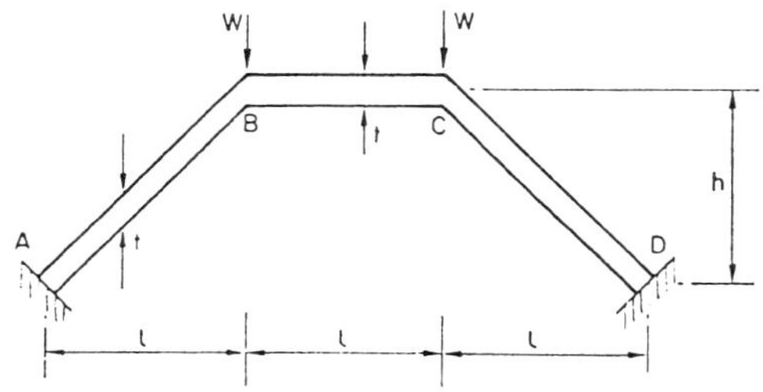

Fig. 6. Idealized arch carrying two point loads.

In practice this "live-load test" could be combined with the geometrical test. For example, a suitable value of point load could be selected, and the minimum thickness of arch determined to contain a thrust line for both the original loading and the test loading. Examples of this type of analysis are given below.

Finally, a strength test could be devised. Although it has been assumed that stress levels are low, it may be necessary to check this for any given design. For very large-span bridges, e.g. Gladesville,[33] it may even be that permitted stresses become the ruling criterion. In the case of the limiting arches of Fig. 5, for which the thrust line approaches the surface of the masonry, stresses will rise locally to very high values even though *mean* stresses are low. However, the concept of a geometrical factor of safety may be adapted to provide an automatic safety margin on stress. That is, if the mean stress levels are satisfactory for the actual loads acting on the theoretically reduced arch, they will be satisfactory *a fortiori* for the actual arch. In this connexion, Yvon Villarceau advocated the use of a safety factor of 10 on stress; the working stresses in the masonry should not exceed one-tenth of the crushing stress.

These tests of safety will be illustrated with some calculations on real arches. However, the ideas may perhaps be made clear by reference to a highly idealized example. Suppose the dead self-weight of an arch to be represented by two concentrated loads W, Fig. 6; the arch has a span $3l$ and a rise h, as shown. The thrust line consists of three linear segments, and it will be assumed that the arch has been built as three straight portions, each of *vertical* thickness t. (This assumption of constant thickness throughout makes the immediate calculations easier; variable thickness will be discussed in the example of the Ponte Mosca below.) The horizontal abutment thrust in the arch has the value $H = Wl/h$, as may be seen from a consideration of Fig. 7.

Assuming the arch material to be infinitely strong, the real arch can be hypothetically reduced in thickness from t to zero; the geometrical factor of safety is infinite. However, suppose now that a travelling point P is imposed over the length BC, Fig. 8. The thrust line for the combined loading will no longer coincide with the centre of the original arch, and a possible position is sketched in Fig. 8. This is, in fact, a special and unique position; the line has been drawn so that its maximum departure from the arch centre-line is e, say.

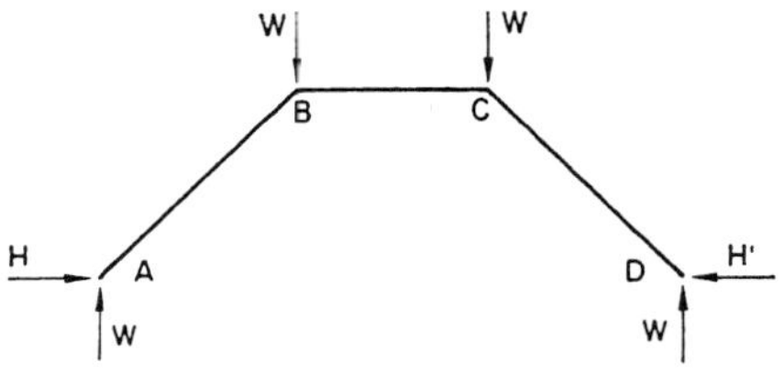

FIG. 7. Centre-line of idealized arch.

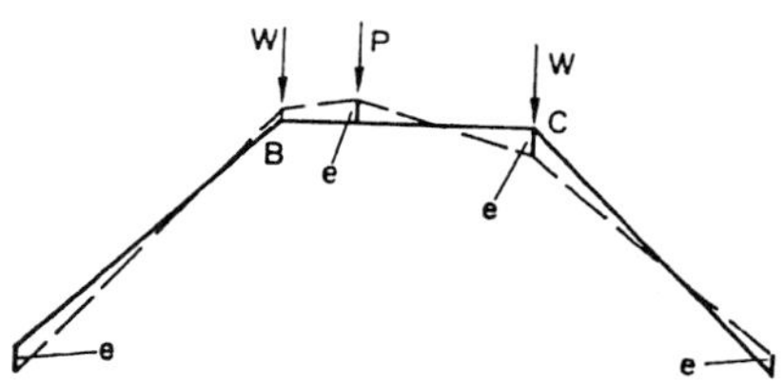

FIG. 8. Line of thrust due to travelling point load.

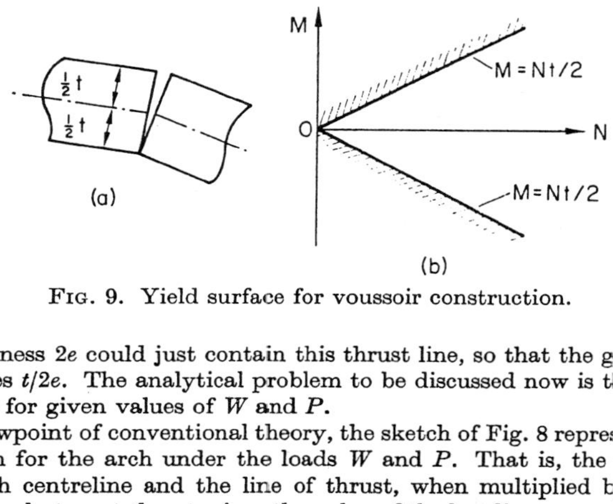

FIG. 9. Yield surface for voussoir construction.

An arch of thickness $2e$ could just contain this thrust line, so that the geometrical factor of safety becomes $t/2e$. The analytical problem to be discussed now is the determination of the value of e for given values of W and P.

From the viewpoint of conventional theory, the sketch of Fig. 8 represents the bending moment diagram for the arch under the loads W and P. That is, the vertical distance between the arch centreline and the line of thrust, when multiplied by the horizontal component of the abutment thrust, gives the value of the bending moment at any section. Now a bending moment in a voussoir arch can be defined by reference to the centre-line of the arch; if the line of thrust is eccentric by an amount e, and the thrust has value N, then the bending moment has value Ne. As long as e is less than $t/2$, Fig. 9(a), then the voussoirs will remain in contact; a hinge will occur for $M = \pm Nt/2$, leading to the *yield surface* of Fig. 9(b) (using a concept familiar from plastic theory).

In the limiting case of Fig. 8, therefore, if the arch thickness is only just sufficient to contain the thrust line, four hinges will form which will turn the arch into a mechanism. It is convenient to use the usual form of calculations for simple plastic theory applied to steel frames. Fig. 10 shows the "steel frame" problem analogous to the masonry problem of Fig. 8; the arch has three redundancies, confirming the necessity for four plastic hinges for a collapse mechanism. The full plastic moment acting at the hinges of Fig. 10 will be denoted M_p; it will be appreciated that, since the values of e have been set equal in Fig. 8, the value of M_p is the same at all four hinges. For an arch of variable thickness, leading to differing values of eccentricity, the corresponding values of M_p will differ from hinge to hinge; see the example of the Ponte Mosca below.

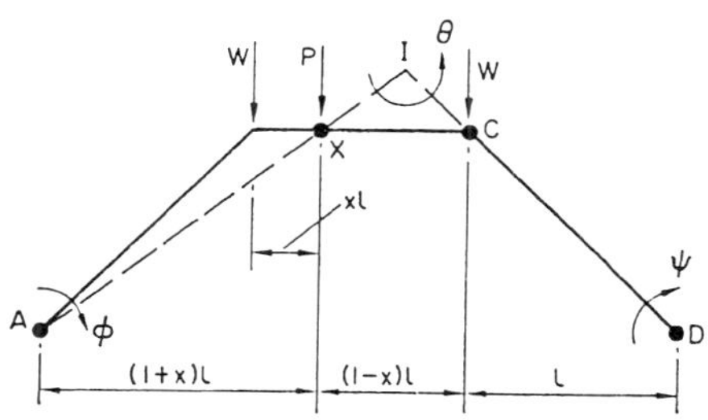

FIG. 10. Plastic collapse mechanism for idealized arch.

The assumption will now be made that the eccentricities e in Fig. 8 are very small compared with the overall dimensions of the arch, so that the mechanism of Fig. 10 may be examined with sufficient accuracy by taking the hinges to lie exactly on the centre-line of the arch. The motion of the mechanism may be referred to a rotation of the portion XC of an amount θ about its instantaneous centre I. It may be confirmed that the rotations ϕ and ψ in Fig. 10 are given by

$$\phi = \psi = \frac{1-x}{2+x}\theta. \tag{4}$$

Thus the work done by the loading on a small motion of the mechanism is

$$P(1+x)\,l\phi + Wl\phi - Wl\psi. \tag{5}$$

By virtue of equation (4), the terms in W cancel in (5). This is not accidental and indeed is of crucial importance for the analytical method developed in this paper. Since the centre-line of the arch coincides with the line of thrust for the dead loads, *any* small inextensional deformation of the centre-line will lead to zero virtual work of those loads. Thus any mechanism of the type sketched in Fig. 10, referred to the dead-load thrust line, will involve work terms only for the additional live loading.

The total hinge rotation in Fig. 10 is $2(\theta+\phi+\psi)$, so that, using equation (4) and equating the work done by the loads [expression (5)] to the work dissipated in the hinges, it is found that

$$M_p = \frac{Pl}{2}\left(\frac{1-x^2}{4-x}\right). \tag{6}$$

The maximum value of M_p occurs for $x = 4-\surd 15 = 0{\cdot}127$ and is equal to

$$M_p = Pl(4-\surd 15) = 0{\cdot}127Pl. \tag{7}$$

It remains to compute the eccentricity e in Fig. 8 corresponding to the value of M_p given by equation (7). For most cases it will be sufficiently accurate to use the *dead-load* abutment thrust in this computation, i.e. the value $H = Wl/h$ in Fig. 7. Since, by definition,

$$M_p = He \tag{8}$$

equation (7) leads to the value

$$e = 0{\cdot}127\frac{P}{W}h. \tag{9}$$

(Inserting some typical values, say $h = 3m$, $W = 10MN$, $P = 1MN$, the value of e is determined as something less than 40 mm. For a ratio e/h of about 1 per cent, little error will be introduced by working from the centre-line, as was done in the calculations corresponding to Fig. 10, instead of locating the hinge points accurately in the extrados and intrados of the fictitious arch of minimum thickness.)

The "work method" just used appears to have some advantages, leaning as it does on established concepts in simple plastic theory. It is not really convenient, however, for arches of more complex geometry, where the solution may be obtained most easily by drawing. For such a graphical method the dead loads may be ignored completely, since they have been shown not to enter into the calculations.

For the problem just considered, therefore, in order to calculate the value of M_p at the hinge points of Fig. 10, it is sufficient to determine the action of the single point load P acting alone. In Fig. 11 the thrust line has been positioned to give four equal bending

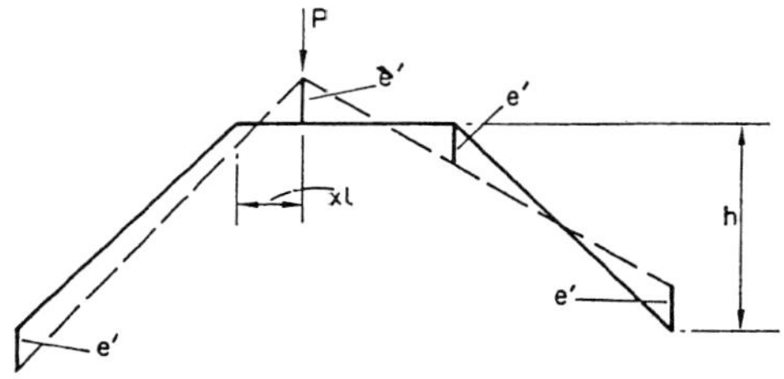

FIG. 11. Thrust line for point load.

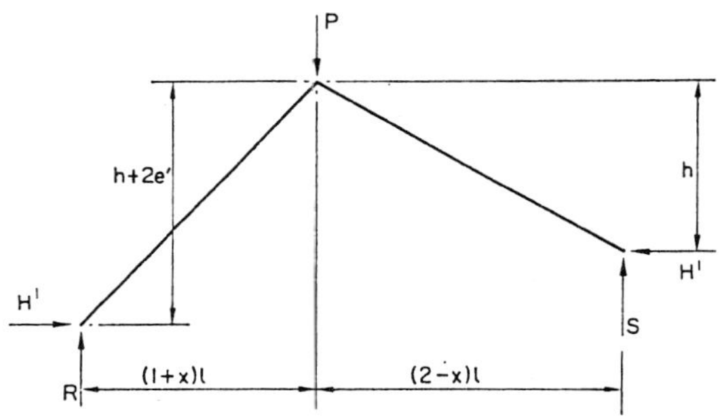

FIG. 12. Geometry of thrust line.

moments, as before. In general this would be done by trial and error; from the simple geometry of Fig. 11, it is found that

$$\frac{2e'}{h} = \frac{1-x}{2-x}. \tag{10}$$

From a consideration of the statics of the thrust line, shown separately for clarity in Fig. 12, it is found that

$$H'h = Pl\frac{(1+x)(2-x)}{(4-x)} \tag{11}$$

so that, from equations (10) and (11),

$$H'e' = \frac{Pl}{2}\left(\frac{1-x^2}{4-x}\right). \tag{12}$$

The quantity $H'e'$ is the required value of M_p, and it will be seen that equations (6) and (12) are identical; the work can now proceed as before.

As a second idealized example the design of a parabolic arch will be considered, and the opportunity will be taken to insert numbers bearing some relation to an actual bridge. The parabolic arch results, of course, from a uniformly distributed horizontal load, and it will be supposed that the arch has span $2l$ and height h, Fig. 13. If now the effect of a superimposed load is considered, the hinges in Fig. 13 indicate the collapse mode; the analysis of the Lincoln arch below indicates another possible mode, but Fig. 13 gives the worst case for the parabolic arch.

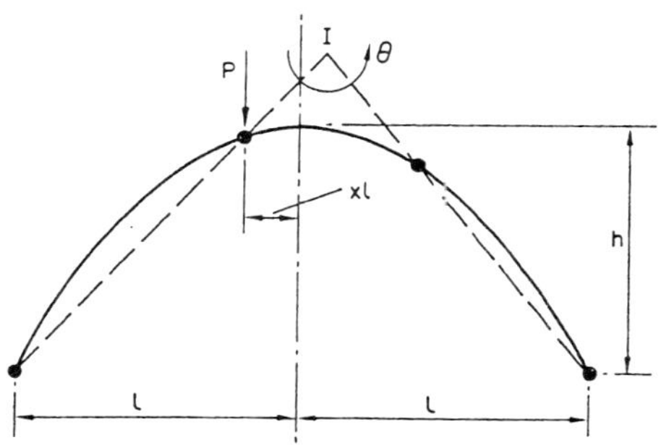

FIG. 13. Parabolic arch.

Treating the problem again as an exercise in simple plastic theory, the value of M_p corresponding to the mechanism of Fig. 13 is given by

$$M_p = \frac{Pl}{2}(1-x)\frac{(1+x)^2}{(3-x)^2}. \tag{13}$$

This expression is a maximum for $x = 5-\sqrt{20} = 0{\cdot}528$, that is, for the load at about quarter span; the maximum value is

$$[M_p]_{\max} = \frac{Pl}{2}(5\sqrt{5}-11) = 0{\cdot}0902Pl. \tag{14}$$

As before, this numerical result could have been obtained by drawing a thrust line similar to that shown in Fig. 11 for the segmental arch.

Suppose a bridge is to be designed to carry a uniformly distributed load of 1500 kN (per metre width of bridge) over a span of 24 m with a rise of 3 m. A point live load of 100 kN (per metre width of bridge) is to be considered. The (uniform) depth of the voussoirs is required, using a geometrical factor of safety of 8 and a factor of safety on stress of 10. (These values are used purely for illustration and are not necessarily recommended as proper to this class of bridge.)

Since the dead-load thrust is parabolic, the arch centre-line will also be made parabolic. The basic value of horizontal thrust is given by $Hh = Wl/4$, i.e. $H = 1500$ kN. From equation (14), the value of M_p is (0·0902) (100) (12) = 108·2 kNm. Thus the eccentricity e of the thrust line is given by $e = 108{\cdot}2/1500 = 0{\cdot}0722$ m; an arch of width 0·144 m would just contain the line of thrust. Using the geometrical factor of safety of 8, the depth of the voussoirs is determined as 1·15 m, if the geometrical criterion governs.

If the voussoirs are made of a medium sandstone having a crushing strength of 40 MN/m² (N/mm²), then the basic thrust would require a voussoir depth of

$$1500/40{,}000 = 0{\cdot}0375 \text{ m}.$$

Using the safety factor on stress of 10, the voussoir depth should be at least 0·375 m, and it is clear that under these circumstances the stress condition is not so severe as the geometrical condition. Indeed, the voussoir depth of 1·15 m would give a mean stress due to the horizontal component of the abutment thrust of 1·50/1·15 = 1·30 MN/m², compared with the crushing strength of 40 MN/m²; the factor of safety is over 30. (The actual thrust at the abutment will, of course, be slightly higher than the horizontal component. In practice, however, the depth of the voussoirs will be increased from the crown towards the springing.)

The ideas developed here will now be applied to two actual arches, of which the first (Lincoln) is especially simple but asymmetrical, and the second (Ponte Mosca at Turin) has readily available data.

THE SAFETY OF TWO BRIDGES

(a) *Lincoln*

Between the western towers of Lincoln Cathedral, and just above the nave vault, there is a thirteenth-century slender stone arch[34, 35] of span 8·54 m, Fig. 14. The northern springing of the arch is some 300 mm lower than the southern, due apparently to settlement, and the rise of the arch, relative to the chord, was determined in 1836 as about 380 mm. The cross-section of the arch has a uniform profile, virtually rectangular, and is about 500 mm wide and 280 mm deep. The spandrels of the arch have stone backings which will be assumed to contribute to the loading but not to carry any thrust.

The function of the arch is obscure;[36] it may have served as a setting-out point for the construction of the nave vault in 1240–50.

The centre-line of the arch is shown to an exaggerated vertical scale (6 ×) in Fig. 15; it will be seen that the arch is two-centred and has a distinct point. Also shown in Fig. 15 is a possible line of thrust for the self-weight of the arch plus backings.

In Fig. 16 is shown an enlarged view of the left-hand end of the arch; the origin of co-ordinates is taken at the springing (of the centre-line of the arch), and the line of thrust passes below the springing by a vertical amount Δ. At any other section of the arch, say A, the vertical distance between the arch centre-line and the line of thrust is given by $\varepsilon_A = y_A - h_A$, where the distances are as marked in Fig. 16.

If the (non-uniform) vertical load acting on the segment OA of the arch has intensity w per unit horizontal length, then statical equilibrium requires that

$$(h_A + \Delta)\, H + \int_0^{x_A} (x_A - z)\, w\, dz = V x_A \tag{15}$$

or

$$h_A = -\Delta + \frac{V}{H} x_A - \frac{1}{H}\int_0^{x_A} (x_A - z)\, w\, dz. \tag{16}$$

Since both H and V/H are, for the moment, unknown quantities, equation (16) can be written for any section x (dropping the suffix A):

$$h = -\Delta + \rho x - \gamma \int_0^{x} (x - z)\, w\, dz. \tag{17}$$

Thus this equation defining the line of thrust is written in terms of three quantities Δ, ρ and γ to be determined; this corresponds to the fact that the arch has three redundancies. By adjusting the values of Δ, ρ and γ, the line of thrust can be shifted with respect to the arch centre-line.

Considering, as before, the thickness of the real arch to be hypothetically reduced until it can only just contain a proper line of thrust, then a solution is sought to the problem of making the two lines of Fig. 15 lie as close together as possible. That is, the distance $\varepsilon = y - h$, Fig. 16, must be minimized for the whole arch. Since there are three variables defining h, equation (17), the minimum will occur when *four* quantities ε are made equal (or, in general, made to bear a fixed ratio to each other, if the arch is of variable depth; see the next example. In the case of Lincoln the vertical depth of the real arch is sensibly constant at 280 mm).

For the purposes of calculation, the integral in equation (17) was replaced by a discrete summation. The arch was divided into 14 segments each of length 0·61 m, whose weights W_r were calculated; for the nth segment, the expression

$$x_n \sum_{r=1}^{n} W_r - \sum_{r=1}^{n} x_r W_r \tag{18}$$

was computed as an approximation to the value of the integral at that segment.

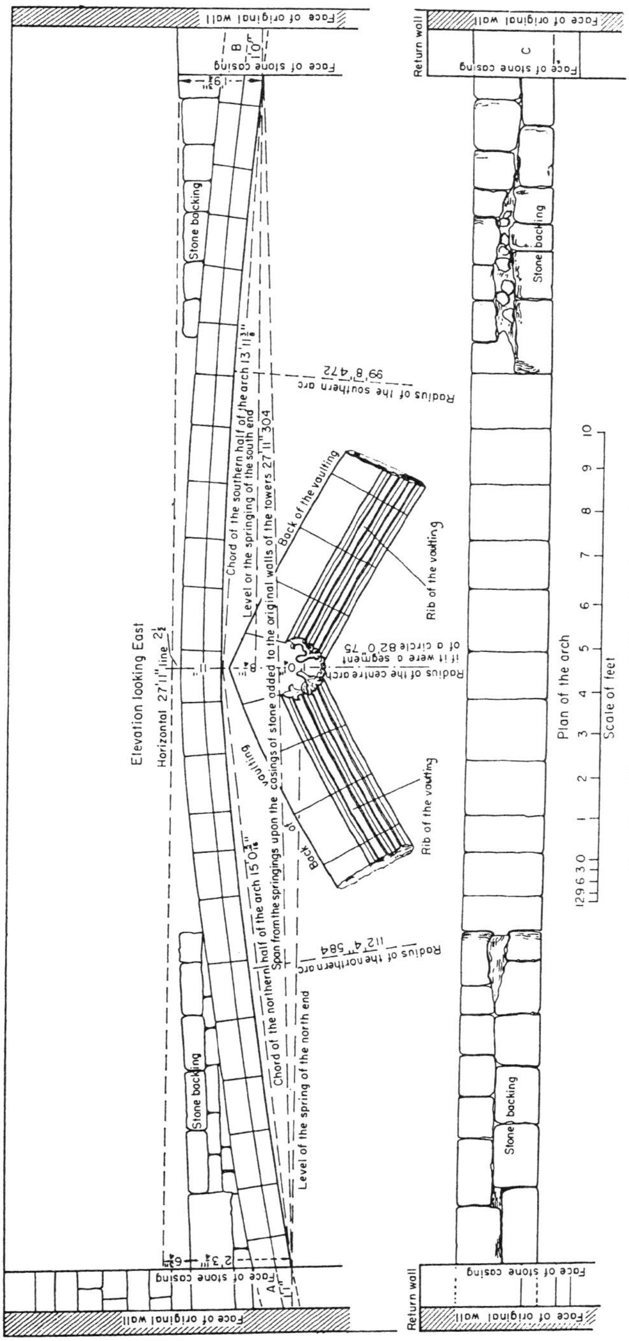

FIG. 14. Stone arch at Lincoln Cathedral.

The analysis then proceeds by trial and error, exactly as for the analogous problem of a plastic frame. The final positions in the arch where the eccentricities of the thrust line are equalized were found to be at the left-hand springing (0) and at segments 3, 7 and 12. The following equations may be written:

$$\left.\begin{aligned} 0:&\quad \varepsilon = \Delta \\ 3:&\quad -\varepsilon = \Delta + 0{\cdot}257 - 1{\cdot}525\rho + 8080\gamma \\ 7:&\quad \varepsilon = \Delta + 0{\cdot}541 - 3{\cdot}965\rho + 47170\gamma \\ 12:&\quad -\varepsilon = \Delta + 0{\cdot}439 - 7{\cdot}015\rho + 122840\gamma \end{aligned}\right\}. \tag{19}$$

In the equation for segment 3, the value 0·257 (m) is the rise of the arch, Fig. 16, the segment is 1·525 m from the origin, and 8080 Nm is the value of expression (18) at the

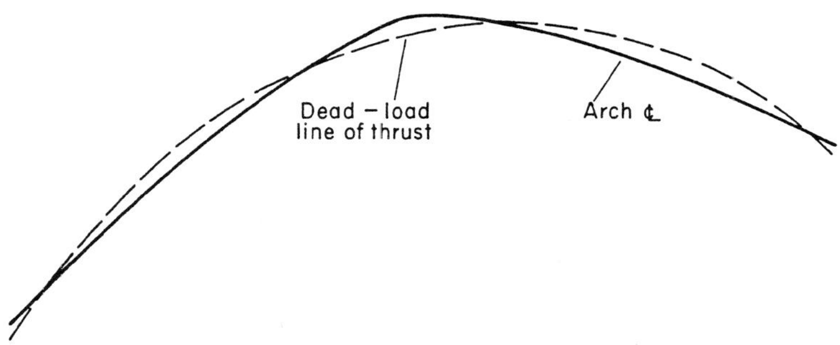

FIG. 15. Centre-line of Lincoln arch; vertical scale × 6.

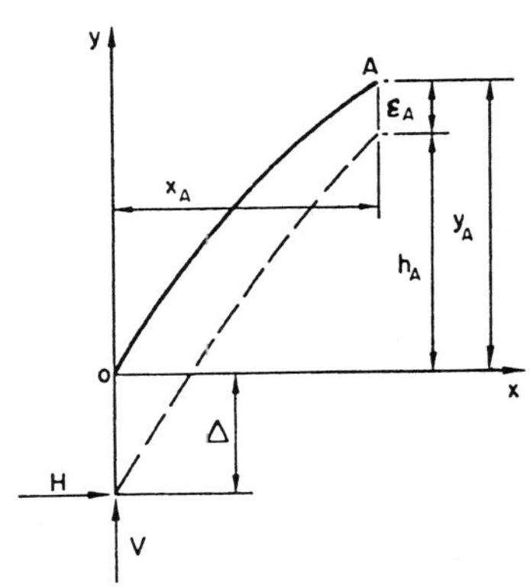

FIG. 16. Centre-line and thrust line at springing.

segment. Note the alternation in sign of ε in equation (19), corresponding to the line of thrust lying below or above the arch centre-line. Equations (19) solve to give:

$$\left.\begin{aligned} \varepsilon &= 0{\cdot}033 \text{ m} \\ H &= 1/\gamma = 87 \text{ kN} \end{aligned}\right\}. \tag{20}$$

The corresponding line of thrust is the one shown (approximately to scale) in Fig. 15; the eccentricity of the line of thrust for the whole arch is shown in Fig. 17.

It may be concluded that an arch of depth 66 mm would just contain the dead-load line of thrust. The geometrical factor of safety of the actual Lincoln arch, against dead load only, is therefore 280/66 = 4·2.

The cross-sectional area of the real arch is about 0·14 m², so that the horizontal component H of the abutment thrust, equation (20), would cause a mean compressive stress of about 0·62 MN/m². The arch is made of oolitic limestone for which the crushing strength (see, for example, Gwilt[37] or Warland[38]) is about 15 MN/m². However, Nicholson[35] states that the mortar beds (about 10 mm) between the voussoirs are friable, so that the crushing strength of the actual stone may not be a reliable guide to the safety factor on stress. Nevertheless, the working value of the mean compressive stress is satisfactorily low.

It remains to assess the stability of the arch under live loading. Nicholson[35] reports (in 1842) that "the arch vibrates perceptibly when jumped upon", and notes that it is "the constant practice of visitors thus to prove its elastic properties" (it still is). The effect of a unit point load placed anywhere on the arch will now be considered.

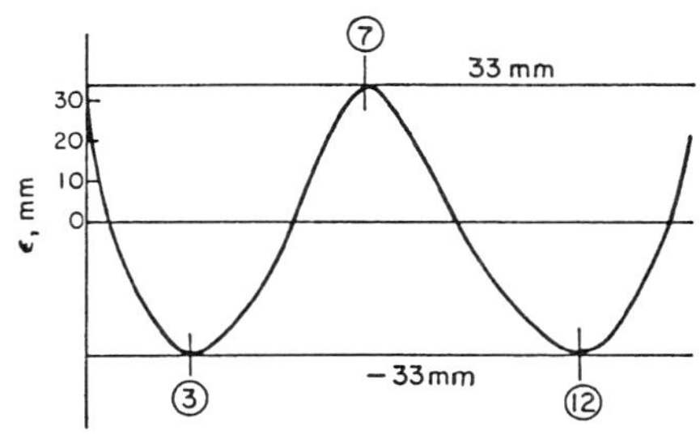

FIG. 17. Eccentricity of dead-load thrust line, Lincoln.

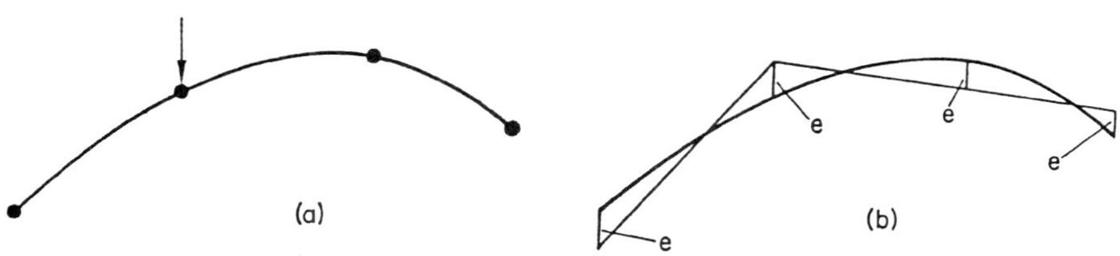

FIG. 18. Collapse mechanism for point load, Lincoln.

For the purposes of this calculation the real profile of the arch will be temporarily discarded; the centre line of the real arch (full line in Fig. 15) will be replaced by the line of thrust calculated from equation (17), in which the integral is replaced by the discrete summations, and the values of Δ, ρ and γ are those resulting from equations (19). Thus the hypothetical arch to be considered has the profile shown dotted in Fig. 15, and the problem to be solved is the following: Given an arch of this profile, what is its minimum (constant) thickness if it is just to contain the thrust line resulting from the dead load acting on the real arch together with a unit live load placed anywhere on the arch?

As was noted above, by referring the calculations to the dead-load line of thrust the dead loads themselves do not appear further in the calculations; the effect of the live load can be investigated in isolation. The calculation is, indeed, similar to that made earlier for the parabolic arch, Fig. 13. Thus the load placed approximately in the position shown in Fig. 18(a) will give rise to the same type of collapse mechanism; the hinges can be located most easily by drawing the thrust line, Fig. 18(b), arranged to give equal eccentricities e at the four hinge points. Fig. 19 shows another possible mechanism when the point load is near the centre of the arch; one hinge has moved away from the springing.

The eccentricity e of the thrust line, when multiplied by the horizontal component T of the abutment thrust, will give the required value of M_p. From the statics of the thrust line, using the notation of Fig. 20, the value of T is given by

$$T = \frac{P}{(h_1/l_1) + (h_2/l_2)}. \tag{21}$$

Calculations were made for the Lincoln arch for 14 positions of the unit live load, and the resulting values of the "full plastic moment" M_p are shown in Fig. 21. The worst position of the load is at about quarter span; the diagram is slightly asymmetrical, and the maximum value of M_p is found to be 0·396 Nm for a unit load.

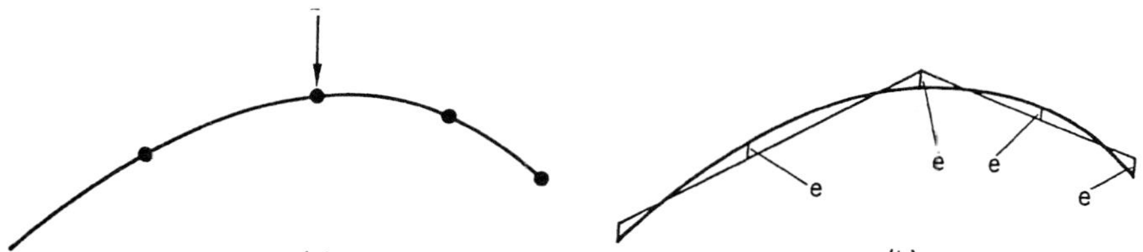

FIG. 19. Alternative collapse mechanism for point load, Lincoln.

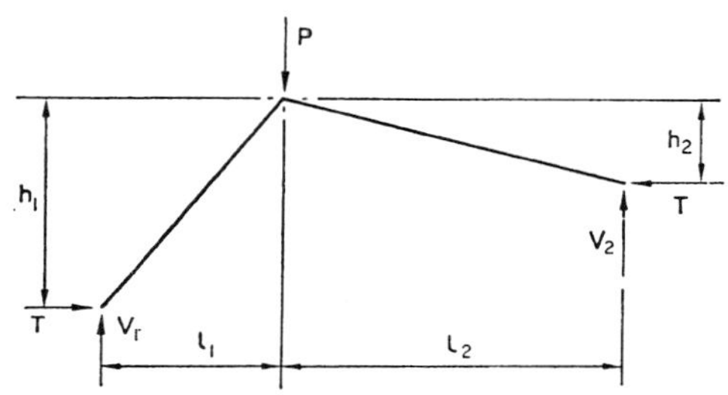

FIG. 20. Statics of thrust line, Lincoln.

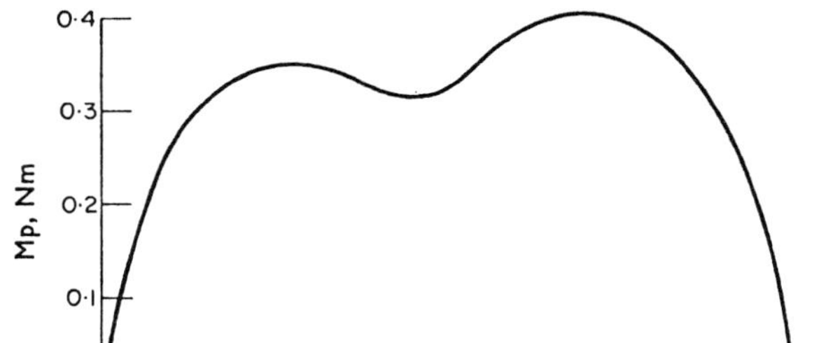

FIG. 21. Required "full plastic moment" for unit point load, Lincoln.

Thus, taking a point load of 2 kN (a heavy man jumping on the arch) and using the basic abutment thrust of 87 kN, equation (20), the effect of the live load is to displace the dead-load line of thrust by a maximum amount ε, where

$$\left.\begin{aligned} (0{\cdot}396)\,(2000) &= (\varepsilon \times 87{,}000) \\ \text{or} \qquad \varepsilon &= 0{\cdot}009 \end{aligned}\right\}. \tag{22}$$

This means that an arch of thickness 18 mm having the *broken line* profile in Fig. 15 would just contain the line of thrust due to the dead load on the Lincoln arch plus the live load of 2 kN in the worst position.

The eccentricities of equations (20) and (22) are simply additive: $33+9 = 42$ mm. That is, an arch of thickness 84 mm having the *full line* profile (i.e. the real profile) of

Fig. 15 would just contain the dead- plus live-load line of thrust. Thus the geometrical factor of safety, which was computed to be 4·2 against dead load only, is reduced by a man jumping in the worst position to 280/84 = 3·3. The arch may vibrate perceptibly, but the geometrical factor of safety is not reduced dangerously.

(b) *Ponte Mosca, Turin*

Castigliano[39] gives details of Mosca's bridge of 1827 over the Dora Riparia, Turin, and he tabulates expressions similar to the summations of (18) above. Castigliano's book is concerned with the exposition of the theorems which bear his name, and he treats the bridge at Turin as an elastic arch having three redundancies, whose values he proceeds to compute.

Castigliano shows that a cracked arch (i.e. an imperfectly elastic arch) may be dealt with by trial and error, and he makes three sets of calculations for the Turin bridge. In the first, full allowance is made for the elasticity of the mortar beds between the voussoirs, and the value of the horizontal abutment thrust is calculated to be 352,990 kgf/m width of bridge. (For a direct comparison with Castigliano's figures, kgf units will be used here instead of Newtons; 1 kgf ≏ 10 N. It will be noticed that Castigliano carries out his computations to an unwarranted number of significant figures.)

A second calculation is made for dry construction (mortar beds of zero thickness), and Castigliano obtains the value 324,710 kgf, a drop of some 8 per cent in the value of the abutment thrust. However, this solution involves the line of thrust falling outside the middle third near the springings, and a final revised calculation allowing for the partially cracked masonry gives a thrust of 333,960 kgf. Castigliano's drawing is reproduced in Fig. 22; the arch is symmetrical, so that one half only need be considered for the purpose of determining the effect of dead load. The radial thickness of the voussoirs increases from 1·50 m at the crown to 2·00 m at the springings; the *vertical* thickness at the springings is 2·23 m. As before, the arch will be imagined to be "shrunk" towards its centre-line until the line of thrust can only just be contained; during the shrinking process, the relative thickness of the arch at the various cross-sections will be maintained constant.

Castigliano divided the arch into six segments, Fig. 22. Taking the origin at the crown (instead of at the springing, cf. Fig. 16), and using the previous notation, the equations for the eccentricities at the critical sections become

$$\left.\begin{array}{lrl} 6\ \text{(crown)} & : & \varepsilon = \Delta \\ 3 & : & -1{\cdot}20\varepsilon = \Delta + 2{\cdot}40 - 1/H(800{,}470) \\ 0\ \text{(springing)} & : & 1{\cdot}49\varepsilon = \Delta + 5{\cdot}36 - 1/H(2{,}021{,}937) \end{array}\right\} . \qquad (23)$$

In the third of these equations, for example, the value $1{\cdot}49\varepsilon$ contains the ratio (2·23/1·50) of the vertical thicknesses of the arch at the springing and the crown; the eccentricity ε of the line of thrust is thus referred to the dimension 1·50 m at the crown. Section 0 lies 5·36 m below section 6 (the origin); the value 2,021,937 kgfm is taken from Castigliano's tables.

Equations (23) solve to give $\varepsilon = \Delta = -0{\cdot}116$ m, $H = 373{,}000$ kgf; this last figure is some 5 per cent higher than Castigliano's elastic value. The geometrical factor of safety against dead load is 1·50/2(0·116) = 6·5.

The greatest mean stress in the arch occurs at the crown, where its value is 373/1·5 = 249 tf/m^2 (the corresponding value at the springings is 438/2·0 = 219 tf/m^2). Ungewitter[40] gives typical strengths of granite in the range 5000–18,000 tf/m^2; the nominal factor of safety on stress appears to be well over 20. Weak mortar may, however, crumble away until the granite voussoirs make contact; Castigliano has some interesting comments on the best time to strike the centering under these conditions.

The dead-load line of thrust is now substituted for the actual centre-line of the arch for the purpose of computing the effect of live load. A graphical technique is again best; a typical position of the thrust line is shown (schematically; vertical scale × 4) in Fig. 23. The eccentricities of this line are not constant; the values e_1, etc. are proportional to the vertical depth of the arch at the corresponding sections. The final results are presented in Fig. 24 (cf. Fig. 21 for Lincoln); the largest value of M_p is 1·72 tfm for a unit (tonne) load.

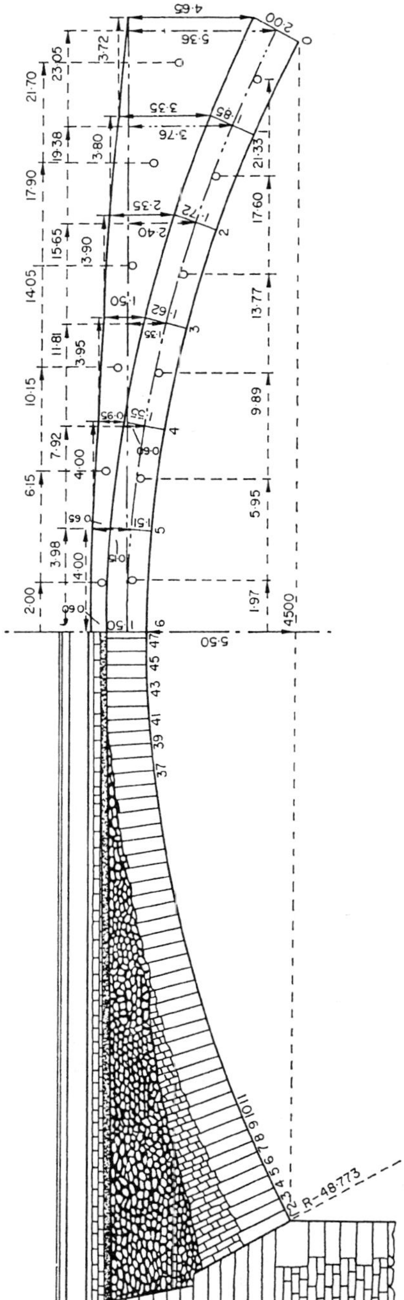

FIG. 22. Ponte Mosca, Turin.

Thus, taking a live load of 10 tf/m width of bridge the value of M_p is 17·2 tfm. The basic abutment thrust is 373 tf, and the live-load eccentricity, referred to the crown depth as before, is therefore 17·2/373 = 0·046 m. Adding in the dead-load eccentricity of

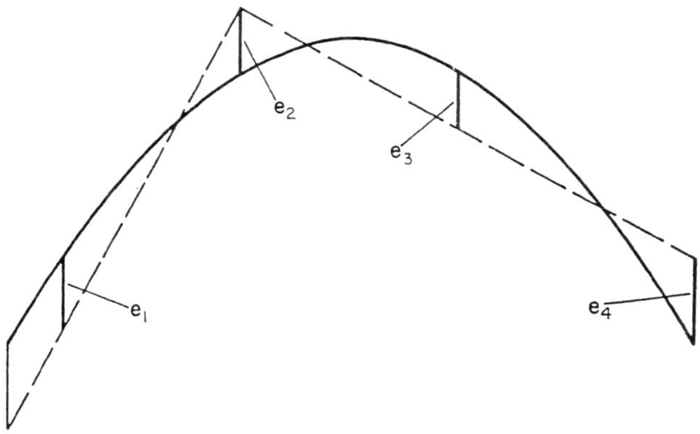

FIG. 23. Thrust line for point load, Turin.

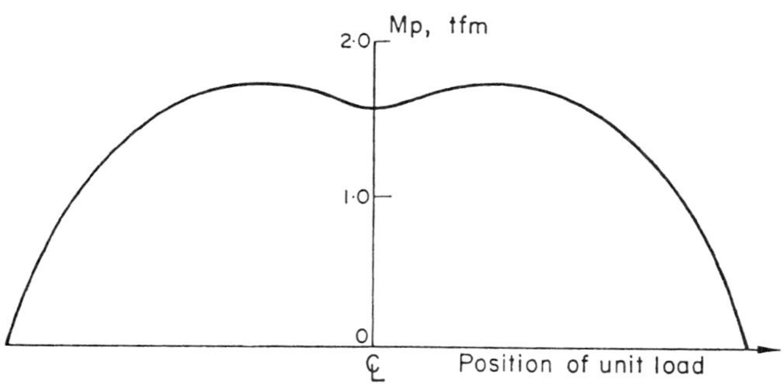

FIG. 24. Required "full plastic moment" for unit point load, Turin.

0·116 m, the geometrical factor of safety is reduced to 1·50/2(0·162) = 4·6 for this particular value of live load.

THE EFFECT OF GROSS DEFORMATIONS

Without precise and detailed information on the foundations it is not possible to discuss the problem of failure of the abutments of a bridge. However, it is of interest to investigate the effect on the masonry arch proper of a limited movement at the springings. That is, it will be assumed that unrestricted movement is not occurring, but that something analogous to consolidation has occurred at one or other foundation; after movement has stopped, it is required to assess the new safety factors of the arch.

It has been noted above, as a corollary of the "safe" theorem, that, providing geometry changes are negligible, any small movements will not affect the ultimate collapse conditions. It is intuitively clear, and this may be confirmed by calculation, that the geometry of the bridge at Turin will not be much changed by small settlement or spreading of the abutments. It may be concluded, therefore, that the safety factors are virtually unchanged. By contrast, a small spread in the abutments (the western towers) of the stone bridge at Lincoln might be expected to produce a marked change in the geometry of such a flat arch. This arch will be examined numerically, and new safety factors calculated.

In Fig. 25(a), a typical arch is supposed to be fitted exactly between its abutments. If the abutments spread, then the arch must crack to accommodate itself to the new span;

in practice, the voussoirs will open as sketched in Fig. 25(b). This phenomenon was investigated extensively by Pippard.[25] Such a cracked state is completely safe; the presence of three hinges does not turn the arch into a mechanism. Indeed, since a real arch will never fit exactly on to its abutments and will be subject to inevitable accidental imperfections, the cracked state might be thought of as the natural state of the arch. Fig. 26 shows the seventeenth-century bridge at Clare College, Cambridge (Grumbold, 1638–40).

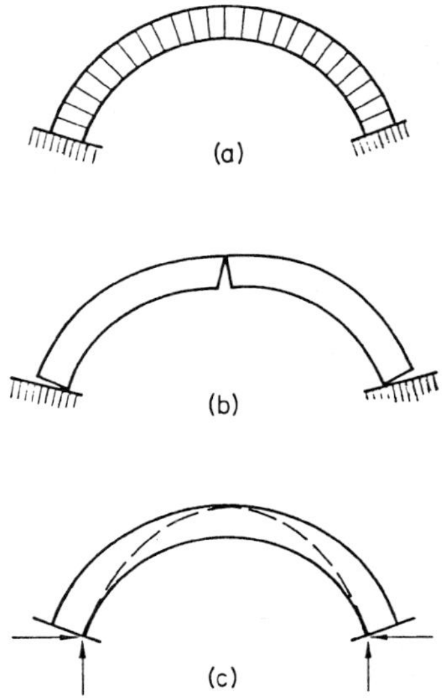

FIG. 25. Cracking of voussoir arch due to spread of abutments.

Since the thrust line must pass through a hinge point, three hinges located as in Fig. 25(b) will define uniquely the position of the thrust line for the arch. This is sketched in Fig. 25(c). Of all possible lines of thrust, this line corresponds to the least value of the abutment thrust. Thus a first paradox of the behaviour of an arch with yielding abutments is that, if the abutments do give way, they will do so under the *minimum* thrust that the arch is capable of delivering. Put another way, if the abutments of an arch start to yield under a certain value of the thrust, then the thrust will immediately drop towards the minimum value corresponding to the line sketched in Fig. 25(c), and the danger is alleviated.

A second paradox is that, if the deformations have a sensible effect on the geometry of the arch, then the arch will in general (but not in every case) be of a "better" shape to resist the applied loading; that is, the geometrical factor of safety will increase. This may be illustrated by a calculation for the Lincoln arch.

Suppose the arch to spread in span by 25 mm. Hinges will form at the springings and at the crown [as in Fig. 25(b)], and compatibility requires that the crown drop 81 mm, which causes an appreciable change in the shape of the arch. Fig. 27 shows (to the same exaggerated scale of Fig. 15), the profile of the centre-line of the arch before and after the spread of 25 mm in the abutments.

Equations similar to (19) may be written for the new profile, and solved to give [cf. equations (20)]:

$$\left.\begin{aligned} \varepsilon &= 0{\cdot}023 \text{ m} \\ H &= 107 \text{ kN} \end{aligned}\right\} \qquad (24)$$

The geometrical factor of safety has been improved by the settlement from the value 4·2 to the value 280/46 = 6·1, some 45 per cent. However, the horizontal thrust has

Fig. 26. Bridge at Clare College, Cambridge.

increased by 23 per cent to 107 kN, so that the factor of safety on stress has been reduced. Nevertheless, the mean compressive stress is still less than 0·8 MN/m².

In the limit the perfectly flat arch (the *plate-bande*) has an infinite geometrical factor of safety, but also, of course, an infinite abutment thrust as the thickness decreases to

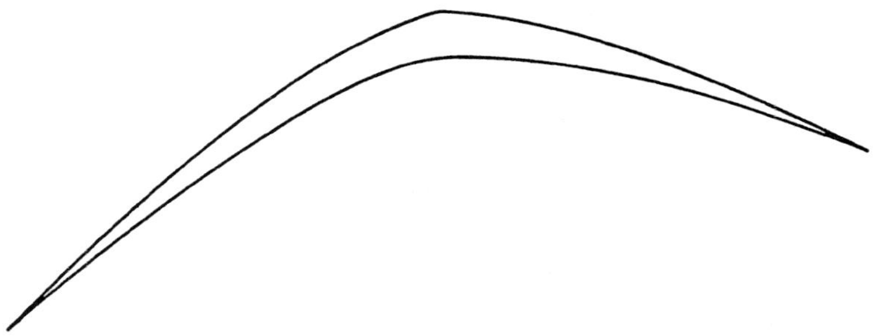

FIG. 27. Deformed centre-line of Lincoln arch due to 25 mm spread in abutments; vertical scale × 6.

zero. Before this condition is reached, "snap-through" of the arch must occur due to the compressibility of the stone, which has been neglected. (There is no danger of snap-through at Lincoln. At a mean stress of 0·8 MN/m² the arch of span 8·54 m will compress, due to thrust, an amount of say 1 mm; a generous allowance for more highly compressible mortar will not double this figure.)

CONCLUSIONS

It is difficult, on the basis of the few bridges analysed here, to make any definite recommendations as to the value of the geometrical factor of safety necessary to give adequate assurance of stability. However, it seems that any practical bridge of reasonable shape has a wide margin of safety; the thrust line will normally lie well within the masonry, working stresses will be low, the bridge can carry comfortably unsymmetrical live load and will not fail due to accidental imperfections such as yielding of the abutments. Indeed, factors of safety against these conditions seem remarkably high, and it requires an exceptionally unskilled and ignorant designer to produce a bridge such as that at Pontypridd which collapsed in 1751.

This conclusion about the essential stability of apparently reasonable designs is supported by the fact that such bridges, once erected and having survived a little while, and given occasional minimum maintenance, do not subsequently collapse, in spite of the fact that they now carry much larger live loads than were envisaged by their designers. The safety of masonry arches under modern traffic conditions has, of course, given rise to some concern, and it is of interest to compare some of the conclusions of Chettoe and Henderson[41] with the implications of the present paper.

Chettoe and Henderson made direct heavy loading tests on a large number of bridges. Although some permanent deformations were observed, the bridges behaved in a sufficiently elastic way for these authors to conclude that an elastic analysis would furnish a proper picture of the response of any particular bridge. The middle-third rule is not adhered to, and Chettoe and Henderson propose the "safe" assumptions that mortar cannot carry tensile stress, and that the structural contribution of any fill should be ignored. These last assumptions are, of course, identical with those made here.

In the absence of any reliable estimate of strengths of old masonry, Chettoe and Henderson propose limiting stresses of 200–250 lbf/in², say 1·5 MN/m². It will be recalled that the mean compressive stress at Lincoln is about 0·6 MN/m², while that for the Ponte Mosca is about 2·5 MN/m²; this latter bridge is made, however, of strong granite.

It is fairly clear that quite rigorous tests of safety, both geometrical and on level of stress, can be passed by a wide range of bridges of usual shapes and spans. Having satisfied himself as to the design of the arch itself, the designer may well conclude that the critical portions of a masonry bridge lie in the abutments. Indeed, the crown deflexions and spread recorded by Chettoe and Henderson were the results very largely of movements of the abutments under load. One of the main problems of the bridge designer is to assign a reasonable value to the abutment thrust so that he can make a proper design of the abutments.

In this connexion the eighteenth-century work seems to have been directed precisely towards the right goal, and it is significant that both theory and such experiments as were made were all concerned with the limit states, that is, with the assessment of the thrust as the arch was on the point of collapse. This is a perfectly proper basis of design, and the basis was changed only when elastic theory was developed sufficiently to enable solutions of the type described by Castigliano in 1879 (although Yvon Villarceau was well aware in 1854 that the arch problem was, in modern terms, statically indeterminate).

This concentration on elastic methods of analysis, with undue weight being given to such concepts as the middle-third rule, has bedevilled structural design for the last century. For the assessment of design values of arch thrust there is, of course, nothing against elastic theory; the application of elastic methods to a structure which is imperfectly elastic will give results which are, if not as "true" as the elastic designer thinks, as least safe in the sense of the plastic theorems. However, elastic methods tend to be lengthy and they rely heavily on a whole range of conventional assumptions; the lack of knowledge of the extent and properties of the mortar beds between voussoirs, for example, make the precise results of an elastic analysis rather meaningless.

By contrast, the sort of equilibrium approach attempted in the present paper does not rely on such assumptions. Moreover, the method is simpler than an elastic analysis, it again gives safe results, and it yields values which are at least as reasonable for design purposes as those given by conventional elastic means.

REFERENCES

1. E. de Maré, *The Bridges of Britain*. Batsford, London (1954).
2. S. Smiles, *Lives of the Engineers*, rev. ed. Vol. 2: *Harbours–Lighthouses–Bridges; Smeaton and Rennie*. Murray, London (1874).
3. H. Shirley Smith, *The World's Great Bridges*. Phoenix House, London (1953).
4. P. C. Lesage, *Recueil de divers mémoires extraits de la Bibliothèque Impériale des Ponts et Chaussées, à l'usage de MM. les ingénieurs*, 2 vols. Didot, Paris (1810).

5. A. A. H. DANYZY, *Histoire de la Societé Royale des Sciences établie à Montpellier*, Vol. 2, p. 40. Lyon (1778).
6. A. F. FRÉZIER, *La théorie et la pratique de la coupe des pierres et des bois pour la construction des voûtes et autres parties des batimens civils et militaires, ou traité de stéréotomic à l'usage de l'architecture*, 3 vols. Doulsseker le fils, Strasbourg and Paris (1737–9).
7. P. DE LA HIRE, *Traité de Mécanique*. Annison, Paris (1695).
8. P. DE LA HIRE, *Mémoires de l'Académie Royale des Sciences*, 1712, p. 69. Paris (1731).
9. A. PARENT, *Essais et recherches de mathematique et de physique*, 3 Vols. Nully, Paris (1713).
10. B. F. DE BÉLIDOR, *La science des ingénieurs dans la conduite des travaux de fortification et d'Architecture civile*. Jombert, Paris (1729).
11. COUPLET, *Histoire de l'Académie Royale des Sciences*, p. 79 (1729); p. 117 (1730).
12. C. A. DE COULOMB, *Mémoires de mathématique et de physique, présentés à l'Académie Royale des Sciences, pars divers savans, et lûs dans ses assemblées*, 1773, p. 343. Paris (1776).
13. C. BOSSUT, *Histoire de l'Académie Royale des Sciences*, 1774, p. 534. Paris (1778).
14. C. BOSSUT, *Histoire de l'Académie Royale des Sciences*, 1776, p. 587. Paris (1779).
15. E. M. GAUTHEY, *Traité de la construction des ponts* (edited by C. L. M. H. NAVIER), 2 Vols. Didot, Paris (1809, 1813).
16. M. G. LAMÉ and E. CLAPEYRON, *Ann. Mines* **8**, 789 (1823).
17. T. LE SEUR, F. JACQUIER and R. G. BOSCOVICH, *Parere di tre mattematici sopra i danni che si sono trovati nella cupola di S. Pietro sul fine dell'anno 1742*. Rome (1743).
18. G. POLENI, *Memorie istoriche della gran cupola del Tempio Vaticano*. Padova (1748).
19. D. GREGORY, *Phil. Trans.* No. 231, p. 637 (1697).
20. A. YVON VILLARCEAU, *Institut de France, Académie des Sciences, Mémoires présentés par divers Savants*, Vol. 12, p. 503 (1854).
21. H. STRAUB, *A History of Civil Engineering*. Leonard Hill, London (1952). [*Die Geschichte der Bauingenieurkunst*. Birkhäuser Verlag, Basle (1949)].
22. S. P. TIMOSHENKO, *History of Strength of Materials*. McGraw-Hill, New York (1953).
23. H. C. F. JENKIN, Article: Bridges. *Encyclopaedia Britannica*, 9th ed. Edinburgh (1876).
24. J. A. EWING, *The Strength of Materials*. Cambridge University Press (1899).
25. A. J. S. PIPPARD, F. TRANTER and L. CHITTY, *J. Instn civ. Engrs* **4**, 281 (1936).
26. A. J. S. PIPPARD and R. J. ASHBY, *J. Instn civ. Engrs* **10**, 383 (1938).
27. A. KOOHARIAN, *Proc. Am. Concr. Inst.* **89**, 317 (1953).
28. W. PRAGER, *An Introduction to Plasticity*. Addison-Wesley, Reading, Mass. (1959).
29. J. HEYMAN, *Int. J. Solids Struct.* **2**, 249 (1966).
30. J. HEYMAN, *Gaz. Beaux-Arts*, **71**, 177 (1968).
31. J. F. BAKER, M. R. HORNE and J. HEYMAN, *The Steel Skeleton*. Vol. 2: *Plastic Behaviour and Design*. Cambridge University Press (1956).
32. C. INGLIS, *Applied Mechanics for Engineers*. Cambridge University Press (1951).
33. J. W. BAXTER, A. F. GEE and H. B. JAMES, *Proc. Instn civ. Engrs* **30**, 489 (1965).
34. *Vetusta Monumenta*, Vol. 3. Society of Antiquaries, London (1796).
35. W. A. NICHOLSON, *Trans. RIBA*, Vol. I, Part II, p. 180. London (1842).
36. J. B. PAPWORTH, *Trans. RIBA*, Vol. I, Part II, p. 184. London (1842).
37. J. GWILT, *An Encyclopaedia of Architecture*, New Impression. Longmans, Green and Co., London (1903).
38. E. G. WARLAND, *Constructional Masonry*. Pitman, London (1946).
39. C. A. P. CASTIGLIANO, *Théorie de l'equilibre des systèmes élastiques et ses applications*. Augusto Federico Negro, Turin (1879). Translated by EWART S. ANDREWS: *Elastic Stresses in Structures*. Scott, Greenwood and Son, London (1919). Republished with an introduction by G. AE. ORAVAS, *The Theory of Equilibrium of Elastic Systems and its Applications*. Dover, New York (1966).
40. G. G. UNGEWITTER, *Lehrbuch der gotischen Konstruktionen*, 2 Vols. Tauchnitz, Leipzig (1901).
41. C. S. CHETTOE and W. HENDERSON, *Proc. Instn civ. Engrs* **7**, 723 (1957).

Name and Place Index

Please note: Page numbers which appear in italics are references to tables or illustrations

Subject Index

Please note: Page numbers which appear in italics are references to tables or illustrations